Functional differential equations: II. C-applications*
Part 1 Equations with continuous coefficients

Pitman Monographs and
Surveys in Pure and Applied Mathematics 94

Functional differential equations: II. C*-applications

Part 1 Equations with continuous coefficients

Anatolij Antonevich
Mikhail Belousov
Andrei Lebedev

LONGMAN

Addison Wesley Longman Limited
Edinburgh Gate, Harlow
Essex CM20 2JE, England
and Associated companies throughout the world.

Published in the United States of America
by Addison Wesley Longman Inc.

First published 1998

AMS Subject Classifications: (Main) 35, 45, 47
(Subsidiary) 22, 46, 58

ISSN 0269-3666

ISBN 0 582 10049 6

British Library Cataloguing in Publication Data

A catalogue record for this book is
available from the British Library

We are grateful to the following for permission to reproduce copyright material:

Reed Consumer Books Ltd / Penguin Putnam Inc. for excerpts from THE HOUSE AT POOH CORNER by A A Milne, originally published by Methuen Children's Books, a division of Reed International Books Ltd / Illustrations by E H Shepard. Copyright 1928 by E P Dutton, renewed © 1955 by A A Milne. Used by permission of Dutton Children's Books, a division of Penguin Putnam Inc. and excerpts from "Introduction" by A A Milne, from NOW WE ARE SIX by A A Milne, originally published by Methuen Children's Books, a division of Reed International Books Ltd / Illustrations by E H Shepard. Copyright 1927 by E P Dutton, renewed © 1955 by A A Milne. Used by permission of Dutton Children's Books, a division of Penguin Putnam Inc.

Printed and bound by Bookcraft (Bath) Ltd

Тем, кого любим — за мученья.

To those whom we love - for the martyrdom.

Contents of Volume II (Part 1)

Preface

This book is an attempt to provide the first thorough and up-to-date exposition of a relatively new field of modern analysis lying in between differential equations and C^*-algebras, and in fact deeply intersecting both of these subjects. The problems considered arise from the investigation of non-local functional differential operators containing transformation operators of an argument.

We have developed a general unified approach to the solution of these problems based on the investigation of C^*-algebras corresponding to the functional differential operators that we examine. The main objects which appear here are C^*-algebras associated with dynamical systems. Attention is principally devoted to the investigation of invertibility and Fredholmity of these operators and their spectral properties. The monograph contains a description of the structure of the C^*-algebras introduced and in particular, as a result of this description, a symbolic calculus for these kinds of problems is obtained. General formulae for the spectral radius and conditions for the invertibility of elements of these algebras are given. The fundamental correlation between the invertibility of elements and hyperbolic decomposition, trajectorial decomposition and approximations of dynamical systems is brought to light. Various applications to equations associated with singular integral operators, Toeplitz operators, pseudodifferential operators (including operators with discontinuous coefficients) and transformation operators of an argument; non-local boundary-value problems, convolution-type equations with oscillating coefficients and other equations generated by different local-type operators and transformation operators of an argument are presented.

In the first part of the book (C^*-theory), Section 1-10 and 17-20 have been written by Antonevich; the rest is due to Lebedev.

At the end of each chapter notes and remarks concerning each section in the chapter are inserted. These short comments are probably incomplete and may contain errors. The authors ask the forgiveness of those mathematicians whose contributions to the subject have not been brought to light because of our oversight or lack of information.

The bibliography contains in essence only those sources directly related to the material considered here and does not reflect the historical development of the subject.

PREFACE

We wish to express our thanks to Professor Alan Jeffrey, the editor of the book, who in fact initiated the writing and has done everything to smooth the publication process.

Anatolij Antonevich Andrei Lebedev

Preface to Volume II

The second volume contains (as its title makes clear) the applications of the C^*-structures and theory already developed in Volume I to the study of solvability conditions of various non-local functional differential equations. The main objects of the investigation here are non-local pseudodifferential, singular integral and Toeplitz operators (with continuous and piecewise continuous coefficients), convolution type operators with oscillating coefficients and shifts and operators associated with non-local boundary value problems containing transformation operators of an argument on the boundary. For all these classes of operators we build the (principal) symbolic calculus, that is we construct the objects (symbols) in terms of which the Fredholmity conditions of these operators are written out. With the help of this symbolic calculus we treat various concrete examples of non-local operators and present the explicit computation of their Fredholmity conditions and the index formulae, and obtain a number of related results (for example Kveselava-Vekua type theorems for non-local singular integral operators with piecewise continuous coefficients and non-local Agranovich-Dynin type theorems).

The volume consists of two parts (Part 1: Equations with continuous coefficients, and Part 2: Equations with discontinuous coefficients and boundary value problems). In Part 1 Chapter 5 and Appendix A were written by Antonevich and Appendix B by Lebedev. In Part 2, Chapter 7 is due to Belousov and Chapter 6 and Appendix C are due to Lebedev.

In fact the volume turned out to be a bit longer than we anticipated when we started writing it and we would like to express once more our thanks to Professor Alan Jeffrey for his attention to the rearrangement of the publication process, which has been very helpful, and to the staff at Addison Wesley Longman, for their patience and understanding of our (time and writing) problems, which allowed us the opportunity to overcome them. We would also like to thank Mike Nugent (the copy-editor) who did his best to turn our English into English-like language and Lynn Kelly (the typist) who heroically typed everything we had written.

Anatolij Antonevich* Mikhail Belousov Andrei Lebedev*

* Partially supported by the Fundamental Research Fund of the Republic of Belarus.

Preface to Volume II

Introduction to Volume II

'... Introduction is really the *er-h'r'm* of the book... . There are some very clever writers who say that it is quite easy not to have an *er-h'r'm*... . I think it is much easier not to have all the rest of the book.' A.A. Milne *Now we are six* (Introduction).

'... сформулируем три бесспорных утверждения:

1. Ни один результат чистой математики не получает никаких практических приложений раньше, чем через 50 лет после его получения.

2. Около 99 процентов результатов, получаемых в чистой математике, бывают полностью забыты через 50 лет' М. Евграфов

'Чего я не понимаю в математиках, – так это почему они стремятся к нулю?' В. Фурсенко*

In this second volume we present various examples of non-local functional differential equations which can be successfully investigated with the help of the C^*-algebra methods described in Chapters 1-4. In reality the historical development of research in this area of mathematics has not proceeded in the order presented in this book (in fact in the reverse order!). The problems we deal with in Chapters 5-7 (a search for the solvability conditions of non-local functional differential equations, the description of their symbolic calculus, the index calculation, etc.) have been the main object of study, while the theory of C^*-algebras generated by dynamical systems has been worked out precisely in the direction necessary for the investigation of this class of problems.

I.1 A rather cursory sketch of the picture of the development of the interrelations between non-local functional differential equations (of the type considered in this book) and the corresponding C^*-structures looks approximately as follows.

One of the first known problems of this sort is the *Carleman problem* which was posed by T. Carleman at the International Mathematical Congress in Zurich in 1932 (see Carleman [1]). The problem was to find a function $Q^+(z)$ which is analytic in a domain $\Omega \subset \mathbb{C}$ and satisfies on its boundary Γ the condition

---

* '...let us state the three undeniable assertions:

1. No result in pure mathematics finds any practical application within 50 years after its derivation.

2. About 99 percent of the results obtained in pure mathematics turn out to be completely forgotten after a lapse of 50 years.' M. Yevgrafov

'What I don't understand about mathematicians is why do they tend to zero?' V. Fursenko

$$Q^{+}(z)+D(z)Q^{+}(\alpha(z))=\varphi(z)$$

where $\alpha:\Gamma\to\Gamma$ is a given mapping and D and φ are given functions (Carleman also assumed the satisfaction of the condition

$$\alpha(\alpha(z))=z$$

which was therefore named the *Carleman condition*).

This problem is equivalent to a certain functional singular integral equation of the form considered in Section 46, and the first advance was obtained precisely for the one-dimensional functional singular integral equations with the shift satisfying the Carleman condition (that is, using the terminology of this book, when the group G consists of two elements) and for the boundary value problems for analytic functions and two-dimensional elliptic equations which can be reduced to these singular integral equations. Here one could mention the pioneering work by D.A. Kveselava after which the theory has been developing in a number of scientific centres of the former USSR (Tbilisi, Rostov-on-Don, Kazan', Minsk, Odessa, Dushanbe). Among the authors of original work in this direction are G.S. Litvinchuk, È.G. Hasabov, R.A. Kordzadze, A.S. Sosunov, È.I. Zverovich, and many others.

The first principal situation which was completely investigated was the class of singular integral operators (SIO) with the shifts satisfying the *generalized Carleman condition* (that is, when the mappings (transformations) of Γ form a finite group G). It was understood that these *non-local* problems could be reduced to *local* problems by means of specially constructed equations (associated with the initial equations) in the spaces of *vector* functions (of dimension $|G|$). The corresponding procedure (as well as a detailed survey on the non-local boundary value problems for analytic functions) is presented in Litvinchuk [2] (in fact the initial idea of this construction goes back to Babbage [1]).

Passing from singular integral operators with a finite group of shifts to those with an infinite group of shifts we should first of all mention the results of the group of Odessa mathematicians headed by G.S. Litvinchuk (Yu.I. Karlovich, V.G. Kravchenko, N.L. Vasilevskij, Yu.D. Latushkin) who have done most in this direction for the last 25 years. The first step here was done by Kravchenko [1] who obtained the Fredholmity condition for a two-term functional SIO in the case when the shift $\alpha:\Gamma\to\Gamma$ is not periodic and has a finite number of periodic points. Another important special case - the situation when the shift α is an irrational rotation of the circle - was investigated by

Antonevich and Ryvkin [1,2]. Almost simultaneously similar results were obtained by another method (under a certain additional constraint on the nature of the irrationality of the rotation) by Karapetyants [1].

Parallel (or in fact along) with the investigation of the one-dimensional situation the theory of multidimensional non-local functional partial differential and pseudodifferential equations was developed. First to be investigated were equations where the non-localness was contained in lower terms and thus in fact this type of non-localness had only a small influence on the properties of the equation. The essential complication of the equations in the situation when the non-localness (transformation of an argument) is contained in their principal part has plagued their successful investigation, which prompted A.D. Myshkis to write in his survey (A. Myshkis [2]) that the study of partial differential equations with transformations of an argument in the principal part needs the development of radically new methods.

These methods have now been found.

Again (as in the one-dimensional situation) the derivation of the theory started from the investigation of functional pseudodifferential operators (PDO) with the shifts forming a *finite* group of diffeomorphisms of a compact manifold (Antonevich [2,5]). The operators considered in these articles include as a particular case the above-mentioned functional SIO generated by an arbitrary finite group of diffeomorphisms of a curve Γ. Axiomatically the C^*-algebras related to this class of operators were described by Antonevich [4] and Antonevich and Brenner [1].

We would like to emphasize that, even in this 'simplest' case, non-local functional PDOs are qualitatively different from non-local functional SIOs. Namely the Lefschetz numbers contained in the index formula (see Theorem 43.10) are all trivial (for $g \in G$, $g \neq e$) in the case of singular integral operators (see Theorem 43.13), while this is not so for the pseudodifferential situation (see Example 43.16).

The first study of functional PDOs containing non-periodic shifts was the paper by Antonevich [6] which considered a two-term functional PDO on the torus generated by a standard shift of the torus. In this work the symbol of the functional PDO under investigation was written out and the conditions for the invertibility of the symbol were obtained (in fact in this paper all the ergodic measures with respect to a standard shift of the torus were written out, though the general theorem on the spectral radius (Theorem 4.4) was not then known). Analogous results were obtained also in the paper by Antonevich [11] for a two-term functional PDO generated by a diffeomorphism which can be included into an action of a compact Lie group.

As a result of this work it was clear that:

(1) the decisive tool for the investigation of functional PDOs is the investigation of functional operators;
(2) the latter operators are poorly known.

This made researchers look for new approaches and as a result the approach based on the construction of a C^*-algebra generated by a dynamical system was proposed. The first C^*-algebraic advance in this direction (obtained precisely when searching for the construction of the symbolic calculus for functional PDOs) was achieved by Lebedev [4,7] (in this work the first 'simplest' variant of the isomorphism theorem (Corollary 12.17) which considered the group $G = \mathbb{Z}$ was proved).

On the basis of this work Antonevich and Lebedev [1] constructed a symbolic calculus for functional PDOs generated by *topologically freely* acting diffeomorphisms of a manifold. Since then the development of research in this area has gone on further in the work of Antonevich, Lebedev, Brenner, Karlovich, and Batkhin.

The C^*-algebraic directions of this development are:

(1) the extension of the class of groups (generating the shifts (automorphisms));
(2) the extension of the class of C^*-algebras which play the role of coefficients;
(3) the extension of the class of group actions.

Now progress in the first and second directions has attained its highest point, namely the isomorphism theorem (Corollary 12.17) has been proved for an *arbitrary C^*-algebra* A of coefficients and an arbitrary *amenable topologically freely acting* group of shifts (automorphisms $\hat{T}_g$, $g \in G$).

If G is not amenable or $\hat{T}_g$ do not act topologically freely then (if we do not presuppose any *specific* interrelations between A and T_g) the results like the isomorphism theorem are no longer true (see Remark 12.18). On the other hand, if we do assume some reasonable interrelation (for the functional PDO situation) between A and T_g then we can achieve further advances in (1) and especially in (3). Namely we can pass to the non-amenable groups such that only some of their singled-out factor groups are amenable (see Corollary B.30) and proceed from the topologically acting groups to the so-called normally acting groups (see Section B.25).

The general C^*-picture of the modern state of achievements in directions (1)-(3) is drawn in Chapter 2 and Appendix B. The principal results here are due to Lebedev.

The extension of the class of coefficients up to arbitrary C^*-algebras (direction (2)) has made it possible to construct successfully the symbolic calculus and to obtain the Fredholmity conditions for various classes of functional SIOs and PDOs with *discontinuous* coefficients. The process of the investigation of such operators is now rapidly developing and here we should mention in the first place the work by Karlovich, Lebedev, Dangh Suan Tkhan, Batkhin, and Böttcher (and, returning to the finite groups actions for SIOs, the work of Karapetyants and Samko [3]).

One more class of non-local functional differential equations treated in this book by means of C^*-algebraic methods is the class of non-local boundary value problems. Again the starting example here could be considered to be the Carleman problem mentioned above. Various boundary value problems with boundary conditions of the form

$$\sum_{i=0}^{m} a_i f(\alpha_i(t)) = \psi(t),\ t \in \Gamma \tag{I.1}$$

(where Γ is the boundary (or a part of the boundary) of a domain $\Omega \subset \mathbb{R}^n$, $\alpha_i : \Gamma \to \Gamma$, $i = \overline{1,m}$ are invertible mappings and a_i, $i = \overline{1,m}$, are some differential or pseudodifferential operators) or boundary conditions which could be represented in this form for various classes of partial differential equations and differential operator equations have been investigated by many researchers, among which are S.A. Aldashev, R. Beals, A.V. Bitsadze, F.F. Browder, V.I. Chesalin, A.A. Dezin, F.I. Frankl', V.L. Il'in, N.I. Ionkin, T.Sh. Kal'menov, V.A. Kordzadze, S.G. Krein, G.I. Laptev, S.K. Kumykova, E.I. Moiseev, A.M. Nakhushev, B.I. Ptashnik, Ya.A. Roitberg, Z.G. Sheftel', V.K. Romanko, G.B. Savchenko, M.S. Salakhitdinov, A.L. Skubachevskij, A.P. Soldatov, M. Schehter, M.H. Shanukov, V.A. Steklov, and others.

With regard to these classes of functional differential operators the first essential advance in the investigation of the Fredholm property of non-local boundary value problems was achieved for the case when the mappings α_i contained in the boundary condition (1) form a finite group of transformations of Γ (Antonevich [1,3], Antonevich and Dobrushkin [1]). Its further development for the case of an infinite group of transformations is also due to Antonevich [20]. In this paper the elliptic partial differential equations considered were assumed to satisfy a certain constraint: namely it was presupposed that the Dirichlet problem for them is Fredholm (in appropriate Sobolev spaces) and this was used as one of the cornerstones in the proofs.

However it could well be that for an elliptic partial differential equation (and even more so for an elliptic system of equations) there does not exist any elliptic boundary value problem with differential boundary conditions. For example, for the Cauchy-Riemann operator

$$\frac{1}{2}\left(\frac{\partial}{\partial x_1}+i\frac{\partial}{\partial x_2}\right)$$

in the circle $\{(x_1,x_2)\in\mathbb{R}^2 : x_1^2+x_2^2<1\}$ one could pose neither an elliptic boundary value problem with differential boundary conditions nor even an elliptic boundary value problem with pseudodifferential conditions. Thus the approach developed in that paper had its natural (and in fact rather rigorous) limits.

This obstacle has been overcome, and moreover fundamental progress in the study of the Fredholm property of non-local boundary value problems has been obtained, in the work of Belousov by means of the construction and investigation of a new class of non-local boundary value problem operators.

The C^*-structures associated with such operators have some qualitatively different features (in comparison with those used for functional SIOs and PDOs) since the shift operators (operators generating automorphisms) are acting here not on the whole of the algebra of coefficients (which in this case is the algebra of the Boutet de Monvel operators) but only on a certain part (ideal) corresponding to the *purely boundary* operators. The approach to their study proposed by Belousov was based on the procedure of reducing of their investigation to the isomorphism theorem (Corollary 12.17). Further development of this idea and general abstract C^*-algebras arising in this way have been worked out by Lebedev and are presented in Appendix C.

This completes the brief life story of the types of functional differential operators and C^*-algebraic objects we are dealing with in this volume. Now let us say a few words about the contents of the volume itself.

I.2 First a note about the terminology. The main classes of functional differential operators (FDOs) under consideration have names like *G*-pseudodifferential operators, *G*-singular integral operators, *G*-Toeplitz operators, etc. In other words their name is *G*-... operators where '...' stands for some well-known class of operators. This means the following. As a rule we are considering operators of the form

$$\sum a_g T_g \tag{I.2}$$

where the 'coefficients' a_g belong to the algebra A of operators of a well-known type (pseudodifferential, singular integral, Toeplitz, etc.) and T_g, $g\in G$, are the 'shift' operators (mostly the operators induced by transformations of an argument) forming a representation of a group G and generating a group of automorphisms of A (precisely the

C^*-algebras generated by the operators (I.2), in an abstract way, were described in Volume I).

Thus in the name *G-... operators*, '...' is used to denote the algebra of coefficients ('local' operators) and G indicates that we generate the 'non-localness' in this class of 'local' operators by adding a group of 'shift' operators T_g, $g \in G$.

I.3 We now begin an overview of this. As the title says we regard the material presented here as the C^*-applications of the C^*-algebraic structures and methods developed in Volume I. In fact, from our point of view it is a collection of various *examples* (though, say, Sections 40-42, 51-55, 61, 63, and 65 clearly represent the full-blooded divisions of the theory of the operators we are considering). The main goal of these examples is to illustrate the *technique* of the usage of the C^*-results obtained and to demonstrate the *fundamental principles* underlying the interrelation between C^* and functional differential objects. At the same time, since the classes of operators described here are 'just examples', the situations we are dealing with are not always the most general ones that could be handled by means of the mathematical apparatus and ideas presented, and a lot of 'examples' are still beyond the scope of the book (this concerns especially Chapters 6 and 7 since an increasing variety of functional differential operators of the kind considered in these chapters is now being constantly involved in the process of thorough investigation). However we hope that we have managed to bring to light the main structure of the C^*-approach to the study of functional differential operators.

I.4 The volume contains three chapters.

Chapter 5 deals with various examples of *G*-... operators (recall the explanation of this term presented above) where the symbols of elements of the algebra of 'coefficients' could be naturally considered as continuous functions over some compact spaces or continuous cross-sections of some vector bundles. The main examples of this sort of operator are *G*-pseudodifferential operators with continuous coefficients which have given their name to the chapter itself. We describe here the symbolic calculus of these operators, calculate the conditions under which they are Fredholm and write out the index formulae.

In Chapter 6 an analogous job is done for *G*-pseudodifferential operators with piecewise continuous coefficients and operators of a similar nature.

Finally in Chapter 7 we present the construction, symbolic calculus and calculation of Fredholmity conditions for a wide class of non-local boundary value problems associated with the Boutet de Monvel operators and such that the non-localness (the 'transformation of an argument' operators) is contained in the boundary conditions operators.

Along with the main functional differential text the volume contains three appendices (A, B and C). These appendices present self-contained expositions of certain purely C^*-algebraic structures and results which were not considered in Volume I, are rather widely used in this volume, and have their own intrinsic C^*-algebraic interest.

Appendix A gives the principal C^*-algebraic scheme of the index calculation for the elements of the algebra $C^*(A, T_g)$ in the case of a finite group G.

Appendix B describes the structure of certain algebras, $C^*(A, T_g)$, of a 'functional differential operator' nature (which is most clear when we apply its results to the investigation of the operators in Chapter 5) in the situation when the 'shifts' T_g, $g \in G$, do not act topologically freely. In fact it contains a rather deep natural generalization of Theorem 16.21.

Appendix C deals with the C^*-structures which have been worked out specially to study the non-local boundary value problems considered in Chapter 7.

I.5 Now we would like to offer a more detailed description of the material of Volumes I and II and sketch the developing 'from on high' C^*-picture of FDOs. In fact the precise drawing of this picture is one of the reasons for writing this book. This picture does not of course solve all the problems arising in the investigation of FDOs and is related first of all to the problems of the Fredholm theory of FDOs.

Let us recall these problems. If we are considering '*local*' (in terms of the 'slang' of Section I.2 above) operators (singular integral operators, pseudodifferential operators, Toeplitz operators, ...) then with respect to these the problems in question are:

(1) the description of the algebras generated by these operators;
(2) the construction of the symbolic calculus (that is the objects in terms of which the Fredholmity conditions of the operators could be expressed);
(3) the evaluation of the Fredholmity conditions;
(4) the construction of a regularizer;
(5) the index calculation.

The theory developed in Volumes I and II is worked out precisely to resolve these problems for the 'non-local' FDOs.

Before drawing the emerging C^*-picture (which will be written out in Section I.12 below) let us explain the distinctions appearing when considering 'local' and 'non-local' objects, that is ... operators and *G*-... operators.

Local situation

In the classical theory of pseudodifferential operators with continuous coefficients on a compact manifold the problem of Fredholmity is resolved according to the following scheme.

1. From a PDO $\mathcal{a}$ by means of the Fourier transform there is constructed a matrix-valued function $\sigma(\mathcal{a})$ (a morphism of the induced vector bundle over the cotangent bundle) which is called the *symbol* of $\mathcal{a}$ (see 2.43). The symbol reflects the main algebraic properties of the operator and moreover *the operator* $\mathcal{a}$ *Fredholm iff* $\sigma(\mathcal{a})$ *is invertible* (Theorems 2.44, 2.46). In other words the Fredholm property of $\mathcal{a}$ reduces to the condition

$$\det \sigma(\mathcal{a})(x,\xi) \neq 0 . \tag{I.3}$$

2. The index of $\mathcal{a}$ is calculated via the homotopic invariants of $\sigma(\mathcal{a})$.

Remarks (1) In fact in this situation one does not need to invoke any serious C^*-algebraic results since the algebra of symbols arising here (= morphisms of a vector-bundle) is the 'simplest' (very thoroughly examined) C^*-algebra and the homotopic properties of its elements are also well known (in *K*-theory).

(2) If one considers PDOs with *discontinuous* coefficients then the invocation of C^*-algebras becomes vital even in the local situation (see Chapter 4 and in particular Sections 31, 33).

Non-local situation

As a basic example let us take here the *G*-PDOs (in a certain functional space *F*(*X*) of functions on *X*) of the form

$$\mathcal{b} = \sum a_k(x,D)U_k \tag{I.4}$$

where $a_k(x,D)$ are differential or pseudodifferential operators (*X* is a domain in $\mathbb{R}^n$ or a manifold) and U_k are the transformation of an argument operators of the form $(U_k f)(x) = f(\alpha_k(x))$, where $\alpha_k : X \to X$ are given mappings.

The more complicated nature of the operators (I.4) (in comparison with those considered above in the local situation) shows itself in particular in that there *does not* exist a matrix-valued function possessing (with respect to the operators (I.4)) the main properties of the classical (i.e., in the sense described above) symbol. As it turns out (see I.7, I.8) the symbols of the operators (I.4) are again certain G-operators and their structure has an *essentially infinite-dimensional* character (related in particular to the cross-products) and operation with such objects is simply impossible without the application of the highly developed C^*-apparatus.

Moreover, as well as obtaining such non-trivial objects as the symbols, in addition to problems (1)-(5) listed above (classical for the 'local' operators) we arrive at a more vital problem: *to obtain the invertibility conditions for the symbols we have constructed.* These conditions could no longer be written out in such a simple way as in (I.3) and to investigate them we were forced to develop a number of qualitatively *new* C^*-algebraic methods (see I.9 below).

Now let us proceed to a cursory description of the ideas and the structure of the emerging C^*-picture of FDOs.

I.6. The main C^*-algebraic objects

The main C^*-algebraic object arising when trying to axiomatize in a unified manner the properties of various classes of G-operators is the C^*-algebra $C^*(A, T_g)$ (the uniform closure of operators of the form (I.2)) which was introduced in Section 12 and thoroughly studied in Chapters 2 and 3 (the corresponding object arising in the situation when $\{T_g\}_{g\in G}$ induce not a representation but only a quasirepresentation of the group G by automorphisms

$$\hat{T}_g = T_g(\cdot)T_g^*,\ g \in G \tag{I.5}$$

of the algebra A is considered in Appendix B).

In fact when dealing with non-local boundary value problems (G-boundary value problem operators) we arrive also at the necessity of introducing a certain generalization of the algebra $C^*(A, T_g)$, namely the C^*-algebra $\underline{C}^*(A, \mathcal{J}, T_g)$ where the operators T_g, $g \in G$, generate the automorphisms not of the whole of A but only of a singled-out ideal $\mathcal{J}$. Such C^*-objects are studied in Appendix C.

Remarks (1) As we mentioned above (see I.2) in situations when we deal with concrete classes of G-... operators the role of A is played by certain classical operator algebras such as the algebra of pseudodifferential operators, the convolution type operators, and the algebra of Boutet de Monvel boundary value problem operators, etc. But the general C^*-picture is drawn for an *arbitrary* C^*-algebra A (and any of its ideal $\mathcal{J}$).

(2) We would like to recall also (see the Introduction to Volume I and Notes and Remarks to Chapter 2) that the C^*-algebras of $C^*(A, T_g)$ type play an important role in the general C^*-theory and many C^*-algebraists have contributed to their study. At the same time the investigation of the problems listed in I.5 for FDOs needs in fact not the application of the classical aspects of C^*-theory of $C^*(A, T_g)$ (presented say in Pedersen [1]) but the consideration of a number of *non-traditional* C^*-algebra problems and primarily the problem of obtaining the invertibility conditions for *particular* elements of $C^*(A, T_g)$. Methods for the solution of these problems have their origin in many ways in other fields of mathematics and thus we see the *synthetic* (C^*-algebras + dynamical systems + ergodic theory + hyperbolicity + ...) nature of mathematical derivation.

I.7. Isomorphism theorems

Among the most important building blocks and methods of the C^*-theory of FDOs described below one should single out the aggregate of the so-called isomorphism theorems. These theorems give easily verifiable conditions (on the action of the automorphisms) under which two algebras of the type mentioned in I.6 are isomorphic.

Theorems of this sort appear for the first time in Volume I in Section 12 (Corollaries 12.9 and 12.17). The 'ideological' value of these results for the Fredholm theory of FDOs was already mentioned in I.1 above. Here we would like to single out Corollary 12.17 which in a way plays the most significant technical role and lies at the heart of the 'philosophy' and proofs of the other results of this nature (which one will find in Sections 15 and 16 and Appendices B and C).

Indeed Corollary 12.17, along with Corollary 12.16, links the algebras of $C^*(A, T_g)$ type and the cross-products $A \underset{\hat{T}}{\times} G$ (it actually establishes their isomorphism); and the 'double value' of this result lies in the fact that on the one hand it provides an opportunity to operate with $C^*(A, T_g)$ with the help of the mathematical apparatus already developed (for the cross-products), and this idea has been heavily exploited to obtain a number of results in Chapter 3; while on the other hand the methods of operation with algebras of $C^*(A, T_g)$ type developed in this book can be considered to be methods of operation with the cross-product-type algebras and in particular various results of Sections 13, 14 and Chapter 3 from this point of view appear to be new for the cross-products.

Remark Apart from their C^*-value the isomorphism theorems in fact play the role of the 'non-commutative Fourier transform' for the non-local operators under consideration (cf. the local situation 1. in I.5) which in reality has made it possible to construct the symbolic calculus for the corresponding classes of *G*-... operators and obtain the conditions of their Fredholmity (see I.8 and I.9).

I.8. Symbol of a FDO

The isomorphism theorems on the one hand suggest the right construction of the symbol for this or that class of 'non-local' FDOs and on the other hand serve as one of the instruments enabling us to affirm the corresponding symbolic calculus.

For example, for the operator $\mathcal{b}$ (I.4) the symbol is a *G*-... operator (acting in the space of functions (of $2n$ variables) on the cotangent bundle of *X*) of the form

$$\sigma(\mathcal{b}) = \sum \sigma(a_k(x_1 D))Q_k$$

where $\sigma(a_k(x,D))$ is the symbol of a PDO $a_k(x,D)$ (see the local situation in I.5) and Q_k is the operator associated with the transformation of an argument generated by the codifferential $\partial\alpha_k$ of the mapping α_k (see Theorem 40.7). The symbols constructed in this way reflect the main algebraic properties of the operators under investigation and *the operator $\mathcal{b}$ is Fredholm iff its symbol $\sigma(\mathcal{b})$ is invertible* (cf. the local situation in I.5). Thus the problem is reduced to the derivation of the invertibility conditions for the symbols (that is, the invertibility conditions of elements in the C^*-algebras considered in I.6).

I.9. The invertibility conditions of elements in $C^*(A, T_g)$

Here we present a list of the *qualitatively distinct* general methods of calculation of the invertibility conditions of elements in $C^*(A, T_g)$ which could be applied according to the situation. The substantination of all these methods essentially exploits the isomorphism theorems.

The figures in parentheses after the name of the method indicate the numbers of the sections where this method is explained.

I. The hyperbolic approach (19,20). In the situation when $G = \mathbb{Z}$ (that is $T_z = T^z$, $z \in \mathbb{Z}$), the invertibility conditions for the operators of the form

$$\sum a_z T^z \qquad \text{(I.6)}$$

turn out to be equivalent to the condition on the linear extension of the dynamical system associated with the action of the automorphism $\hat{T}$ (I.5) on the spectrum of A to be hyperbolic.

Using this property we obtain, in particular, the infinite-dimensional C^*-algebraic analogues and generalizations of the corresponding finite dimensional results of Mather [1] and Sacker and Sell [1,2].

With the help of this method (and the Kveselava-Vekua type operators (see I.10) one can also calculate the index of a FDO.

II. The trajectorial approach (21). In this method the invertibility conditions of the *non-discrete* operators of the form (I.2) are calculated by means of the *discrete* operators associated with the trajectories of points of the spectrum of A (under the action of the automorphisms $\hat{T}_g$ (I.5)).

With the help of this method we obtain in particular the (discrete) analogy to the classical Fredholm alternative for integral equations (see Remarks 21.7) and establish the interrelation between the hyperbolicity (see I) and the 'pointwise' exponential dichotomy.

III. Ergodic measures and the Lyapunov exponents (4,7,17,25). In the case when A is a commutative C^*-algebra for the two-term operators of the form $b = a_0 + a_1 T$ the hyperbolic approach presented above could be developed further by using the explicit calculation of the spectral characteristics of the operator via the integrals along the ergodic measures (with respect to the dynamical system considered).

The generalization of this idea to the situation when the coefficients a_0 and a_1 are continuous operator functions with compact values leads us to the calculation of the spectral characteristics via the corresponding Lyapunov exponents.

IV. The approximation approach (27). In this method the dynamical system, generated by the (shift) operators under investigation, is substituted (approximated) by another one and the invertibility conditions of the initial operators are calculated via the invertibility conditions of the resulting approximation operators.

In particular if we could approximate the initial dynamical system by finite (periodic) dynamical systems then for these operators this approach enables us to apply the invertibility conditions of elements in C^*-algebras associated with *finite* dynamical systems (see V) which means that this method could be considered to be the C^*-dynamical analogy to the procedure of the approximation of almost-periodic functions by periodic ones.

By means of this method the spectral stability of these operators with respect to strong perturbations (caused by a change of the dynamical system) is also proved.

It is also worth mentioning that one of the principal instruments exploited in the proof of this method is the Wiener-type theorems on a given form of the inverse operator (derived in Section 26).

V. The finite group of automorphisms case (15). In the situation when the group $\{T_g\}_{g\in G}$ is finite the invertibility conditions of elements in $C^*(A,T_g)$ could be calculated via the invertibility conditions of a specially constructed matrix-function (with the coefficients from A) thus reducing to the conditions of invertibility of 'local' operators.

Remark We shall not discuss here a number of other ideas and results related to the calculation of invertibility conditions (which the reader can find in Chapter 3) since from our point of view they do not have such a universal character as those listed above.

I.10. Index calculation

In the case when $\{T_g\}_{g\in G}$ is a finite group the index calculation for elements in $C^*(A,T_g)$ could be carried out successfully by means of the Lefschetz numbers (Theorem A.23). If $\{T_g\}_{g\in G}$ is infinite then the most accessible situation arise when $G=\mathbb{Z}$ and is related to the hyperbolic approach (see I in I.9). In this case by means of Theorem A.36 and Theorem 19.35 the problem could be reduced to the index calculation for an operator of the form

$$a_0P\oplus a_1T(I-P)$$

where P is a certain projection (canonically constructed in Lemma 19.18), and the index calculation of the latter operator is closely related to the index calculation of an operator of the form

$$P+T(I-P)$$

which (in the situation when we are dealing with G-singular integral operators) is known as the *Kveselava-Vekua operator*, and its index (along with the indexes of operators of a similar nature) is calculated (for various types of T) in Sections 46 and 56 (and in all the situations we consider it turns out to be equal to zero). These index calculations for G-

SIOs and G-Toeplitz continuous and piecewise continuous coefficients are presented in Sections 47, 48 and 56.

I.11. C^*-localization

If the algebra A (playing the role of coefficients for G-... operators) arises from operators with 'singularities' (for example PDOs with discontinuous coefficients or PDOs in functional spaces on manifolds with corners, and so on) then among the powerful tools of operating with such objects and revealing their microstructure is the C^*-localization procedure which is described in Chapter 4 and which is naturally also exploited in this case as an auxiliary instrument when passing from A to $C^*(A, T_g)$.

I.12. The structure of the C^*-theory of FDOs

The scheme shown below illustrates the structure of the C^*-theory developed for FDOs. The figures at the arrows indicate the numbers of the foregoing subsections (thus 7 stands for I.7, etc.) where the method exploited to derive the corresponding results is presented.

In the block labelled 'The symbol invertibility conditions...' the Roman numerals I, II,... indicate the corresponding methods listed in I.9.

Needless to say, when passing from the block 'FDOs under investigation' to the block 'C^*-theory' one has to apply not only the pure C^*-algebraic methods but also the methods of operation with the 'algebra of coefficients' of the G-... operators under study (that is to say, if we are dealing with G-PDOs we have naturally to exploit PDO methods, etc.).

I.13. To complete this introduction we would also like to say a few words about the bibliography and list of notations. The bibliography of this volume is an extension of the bibliography contained in the first volume.

The list of notations is in fact very far from being complete since we cannot see a way (and maybe the sense) of deriving unified notations throughout the whole of the book, as it involves a lot of objects having different origins and natures. Therefore the list of notations contains:

(1) the notations which have a *standard* meaning throughout the whole of the book, such as say $C^*(A, T_g)$ (12.1, B.5), $A \underset{\alpha}{\times} G$ (11.1), $\hat{T}$ (4.1, 12.1);

(2) some special notations such as $\overline{\alpha}_g$ (61.1), $\pi(m, \varphi, \theta)$ (33.13);

(3) some notations which, though not very frequently used, have a certain specific '*ideological*' sense, such as say $\underline{C}^*(A, T_g)$ (C.3), $\underline{C}^*(A, \mathcal{J}, T_g)$ (C.13).

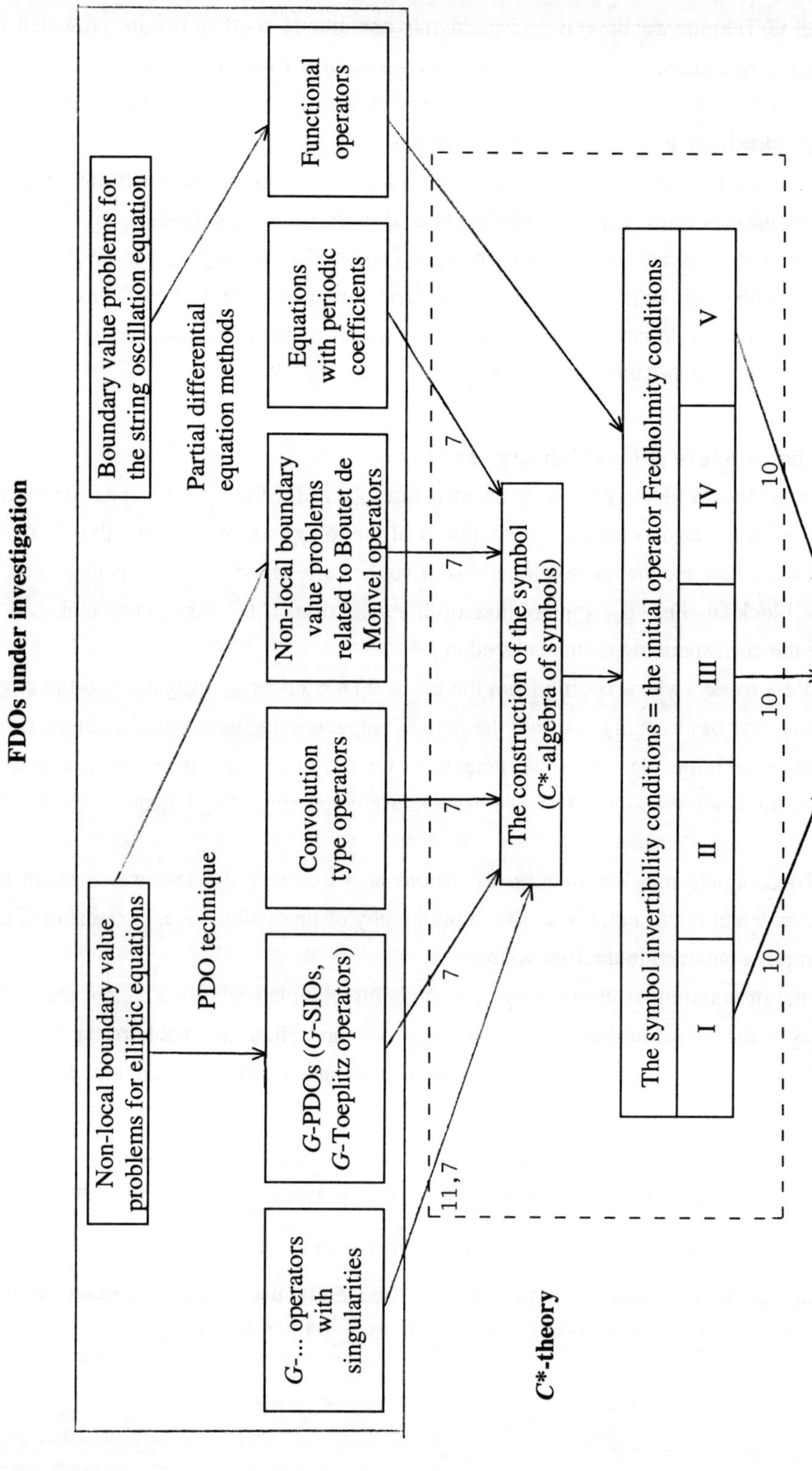
FDOs under investigation
Non-local boundary value problems for elliptic equations
PDO technique
Boundary value problems for the string oscillation equation
Partial differential equation methods
G-... operators with singularities
G-PDOs (G-SIOs, G-Toeplitz operators)
Convolution type operators
Non-local boundary value problems related to Boutet de Monvel operators
Equations with periodic coefficients
Functional operators
11, 7
7
7
7
7
The construction of the symbol (C*-algebra of symbols)
The symbol invertibility conditions = the initial operator Fredholmity conditions
I
II
III
IV
V
10
10
10
Index calculation
C*-theory

At the same time we have not inserted into the list of notations a lot of standard mathematical notations, the meaning of which is quite clear from the context, neither have we included the notations which are used only within the limits of a certain singled-out section and do not have a specific 'ideological' sense.

Finally it goes without saying that theorem (etc.) numbers lower than 36 (the number of the first section of this Volume) refer to those theorems (etc.) in Volume I.

Chapter 5
G:pseudodifferential operators with continuous coefficients

'I have forgotten to explain one very important thing to you. The World could be changed only *in* our mind, nothing else can solve this problem.' Yukio Mishima *The Temple of the Golden Pavilion.*

'Ты мне говоришь – раз Закон известен, значит он существует! А я Тебя спрошу – а существовал ли Закон до того, как он был открыт?' А. Киселев 'Суть № I'

'Различие между Математикой и Реальностью того же сорта, что и между Демократией и Справедливостью (...или наоборот?)...' О. Слин 'Другая вода'

'Плохих баб нет, только леммы одни бывают плохие...' Венедикт Ерофеев 'Москва–Петушки'*

The basic property which unifies the non-local (in this or that way) operators studied in this chapter is that the symbols of elements of their 'coefficients' algebra can be represented as continuous functions (or cross-sections) on some compact space (of some vector bundles over a compact space). The typical model operators of this sort are the operators of the form

$$\sum a_k U_{g_k}$$

where a_k are pseudodifferential operators (with continuous coefficients) on a manifold M and

$$\left(U_g f\right)(x) = f\left(\alpha_g^{-1}(x)\right)$$

---

* 'You are saying - if the Law is known it does exist! And I am asking - had it existed before it was discovered?' A. Kiselev *The essence N 1*

'The difference between Mathematics and Reality is of the same sort as between Democracy and Fairness (... or vice versa?)...' O. Slin *Different water*

'All chicks are cool whereas some lemmas are not...' Venedikt Yerofeyev *Moscow - Petushki*

where $\{\alpha_g\}_{g\in G}$ is a representation of a group G by diffeomorphisms of M.

We call these operators G-PDOs with continuous coefficients (for the explanation of this name see the Introduction to this volume) and the chapter is named 'in honour' of them. It is known that an important role in the investigation of differential and pseudodifferential operators is played by their total and principal symbols which in the case of differential operators are simply their characteristic polynomials and the principal parts of these polynomials respectively. The results are formulated as a rule in terms of the symbols. One of the central results in the theory of PDOs states that

A PDO on a compact manifold is Fredholm in the corresponding Sobolev spaces iff it is elliptic, that is its principal symbol is invertible.

In this chapter we consider problems of this sort related to G-PDOs, namely; we are looking for necessary and sufficient conditions for a G-PDO to be Fredholm.

The first step in the solution of this problem is the derivation of the analogue of the principal symbol for G-PDOs. The fundamental obstacle to the construction of the symbol for G-PDOs is the essential non-commutativity of the algebra generated by G-PDOs. In particular it follows that the symbol of a G-PDO cannot be a scalar function or even (in the case of an infinite group G) a matrix-valued function since the symbols of G-PDOs *should* generate a rather complicated non-commutative algebra.

Luckily the structure of the symbols of G-PDOs is naturally suggested by the isomorphism theorem (Corollary 12.17) and following this suggestion the symbolic calculus for G-PDOs is developed in Sections 40-42.

These sections are preceded by the 'preparatory' Sections 36, 37, the first of which contains the 'elementary symbolic ideology' and some general simple results concerning compact and Fredholm operators (in pairs of spaces) among which we would like to distinguish Lemma 36.5, the benefit of which will be clearly demonstrated in Chapter 7 (Sections 63, 65). Section 37 deals with G-PDOs on $\mathbb{R}^n$ with linear and affine shifts - the situation where the corresponding symbolic calculus looks 'rather evident'.

Since the symbols of G-PDOs turn out to be functional operators (i.e. elements of algebras of the type $C^*(A,T_g)$ which were thoroughly studied in Volume I) many aspects of the theory of their invertibility are in fact already resolved by means of the results of Volume I and concrete examples presenting the total description of the Fredholmity conditions of certain G-PDOs are written out in Sections 39, 44, 45. In particular since the invertibility conditions of a two-term operator $a_0 + a_1T$ are linked with the invariant measures (with respect to the corresponding transformation of a maximal ideal space; see Theorem 17.3) we give, in Section 38, a description of the latter measures (interesting in its own right) for the mappings of S^{n-1} induced by linear transformations of $\mathbb{R}^n$.

In Section 43 we present the index formula for G-PDOs in the case when G is a finite group. In fact the derivation of this formula is based on general C^*-algebraic ideas which are written out in Appendix A.

In Sections 46, 47 we pass to the index computation of G-PDOs (mainly G-SIOs) for $G=\mathbb{Z}$.

The rest of the sections of this chapter could be regarded as 'particular examples' of structures of a G-PDO nature.

In Section 48 the results of the previous sections are 'lowered' to the level of G-Toeplitz operators with continuous coefficients.

Section 49 deals with boundary value problems with strong non-localness, the investigation of which is transformed by means of the 'reduction onto the boundary' into that of G-PDOs.

Finally in Section 50 we write out the symbolic calculus for the (one-dimensional and multidimensional) convolution operators with oscillating coefficients and shifts.

36. Symbolic calculus

36.1 The algebraic sense of the symbolic calculus in general is quite simple. Its main idea is the following. Let $\mathcal{B}$ be some C^*-subalgebra of the algebra $L(H)$ of operators acting in the Hilbert space H and let $\mathcal{K}$ be certain closed two-sided ideal in $L(H)$ (in practice $\mathcal{K}$ is a rule the ideal of compact operators).

Consider the canonical isomorphism

$$\mathcal{B}+\mathcal{K}/\mathcal{K} \cong \mathcal{B}/\mathcal{B}\cap\mathcal{K} \tag{36.1}$$

(see Corollary 2.77). If we denote the right-hand part of (36.1) by B and $\mathcal{B}+\mathcal{K}$ by $\overline{\mathcal{B}}$ then the canonical isomorphism defines the mapping

$$\sigma:\overline{\mathcal{B}} \to B$$

with the properties

(1) $\sigma(b_1+b_2)=\sigma(b_1)+\sigma(b_2)$, $b_i \in \overline{\mathcal{B}}$;
(2) $\sigma(b_1 \cdot b_2)=\sigma(b_1)\cdot\sigma(b_2)$;
(3) $\sigma(b^*)=[\sigma(b)]^*$, $b \in \overline{\mathcal{B}}$;

(4) $\sigma(b) = 0$ iff $b \in \mathcal{K}$;

(5) an element $b \in \overline{\mathcal{B}}$ is $\mathcal{K}$-Fredholm iff $\sigma(b)$ is invertible.

(Recall (see 32.9) that we say that b is $\mathcal{K}$-Fredholm if there exist elements $R_r, R_l \in L(H)$ such that $R_l b \sim 1$ and $bR_r \sim 1$ where 1 is the identity operator in $L(H)$ and the sign $\sim$ means that the right-hand and the left-hand parts of the expression differ by an element $k \in \mathcal{K}$. If $\mathcal{K}$ is the ideal of compact operators then $\mathcal{K}$-Fredholmity coincides with the classical definition of Fredholmity).

Observe also that since in (36.1) it is always necessary to consider in the left-hand part the algebra $\overline{\mathcal{B}} = \mathcal{B} + \mathcal{K}$ it is reasonable to presuppose from the very beginning that $\mathcal{K} \subset \mathcal{B}$ (that is $\overline{\mathcal{B}} = \mathcal{B}$). Thus *henceforth here we assume that* $\mathcal{K} \subset \mathcal{B}$.

The above-mentioned properties (1)-(5) of the mapping σ are just the properties which should possess the so-called (principal) symbol of an element $b \in \mathcal{B}$ (the principal symbol therefore is the mapping which preserves the C^*-algebraic properties and solves the Fredholmity problem for elements of $\mathcal{B}$).

Now if σ_1 is any other mapping from $\mathcal{B}$ onto a C^*-algebra B_1 possessing the same properties (1)-(5) then (by Corollary 2.76) clearly

$$B = \mathcal{B}/\mathcal{K} \overset{\psi}{\cong} B_1$$

and under the natural isomorphism the mapping σ is transformed into σ_1, that is the following diagram is commutative

$$\begin{array}{ccc} & \xrightarrow{\ \sigma\ } & \mathcal{B}/\mathcal{K} = B \\ \mathcal{B} & & \downarrow \psi \\ & \xrightarrow[\ \sigma_1\ ]{} & B_1 \end{array}$$

Moreover obviously any isomorphism ψ of the algebra $\mathcal{B}/\mathcal{K}$ with some algebra B_1 induces the mapping $\sigma_1 = \psi \circ \sigma$ with the properties (1)-(5) of the symbol.

Thus the problem of the construction of the *symbolic calculus* is in fact equivalent to the *explicit* presentation of some algebra (B_1) isomorphic to $\mathcal{B}/\mathcal{K}$, that is the symbolic calculus is simply the explicit form of some faithful representation of $\mathcal{B}/\mathcal{K}$.

36.2 If we consider the algebra of pseudodifferential operators then Theorem 2.46 tells us that we can take $B_1 = \mathrm{HOM}(E_2)$ (E_2 is defined in 2.45) and this is what we mean by writing $\sigma(b) \in HOM(E_2)$. Theorem 2.46 here is the *explicit* record of the symbolic calculus for PDOs.

The classes of operators considered in this and subsequent chapters (various G-operators) differ essentially in nature from PDOs and are related to C^*-algebras associated with automorphisms. That is why Volume I (C^*-theory) forms the necessary C^*-algebraic basis to be exploited in the Fredholm theory of G-operators.

In this chapter we shall mainly use it for the following purposes:

(1) The construction of the symbolic calculus of G-operators. Here we shall apply the ideas developed in Chapter 2 (the isomorphism theorems obtained in Sections 12, 15, 16).

(2) The derivation of the explicit conditions of Fredholmity (that is, the derivation of the explicit conditions of the *invertibility of the constructed symbol*) and the index calculation. Here we shall apply the material of Chapter 3.

Remark We would like to emphasize immediately that the latter problem (2) for G-PDOs is quite different from the corresponding problem in the PDO case. Even in the rather 'simple' examples of concrete G-operators presented in this chapter the invertibility conditions for their symbols have a very non-elementary character.

36.3 Recall that as a rule, PDOs are considered as acting not from a space E into itself but from E into some other space F, that is they are elements of $L(E,F)$. Thus they are not the elements of a C^*-algebra. In any case, it is well known that all the calculations necessary for our purposes (for example, the invertibility conditions of the symbol) could be reduced to the C^*-algebra situation.

Indeed let $b \in L(E,F)$ where E and F are Hilbert spaces. Consider the space

$$H = E \oplus F$$

and the operator $\tilde{b}$ given by the operator-matrix

$$\tilde{b} = \begin{bmatrix} 0 & b^* \\ b & 0 \end{bmatrix}$$

where we assume that $b^* \in L(F,E)$ is the operator determined by the equation

$$\langle bx, y\rangle_F = \left\langle x, b^* y\right\rangle_E . \tag{36.2}$$

The operator $\tilde{b}$ is an element of a C^*-algebra $L(H)$ and b is invertible iff $\tilde{b}$ is invertible (the inverse of $\tilde{b}$ has the same form

$$\tilde{b}^{-1} = \begin{bmatrix} 0 & b^{-1} \\ b^{*-1} & 0 \end{bmatrix}).$$

We shall also present certain results on compact and Fredholm operators showing that we can always confine ourselves to the case $b \in L(H)$.

Lemma 36.4 *Let $b \in L(E,F)$ and $b^* \in L(F,E)$ be operators linked by equation (36.2). Then*

(i) *b is compact iff both the operators b^*b and bb^* are compact;*
(ii) *b is Fredholm iff both the operators b^*b and bb^* are Fredholm.*

Proof Let us prove (i) first.

Clearly if b is compact then b^*b and bb^* are both compact so we have to check only the inverse statement. Let, for example, b^*b be compact while b is not. Since b is not compact it follows (Rudin [1], Chapter 4, Exercise 18) that there exists a sequence $\{\xi_n\}$, $\xi_n \in E$ and $\delta > 0$, such that

$$\xi_n \xrightarrow{\text{weakly}} 0 \tag{36.3}$$

and

$$\|b\xi_n\| \geq \delta . \tag{36.4}$$

But (36.4) and (36.2) imply

$$\langle b\xi_n, b\xi_n\rangle = \left\langle \xi_n, b^* b\xi_n\right\rangle \geq \delta^2 . \tag{36.5}$$

On the other hand, as b^*b is compact (36.3) implies

$$b^* b\xi_n \to 0$$

and again in view of (36.3) and by Rudin [1], Chapter 4, Exercise 18, we have

$$\sup\|\xi_n\| < \infty$$

which, along with the foregoing, means that

$$\langle \xi_n, b^* b\xi_n \rangle \to 0$$

and we arrive at a contradiction to (36.5).

The operator b^*b could be considered using the same argument by replacing b by b^*. So (i) is proved.

(ii) Note first that b is Fredholm iff b^* is Fredholm (see, for example, Treves [1], Chapter II, 2.1) and thus, in view of 2.8, it follows that b^*b and bb^* are both Fredholm.

On the other hand, if b^*b and bb^* are Fredholm then they have left and right regularizers, that is there exist operators c and c' such that

$$cb^*b \sim I_E \text{ and } bb^*c' \sim I_F$$

which means that the operators cb^* and b^*c' are left and right regularizes of b. Thus b is Fredholm. □

Lemma 36.5 *Let $b \in L(E,F)$ and $b^* \in L(F,E)$ be operators linked by equation (36.2). Then*

$$|||b|||^2_{\mathcal{K}(E,F)} = |||b^*|||^2_{\mathcal{K}(F,E)} = |||bb^*|||_{\mathcal{K}(F,F)} = |||b^*b|||_{\mathcal{K}(E,E)}$$

where $\mathcal{K}(\cdot,\cdot)$ are sets of compact operators in the corresponding spaces and $|||\cdot|||$ is the essential norm sign (that is, the factor norm with respect to $\mathcal{K}(\cdot,\cdot)$).

Proof Let us show first that

$$|||b|||^2_{\mathcal{K}(E,F)} \geq |||b^*b|||_{\mathcal{K}(E,E)}. \tag{36.6}$$

Indeed, we have

$$|||b|||^2_{\mathcal{K}(E,F)} = \inf_{k\in\mathcal{K}(E,F)} \|b+k\|^2 = \inf_{\mathcal{K}(E,F)} \sup_{\|\xi\|_E=1} \left|\langle (b+k)\xi, (b+k)\xi\rangle_F\right|$$

$$= \inf_{\mathcal{K}(E,F)} \sup_{\|\xi\|_E=1} \left|\langle \xi, [b^*b+(b^*k+k^*b+k^*k)]\xi\rangle_E\right| . \tag{36.7}$$

Observe that the operator

$$\tilde{k} = b^*k + k^*b + k^*k \tag{36.8}$$

is a compact self-adjoint operator in E and b^*b is a self-adjoint operator. Thus we have

$$\sup_{\|\xi\|_E=1} \left|\langle \xi, b^*b+\tilde{k}\rangle\right| = \|b^*b+\tilde{k}\| \tag{36.9}$$

and (36.7), (36.8) and (36.9) imply

$$|||b|||^2_{\mathcal{K}(E,F)} = \inf_{k\in\mathcal{K}(E,F)} \|b^*b+\tilde{k}\| \geq \inf_{k\in\mathcal{K}(E,E)} \|b^*b+k\| = |||b^*b|||_{\mathcal{K}(E,E)}$$

thus proving (36.6).

On the other hand

$$|||b|||^2_{\mathcal{K}(E,F)} = \inf_{\mathcal{K}(E,F)} \|b+k\|^2 = \inf_{\mathcal{K}(E,F)} \|b^*+k^*\|^2$$

$$= \inf_{k\in\mathcal{K}(F,E)} \|b^*+k\|^2 = |||b^*|||^2_{\mathcal{K}(F,E)}$$

which, along with (36.6), means that

$$|||b|||^2_{\mathcal{K}(E,F)} = |||b^*|||_{\mathcal{K}(F,E)} \geq |||bb^*|||_{\mathcal{K}(F,F)} . \tag{36.10}$$

Now let $\{p_\lambda\}$ be the set of all orthogonal finite dimensional projections in F with the natural partial order (that is $\lambda_1 < \lambda_2$ iff $P_{\lambda_1} \leq P_{\lambda_2}$). As is well known, for every $k \in \mathcal{K}(E,F)$ we have

$$\|k - p_\lambda k\| \underset{\lambda}{\to} 0 . \tag{36.11}$$

Note that

$$|||b|||_{\mathcal{K}(E,F)} = \lim_{\lambda} \|b - p_\lambda b\| . \tag{36.12}$$

Indeed in view of (36.11) we conclude that for every $k \in \mathcal{K}(E,F)$

$$\begin{aligned}\overline{\lim_{\lambda}}\|b - p_\lambda b\| &= \overline{\lim_{\lambda}}\|b - p_\lambda b + k - p_\lambda k\| \\ &= \overline{\lim_{\lambda}}\|(I_F - p_\lambda)(b+k)\| \le \|b+k\|\end{aligned}$$

and thus (bearing in mind that $p_\lambda b \in \mathcal{K}(E,F)$) we obtain

$$\begin{aligned}|||b|||_{\mathcal{K}(E,F)} &\ge \overline{\lim_{\lambda}}\|b - p_\lambda b\| \ge \underline{\lim_{\lambda}}\|b - p_\lambda b\| \\ &\ge \inf_{k \in \mathcal{K}(E,F)} \|b+k\| = |||b|||_{\mathcal{K}(E,F)}\end{aligned}$$

which proves (36.12).

Equation (36.12) in turn implies

$$|||b|||_{\mathcal{K}(E,F)} = \lim_{\lambda} \|b^* - b^* p_\lambda\| . \tag{36.13}$$

Now let us verify the inequality

$$|||b|||^2_{\mathcal{K}(E,F)} \le |||bb^*||| . \tag{36.14}$$

Applying (36.13) and (36.11) we obtain (for any $k \in \mathcal{K}(F,F)$)

$$\begin{aligned}|||b|||^2_{\mathcal{K}(E,F)} &= \lim_{\lambda}\|b^* - b^* p_\lambda\|^2 = \lim_{\lambda}\|(b - p_\lambda b)(b^* - b^* p_\lambda)\| \\ &= \lim_{\lambda}\|bb^* - p_\lambda bb^* - bb^* p_\lambda + p_\lambda bb^* p_\lambda\|\end{aligned}$$

$$= \lim_{\lambda} \left\| bb^* + k - p_\lambda k - p_\lambda bb^* - bb^* p_\lambda - kp_\lambda + p_\lambda kp_\lambda + p_\lambda bb^* p_\lambda \right\|$$

$$= \lim_{\lambda} \left\| (I_F - p_\lambda)(bb^* + k)(I_F - p_\lambda) \right\|$$

$$\leq \left\| bb^* + k \right\|$$

which means that (36.14) is true.

Inequality (36.14), along with (36.10), means that

$$\|b\|^2_{\mathfrak{X}(E,F)} = \left\| b^* \right\|^2_{\mathfrak{X}(F,E)} = \left\| bb^* \right\|_{\mathfrak{X}(F,F)} \tag{36.15}$$

and since in this situation

$$b^{**} = (b^*)^* = b$$

(36.15) implies

$$\left\| b^* \right\|^2_{\mathfrak{X}(F,E)} = \left\| b^* \cdot (b^*)^* \right\|_{\mathfrak{X}(E,E)} = \left\| b^* b \right\|_{\mathfrak{X}(E,E)}$$

thus completing the proof. □

37. *G*-pseudodifferential operators on $\mathbb{R}^n$. Pseudodifferential difference operators and pseudodifferential operators with affine shifts

This section should be considered as preparation for the main (theoretical) Sections 40-42 of this chapter. Here we confine ourselves to the 'simplest' situation - the investigation of the objects generated be the classical PDOs (of zero order) in $L^2(\mathbb{R}^n)$ and the shift operators generated by 'linear' transformations of an argument. In any case the separate presentation of these topics has its own value. On the one hand, we have the possibility of writing out all the basic operators and constructions in explicit form and, on the other hand, as the reader will discover even for these 'simplest' objects we obtain a natural and essential difference between the theory of pseudodifferential difference operators (37.2-37.15) and pseudodifferential operators with affine shifts (37.16-37.23). When investigating the latter operators we arrive at a number of general ideas and difficulties (see Remark 37.23), the consideration of which will be developed further in Sections 40-42.

37.1 To begin with for the reader's convenience we recall the construction of the algebra of pseudodifferential operators of order 0 in $L^2(\mathbb{R}^n)$. Let $a(x,\xi)$ be a smooth function on $\mathbb{R}^n \times \{\mathbb{R}^n \setminus 0\}$ which is zero-degree positive homogeneous with respect to ξ and suppose there exists a constant C such that $a(x,\xi)$ does not depend on x when $|x|> C$, that is

$$a(x,\xi) = a(\infty,\xi),\ |x|> C .$$

With the function $a(x,\xi)$ we associate an operator $\mathfrak{a} = a(x,D)$ acting from $C_0^\infty(\mathbb{R}^n)$ into $C^\infty(\mathbb{R}^n)$ via the formula

$$\mathfrak{a}u = a(x,D)u = F^{-1}a(x,\xi)Fu \tag{37.1}$$

where F is the Fourier transform on $\mathbb{R}^n$. For every $s \in \mathbb{R}$ the operator $a(x,D)$ could be extended by continuity up to a bounded operator acting from $H^s(\mathbb{R}^n)$ into itself.

Let $\mathfrak{A}$ be the uniform closure of the algebra of operators in $L^2(\mathbb{R}^n)$ generated by the operators of the form (37.1) and the set K of compact operators in $L^2(\mathbb{R}^n)$. Let $\overline{\mathbb{R}}^n = \mathbb{R}^n \cup \{\infty\}$ be the one-point compactification of $\mathbb{R}^n$, S^{n-1} be the unit sphere and $Q = \overline{\mathbb{R}}^n \times S^{n-1}$.

For every $\mathfrak{a} \in \mathfrak{A}$ there is defined the *(principal) symbol* $\sigma(\mathfrak{a})$ which is a continuous function on Q. In particular, for the operator of the form (37.1) we have

$$\sigma(a(x,D)) = a(x,\xi) .$$

The algebra of the symbols of the elements of $\mathfrak{A}$ will be denoted by A.

Pseudodifferential difference operators

37.2 We now consider the algebra generated by the pseudodifferential operators and the transformation operators of an argument. We begin with the simplest case - the pseudodifferential difference operators, that is the case when the transformation of an argument (the shift) is generated by the group operation (addition of vectors) on $\mathbb{R}^n$; namely, let us consider in the space $L^2(\mathbb{R}^n)$ the operator $\mathfrak{b}$ defined via the formula

$$\mathscr{b} = \sum_{k=1}^{m} a_k(x,D) T_{h_k} \tag{37.2}$$

where $a_k(x,D) \in \mathscr{A}$ and $T_h\left(h \in \mathbb{R}^n\right)$ is the shift operator

$$(T_h u)(x) = u(x-h), \; u \in L^2\left(\mathbb{R}^n\right). \tag{37.3}$$

Clearly $\{T_h\}_{h \in \mathbb{R}^n}$ is the group of unitary operators in $L^2\left(\mathbb{R}^n\right)$.

We denote by

$$\mathscr{B} = C^*\left(\mathscr{A}, T_h\right) \tag{37.4}$$

the C^*-algebra (of *pseudodifferential-difference operators*) generated by $\mathscr{A}$ and $\{T_h\}_{h \in \mathbb{R}^n}$.

37.3 We define the *symbol* of the operator $\mathscr{b}$ of the form (37.2) to be the functional operator $\sigma(b)$ in $L^2(M)$ (where $M = \mathbb{R}^n \times S^{n-1}$) described by the formula

$$\sigma(b) = \sum_{k=1}^{m} \sigma(a_k) V_{h_k} \tag{37.5}$$

where $\sigma(a_k) = a_k(x,\xi)$ denotes the operator of multiplication by the function $a_k(x,\xi)$ (the symbol of $a_k(x,D)$) and $V_h \left(h \in \mathbb{R}^n\right)$ is the unitary operator in $L^2(M)$ of the form

$$(V_h w)(x,\xi) = w(x-h,\xi), \; w \in L^2(M). \tag{37.6}$$

We denote by

$$B = C^*\left(A, V_h\right)$$

the C^*-algebra generated by the algebra A (the symbols of the elements of $\mathscr{A}$) and $\{V_h\}_{h \in \mathbb{R}^n}$.

The next theorem says that the algebra B is indeed the algebra of the symbols of elements of $\mathscr{B}$ (see Section 36).

Theorem 37.4 *The mapping*

$$\sigma : \mathcal{b} \to \sigma(\mathcal{b})$$

*(see (37.2) and (37.5)) can be extended up to the *-epimorphism*

$$\sigma : \mathcal{B} \to B$$

the kernel of which is the ideal K of compact operators.

In particular this mapping possesses all the main properties of the symbol of PDOs (cf. 2.41)

(1) $\sigma(\mathcal{b}_1 + \mathcal{b}_2) = \sigma(\mathcal{b}_1) + \sigma(\mathcal{b}_2)$*;*
(2) $\sigma(\mathcal{b}_1 \cdot \mathcal{b}_2) = \sigma(\mathcal{b}_1) \cdot \sigma(\mathcal{b}_2)$*;*
(3) $\sigma(\mathcal{b}^*) = \sigma(\mathcal{b})^*$*;*
(4) $\sigma(\mathcal{b}) = 0$ *iff* $\mathcal{b}$ *is compact;*
(5) *an operator* $\mathcal{b}$ *is Fredholm iff* $\sigma(\mathcal{b})$ *is invertible.*

Proof Let

$$[\mathcal{B}] = \mathcal{B}/K .$$

To prove the theorem it is enough to check that the mapping σ defined by (37.5) establishes the isomorphism between $[\mathcal{B}]$ and B. So let us do this.

Recall that we have (see 37.1)

$$[\mathcal{A}] = \mathcal{A}/K \cong A = \sigma(\mathcal{A}) \tag{37.7}$$

and for every $\alpha = a(x, D)$ of the form (37.1)

$$T_h a(x, D) T_h^* = T_h a(x, D) T_{h^{-1}} = a(x - h, D)$$

which implies for every $\alpha \in \mathcal{A}$

$$\sigma(T_h \alpha T_h^*) = V_h \sigma(\alpha) V_h^* \tag{37.8}$$

(here V_h is defined by (37.6)).

Observe that on the space

$$\operatorname{Prim} A = Q = \overline{\mathbb{R}}^n \times S^{n-1}$$

the group of its homeomorphisms $\tilde{h}$, $h \in \mathbb{R}^n$,

$$\tilde{h}(x,\xi) = \begin{cases} (x+h,\xi), & x \in \mathbb{R}^n \\ (\infty,\xi), & x = \infty \end{cases}$$

generated by the automorphisms $V_h(\cdot)V_h^*$ acts topologically freely (see 12.13).

The group $\mathbb{R}^n$ is commutative and thus amenable.

All this, along with (37.8) and the isomorphism theorem (Corollary 12.17), means that

$$[\mathcal{B}] = C^*([\mathcal{A}],[T_h]) \overset{\sigma}{\cong} C^*(A,V_h) = B$$

and the proof is complete. □

37.5 Passing to the conditions of the Fredholmity of elements of $\mathcal{B}$ (the invertibility of the elements of B) one can use various ideas and approaches (see Chapter 3). Let us consider now the set of representations of $\mathcal{B}$ related not to the points of $\operatorname{Prim} A = Q$ (cf. Section 21) but to a certain localization of $\mathcal{B}$ (see Chapter 4).

Namely let us identify the space $L^2(M)$ with the space $L^2\left(S^{n-1}, L^2(\mathbb{R}^n)\right)$ and observe that all the operators V_h $\left(h \in \mathbb{R}^n\right)$ given by (37.6) commute with the operators of multiplication by the functions from $L^\infty\left(S^{n-1}\right)$, which implies (since the elements $a \in A$ also commute with such operators) that every element $b \in B = C^*\left(A, V_h\right)$ possesses this property as well, and consequently every $b \in B$ can be expressed as the operator-valued function

$$b(\xi) = \sum a_k(\cdot,\xi)T_{h_k} \in L\left(L^2\left(\mathbb{R}^n\right)\right), \quad \xi \in S^{n-1}.$$

Remarks (1) In fact the values $b(\xi)$ are simply the local representatives of b with respect to the subalgebra of A (and thus of B) generated by the symbols of PDOs of zero order with constant coefficients (that is not depending on x) (see Section 28); and S^{n-1} is the maximal ideal space of the algebra of symbols of such PDOs.

(2) The passage to the consideration of the operators $b(\xi)$ is useful for a number of reasons. Two of them are:

(i) the operators $b(\xi)$ are simpler in a sense than b;

(ii) the construction of $b(\xi)$ is extremely natural - in fact we simply exchange $a_k(x,D)$ with $a_k(x,\xi)$ and then consider the family of operators depending on the parameter ξ.

Remark (1) here leads in turn to

Lemma 37.6 *The element $b \in B$ is invertible iff all the operators $b(\xi)$, $\xi \in S^{n-1}$, are invertible.*

Proof This lemma is a particular case of Theorem 28.9, but in the simple situation under consideration we would like to present an independent proof illuminating a certain useful idea previously exploited in Example 21.10; namely observe that $b(\xi)$ is a continuous (on $\xi \in S^{n-1}$) operator-valued function. Indeed for the elements $b = \sigma(\ell)$ of the form (37.5) this is obvious (since $\sigma(a_k)$ are operators of multiplication by uniformly continuous functions) and any element $b \in B$ is a (uniform) limit of such operators.

If b is invertible then (since, for every $\xi \in S^{n-1}$, $b(\xi)$ is a representation of b) all the operators $b(\xi)$ are invertible.

Now let all the operators $b(\xi)$, $\xi \in S^{n-1}$, be invertible. In view of the continuity of the function $b(\xi)$ and as S^{n-1} is compact it follows that the set

$$\sum = \{b(\xi)\}_{\xi \in S^{n-1}} \subset L\left(L^2\left(\mathbf{R}^n\right)\right)$$

is compact. Since the passage to the inverse is a continuous function on the group of invertible operators the set

$$\sum^{-1} = \left\{b^{-1}(\xi)\right\}_{\xi \in S^{n-1}}$$

is also compact and consequently is bounded.

Thus the operator-valued function $b^{-1}(\xi)$ defines the inverse b^{-1} of b, and the proof is complete. □

Remarks 37.7. (1) We would like to emphasize that the existence of $b^{-1}(\xi)$ for every ξ is not sufficient in general (for an arbitrary operator-valued function $b(\xi)$ for the invertibility of the operator b defined by this operator-valued function $b(\xi)$, since if b^{-1} does exist its norm satisfies the equality

$$\left\|b^{-1}\right\| = \operatorname{ess\,sup}_{\xi} \left\|b^{-1}(\xi)\right\|,$$

and thus $\left\|b^{-1}(\xi)\right\|$ should be essentially bounded.

In our case we in fact deduced the bounds of $\left\|b^{-1}(\xi)\right\|_{\xi\in S^{n-1}}$ from the compactness of the set $\{b(\xi)\}_{\xi\in S^{n-1}}$.

(2) Since for every ξ the algebra generated by $b(\xi)$, $b \in B$, has a structure analogous to $B = C^*(A, V_h)$ (in fact it has the form

$$B(\xi) = C^*(A(\xi), V_h(\xi))$$

where $V_h(\xi) \in L(L^2(\mathbb{R}^n))$ is the unitary operator given by

$$([V_h(\xi)]f)(x) = f(x-h) \tag{37.9}$$

the conditions for invertibility of the operators $b(\xi)$ can be investigated with the help of the methods and ideas already developed in Chapters 3 and 4.

Now let us consider certain examples of the operators of $\mathcal{B}$ and B.

Example 37.8 Let the symbols of the operators $\alpha_k \in \mathcal{A}$, $k = \overline{1,m}$, be independent of x (that is $\sigma(\alpha_k) = a_k(\xi)$). In this case, for every $\xi \in S^{n-1}$ the operators

$$b(\xi) = \sigma(\mathcal{b})(\xi)$$

(see (37.5)) are the difference operators with constant coefficients.

Under the Fourier transform $F_{x\to\eta}$ the operator $b(\xi)$ is transformed into the operator of multiplication by the continuous function

$$\tilde{b}(\eta)=\sum_{k=1}^{m}a_k(\xi)e^{i\langle h_k,\eta\rangle}.$$

Thus (in view of Lemma 37.6) the operator b is invertible iff for every ξ the condition

$$\inf_{\eta\in\mathbf{R}^n}\left|\sum_{k=1}^{m}a_k(\xi)e^{i\langle h_k,\eta\rangle}\right|>0 \tag{37.10}$$

is satisfied.

Remark The methods developed in Chapters 3 and 4 are oriented towards the investigation of more complicated objects (in particular operators with symbols depending on x) and of course in general the explicit conditions for invertibility cannot be written out in such a simple way as in (37.10).

Example 37.9 In this example we confine ourselves to the case of PDOs on $\mathbf{R}^n$, $n\geq 2$ (the case $n=1$ - singular integral operators - will be considered separately in Sections 46, 47). Let $\mathcal{b}\in\mathcal{B}$ be an operator of the form

$$\mathcal{b}=\alpha_0+\alpha_1 T_h \tag{37.11}$$

where $\alpha_i\in\mathcal{A}$, $i=\overline{0,1}$, and

$$\sigma(\alpha_i)=a_i(x,\xi),\quad i=\overline{0,1},\quad \left((x,\xi)\in Q=\overline{\mathbf{R}}^n\times S^{n-1}\right).$$

The conditions of Fredholmity of this operator are described by the next statement.

The operator $\mathcal{b}$ of the form (37.11) is Fredholm in $L^2(\mathbf{R}^n)$ iff one of the following two conditions (i) or (ii) is satisfied:

(i) $a_0(x,\xi)\neq 0$ *for every* $(x,\xi)\in Q$

and
$|a_0(\infty,\xi)|>|a_1(\infty,\xi)|$, $\xi\in S^{n-1}$;

(ii) $a_1(x,\xi) \neq 0$ *for every* $(x,\xi) \in Q$

and

$$|a_0(\infty,\xi)| < |a_1(\infty,\xi)|, \ \xi \in S^{n-1} .$$

Proof In view of Theorem 37.4 and Lemma 37.6 the operator $\mathcal{b}$ is Fredholm iff for every $\xi_0 \in S^{n-1}$ the operator $b(\xi_0)$ given by

$$b(\xi_0) = a_0(x,\xi_0) + a_1(x,\xi_0)W_h \tag{37.12}$$

(where $a_i(x,\xi_0)$, $i = \overline{0,1}$, $(x \in \mathbb{R}^n)$ are the operators of multiplication (in $L^2(\mathbb{R}^n)$) by the functions $a_i(x,\xi_0)$ and $W_h = V_h(\xi)$ is given by (37.9) is invertible.

The conditions for the invertibility of the operator $b(\xi_0)$ can be written out with the help of Theorem 17.3. Indeed recall that the functions $a_i(x,\xi_0)$, $i = \overline{0,1}$, both have limits

$$\lim_{|x| \to \infty} a_i(x,\xi) = a_i(\infty,\xi)$$

and thus can be considered to belong to the algebra $C(\overline{\mathbb{R}}^n)$ having maximal ideal space equal to $\overline{\mathbb{R}}^n = \mathbb{R}^n \cup \{\infty\}$.

The operator W_h generates the automorphism $W_h(\cdot)W_h^*$ of this algebra, and the corresponding homeomorphism $\tilde{h}$ of $\overline{\mathbb{R}}^n$ is given by

$$\tilde{h}(x) = \begin{cases} x+h, & x \in \mathbb{R}^n \\ \infty, & x = \infty . \end{cases}$$

This homeomorphism admits only one ergodic measure - the Dirac measure $\nu = \delta_\infty$ which implies (by Theorem 17.3) that $b(\xi_0)$ is invertible iff one of the following two conditions (i′) or (ii′) is satisfied:

(i′) $a_0(x,\xi_0) \neq 0$ for every $x \in \mathbb{R}^n$

and

$$|a_0(\infty,\xi_0)| > |a_1(\infty,\xi_0)| ;$$

(ii′) $a_1(x,\xi_0)\neq 0$ for every $x\in\mathbb{R}^n$

and

$$|a_0(\infty,\xi_0)|<|a_1(\infty,\xi_0)|.$$

In view of the connectedness of S^{n-1} and the continuity of $a_i(\infty,\xi)$ the implementation of one of the conditions (i′) or (ii′) at every $\xi\in S^{n-1}$ implies in fact the implementation of (i) or (ii), and the proof is complete. □

Example 37.10 Let the vectors $h_1,\dots,h_m$ in formula (37.2) form a convex and rationally independent system (see Section 24). For convenience we recall here the definitions 24.7 and 24.9; namely the system $\{h_1,\dots,h_m\}$ is called *convex* if h_i, $i=\overline{1,m}$, are vertices of a convex polygon in $\mathbb{R}^n$. This system is called *rationally independent* if the system of $m-1$ vectors $\{h_2-h_1,h_3-h_1,\dots,h_m-h_1\}$ is linearly independent with respect to the field of rational numbers.

Now the results of Section 24 lead to the following statement:

Let $\mathfrak{b}$ be an operator of the form (37.2) and $\{h_1,\dots,h_m\}$ be a convex and rationally independent system of vectors. Then the operator $\mathfrak{b}$ is Fredholm iff there exists a number $k_0\in\overline{1,m}$ such that

(i) $a_{k_0}(x,\xi)\neq 0$ *for every* $(x,\xi)\in Q$ (37.13)

and

(ii) $$|a_{k_0}(\infty,\xi)|>\sum_{k\neq k_0}|a_k(\infty,\xi)|. \tag{37.14}$$

Proof To obtain this result (having in mind Theorem 37.4 and Lemma 37.6) just apply Theorem 24.11 to the operator $b(\xi_0)$, $\xi_0\in S^{n-1}$, and observe that the number k_0 does not depend on $\xi_0\in S^{n-1}$ (since otherwise there exists a point ξ where the second condition is violated). □

Remark 37.11 Observe that $\mathbb{R}^n$ as a vector space with respect the field $\mathbb{Q}$ of rational numbers is infinite dimensional and there exist countable convex rationally independent systems of vectors in $\mathbb{R}^n$. (The convexity of a countable system $h_1,\dots,h_k,\dots$ means that for the convex hull of $\{h_1,\dots,h_k,\dots\}$ all the points h_k, $k=\overline{1,\infty}$, are extreme points, that is there exist linear functionals f_k such that $f_k(h_k) > f_k(h_j)$, $k \neq j$.) The analogue to the statement considered in 37.10 could be obtained for this systems of vectors as well.

37.12 The conditions for the invertibility of the operators $b(\xi)$ can be investigated by a number of methods related to the trajectorial approach (see Section 21).

For example let G be the subgroup of $\mathbb{R}^n$ generated by the vectors $h_1,\ldots,h_m$ (used in formula (37.2)). G is an abelian group and is isomorphic to $\mathbb{Z}^l$ $(l \leq m)$. According to the general construction used in the trajectorial approach (see 21.1), with an operator $b(\xi)$ and every point $x \in \overline{\mathbb{R}}^n$ we can associate the operator

$$b(\xi)_x = \pi_x(b(\xi)) \tag{37.15}$$

in $l^2(\mathbb{Z}^l)$.

Let $\tau_1,\ldots,\tau_l$ be the generators of G (here by the generators of $\mathbb{Z}^l$ we mean the elements of the form $(0,\ldots,0,1,0,\ldots,0)$). Then for every h_k, $k = \overline{1,m}$, we have

$$h_k = \sum_{i=1}^{l} n_k^i \tau_k, \quad n_k^i \in \mathbb{Z} \tag{37.16}$$

and the operator b (given by (37.2)) can be expressed as a finite sum

$$b = \sum a_z T_z \tag{37.17}$$

where $z = (n_1,\ldots,n_l) \in \mathbb{Z}^l$ and $T_z = T_{\tau_1}^{n_1} \circ \ldots \circ T_{\tau_l}^{n_l}$. Now if we fix $(x,\xi) \in Q$ then the operator $b(\xi)_x$ (defined by (37.15)) is given (in view of (37.17) by

$$b(\xi)_x = \sum a_{z,x} U_z \tag{37.18}$$

where

$$(U_z f)(k_1,\ldots,k_l) = f(k_1 - n_1,\ldots,k_l - n_l), \quad f \in l^2(\mathbb{Z}^l) \tag{37.19}$$

and $a_{z,x}$ is the operator of multiplication by the function

$$a_{z,x}(k_1,\ldots,k_n) = a_z(x - k_1\tau_1 - \ldots - k_l\tau_l, \xi) \tag{37.20}$$

(here $a_z(x,\xi)=\sigma(a_z)$ is the symbol of a_z).

The operators of the form (37.18) in a general situation are rather complicated and their properties depend in essence on the behaviour of the coefficients $a_{z,x}(k)$ $(k=(k_1,\ldots,k_l)\in\mathbb{Z}^l)$ at infinity ($k\to\infty$ means $|k|=\sum_{i=1}^{l}|k_i|\to\infty$).

Remark Observe that if $x=\infty$ then

$$a_{z,\infty}(k)=a_z(\infty,\xi)\ .$$

Thus the operator $b(\xi,\infty)$ is the operator with constant coefficients. This operator is Fredholm iff it is invertible, which in turn is equivalent to the condition

$$\sum a_z(\infty,\xi)\lambda^z\neq 0 \tag{37.21}$$

where $\lambda=(\lambda_1,\ldots,\lambda_l)\in\mathbb{T}^l=\hat{\mathbb{Z}}^l$ ($\mathbb{T}^l$ is the l-dimensional torus, that is $\lambda_i\in\mathbb{C}$ and $|\lambda_i|=1$) and $\lambda^z=\lambda_1^{z_1}\ldots\lambda_l^{z_l}$.

The condition (37.21) is equivalent to the condition

$$\inf_{\eta\in\mathbb{R}^n}\sum_k a_k(\infty,\xi)e^{i\langle h_k,\eta\rangle}$$

but in a sense (37.21) is preferable since it does not contain the calculation of the infinum over $\mathbb{R}^n$.

37.13 Bearing in mind the operators $b(\xi)$ considered in Example 37.10 it is reasonable to single out the case when all the coefficients $a_{z,x}(k)$ defined by (37.20) have limits at infinity (when $|k|\to\infty$). This condition will be fulfilled when G is a discrete group (in the topology induced from $\mathbb{R}^n$). Indeed in this case the sequence $x(k)=(x-k_1\tau_1-\ldots-k_l\tau_l)$ $(k=(k_1,\ldots,k_l)\in\mathbb{Z}^l)$ tends to infinity when k tends to infinity and thus

$$\lim_{k\to\infty}a_{z,x}(k)=a_z(\infty,\xi)\ . \tag{37.22}$$

The operator $b(\xi)_x$ given by (37.18) is a multidimensional discrete convolution operator. These operators with coefficients having the above-described behaviour at infinity were investigated by Antonevich [10], Gohberg, Feldman [1], Karapetyants, Samko [2], and Simonenko [1,2].

In particular $b(\xi)_x$ is Fredholm iff

$$\sum a_z(\infty,\xi)\lambda^z \neq 0, \quad \lambda \in \mathbb{T}^l$$

(that is when the condition (37.21) is fulfilled) and if this condition is satisfied then

$$\operatorname{ind} b(\xi)_x = 0 .$$

Thus summarizing all these facts we obtain

Theorem 37.14 *Let the group G generated by the vectors $h_1,\dots,h_m$ (in (37.2)) be discrete (in the topology induced from $\mathbb{R}^n$). Then the operator b is Fredholm iff the following two conditions are satisfied:*

(1) $\sum a_z(\infty,\xi)\lambda^z \neq 0$ *for every* $\xi \in S^{n-1}$ *and* $\lambda \in \mathbb{T}^l = \hat{\mathbb{Z}}^l$;

(2) $\operatorname{Ker} b(\xi)_x = 0$ *for every* $(x,\xi) \in Q$.

Remark 37.15 Recalling the proofs of Theorems 24.4 and 24.5 we note that the statement of Theorem 37.14 in the situation when a group is generated by a convex and rationally independent system of vectors is equivalent to the result stated in Example 37.10; namely, observe that if $\{h_1,\dots,h_m\}$ is a convex and rationally independent system of vectors then the vectors $\{h_2-h_1,\dots,h_m-h_1\}$ generate a group G isomorphic to $\mathbb{Z}^{m-1}$ and discrete (in the topology induced from $\mathbb{R}^n$).

Indeed this group is isomorphic to $\mathbb{Z}^{m-1}$ in view of the rational independence condition (see Example 37.10) and since the system $\{h_1,\dots,h_m\}$ is convex the system $\{0,h_2-h_1,\dots,h_m-h_1\}$ is convex as well, which implies the discreteness of the group G.

The operator b (given by (37.2)) is Fredholm iff the operator

$$b_1 = a_1(x,D) + \sum_{k=2}^{m} a_k(x,D)T_{h_k-h_1}$$

is Fredholm.

For this operator the condition (1) of Theorem 37.14 (in view of the foregoing observations) means that

$$a_1(\infty,\xi)+\sum_{k=2}^{m} a_k(\infty,\xi)\lambda_k \neq 0$$

for any $\lambda_k \in \mathbf{C}, |\lambda_k|=1$, which is equivalent to the condition

$$\sum_{k=1}^{m} a_k(\infty,\xi)\lambda_k \neq 0 \tag{37.23}$$

for any $\lambda_k, |\lambda_k|=1$.

Condition (37.23) in turn (see the proof of Theorem 24.4) implies the satisfaction of condition (37.14).

On the other hand, condition (37.13) is equivalent to condition (2) of Theorem 37.14 (see the proof of Theorem 24.5).

Pseudodifferential operators with affine shifts

37.16 Our next step is the generalization of the group of transformation operators of an argument. Let $\alpha_g:\mathbf{R}^n \to \mathbf{R}^n$, $g \in G$, be a certain group of diffeomorphisms of $\mathbf{R}^n$. Denote by U_g the operators

$$(U_g v)(x) = v\left(\alpha_g^{-1}(x)\right) \tag{37.24}$$

generated by these diffeomorphisms.

In general the operator U_g is not necessarily bounded in $L^2(\mathbf{R}^n)$; in fact it is bounded iff the function $\left|\alpha_g^{-1\prime}(x)\right|^{-\frac{1}{2}}$ ($\left|\alpha_g^{-1\prime}\right|$ is the modulus of the Jacobian of α_g^{-1}) is bounded, and if it is then

$$\left\|U_g\right\|_{L^2} = \sup_{\mathbf{R}^n}\left|\alpha_g^{-1\prime}(x)\right|^{-\frac{1}{2}}.$$

In view of this we shall, along with U_g, consider the operator T_g of the form

$$\left(T_g v\right)(x) = \left|\alpha_g^{-1'}(x)\right|^{\frac{1}{2}} v\left(\alpha_g^{-1}(x)\right) \tag{37.25}$$

which is a unitary operator in $L^2\left(\mathbf{R}^n\right)$.

37.17 Besides the boundness considerations in 37.16 we have to consider also the conditions on α_g under which the operator T_g generates the automorphism $T_g(\cdot)T_g^*$ of the algebra $\mathcal{A}$ of pseudodifferential operators described in 37.1. Evidently the necessary condition on α_g to possess this property is the existence of an $n \times n$ matrix L and a matrix function $A(x)$, $x \in \mathbf{R}^n$, such that

$$\alpha_g'(x) = A(x)L$$

and

$$A(x) = \left[a_{ij}(x)\right]$$

possesses the properties

$$\begin{cases} a_{ij}(x) \to 0 \text{ when } x \to \infty \text{ and } i \neq j \\ \left(a_{ii}(x) - a_{jj}(x)\right) \to 0 \text{ when } x \to \infty . \end{cases}$$

Certainly, a diffeomorphism which is affine for sufficiently large x, that is has the form

$$\alpha_g(x) = Lx - h \left(h \in \mathbf{R}^n\right)$$

(for sufficiently large x), generates the automorphism of $\mathcal{A}$.

Throughout the rest of this section we shall confine ourselves to the consideration of the shifts generated by an affine transformation of an argument, that is

$$\alpha(x) = Lx - h .$$

For these shifts the corresponding automorphism of $\mathcal{A}$ is described by the following

Lemma 37.18 *Let L be a non-degenerate matrix and $h \in \mathbf{R}^n$ be a certain vector. Consider the affine mapping $\alpha_g(x) = Lx - h$ and the operators*

$$\left(U_g v\right)(x)=v\left(L^{-1}(x+h)\right)$$

$$\left(T_g v\right)(x)=|L|^{-\frac{1}{2}} v\left(L^{-1}(x+h)\right)$$

(see (37.24) and (37.25)).

Then for any operator $\mathfrak{a}$ of the form (37.1) the operator

$$\mathfrak{a}_1=U_g \mathfrak{a} U_g^{-1}=T_g \mathfrak{a} T_g^* \tag{37.26}$$

has a form analogous to $\mathfrak{a}$, namely

$$\mathfrak{a}_1=a_1(x, D)=F^{-1} a_1(x, \xi) F \tag{37.27}$$

where $a_1(x,\xi)=a\left(L^{-1}(x+h), L'\xi\right)$ (L' is the transpose of L).

Proof Clearly $U_g \mathfrak{a} U_g^{-1}=T_g \mathfrak{a} T_g^*$ since U_g and T_g differ by a constant multiplier and T_g is unitary. Thus it is enough to consider the operator

$$\mathfrak{a}_1=U_g \mathfrak{a} U_g^{-1}.$$

For any $u \in L^2\left(\mathbb{R}^n\right)$ we have

$$\left[\mathfrak{a}_1 u\right](x)=\frac{1}{(2\pi)^n} \int \mathrm{e}^{\mathrm{i}\left\langle L^{-1}(x+h), \xi\right\rangle} a\left(L^{-1}(x+h), \xi\right) \int \mathrm{e}^{-\mathrm{i}\langle y, \xi\rangle} u(Ly-h)\, \mathrm{d}y\, \mathrm{d}\xi .$$

Substituting the argument $z=Ly-h$ we obtain

$$\left[\mathfrak{a}_1 u\right](x)=\frac{1}{(2\pi)^n} \int \mathrm{e}^{\mathrm{i}\left\langle L^{-1}(x+h), \xi\right\rangle} a\left(L^{-1}(x+h), \xi\right) \int \mathrm{e}^{-\mathrm{i}\left\langle L^{-1}(z+h) \xi\right\rangle} u(z)\left|L^{-1}\right| \mathrm{d}z\, \mathrm{d}\xi$$

$$=\frac{1}{(2\pi)^n} \int \mathrm{e}^{\mathrm{i}\left\langle x, L^{-1'} \xi\right\rangle} \cdot \mathrm{e}^{\mathrm{i}\left\langle L^{-1} h, \xi\right\rangle} a\left(L^{-1}(x+h), \xi\right) \int \mathrm{e}^{-\mathrm{i}\left\langle z, L^{-1'} \xi\right\rangle} \cdot \mathrm{e}^{-\mathrm{i}\left\langle L^{-1} h, \xi\right\rangle} u(z)\left|L^{-1}\right| \mathrm{d}z\, \mathrm{d}\xi. \tag{37.28}$$

In (37.28) substituting the argument $\eta=L^{-1}\xi$ we find

$$(37.28) = \frac{1}{(2\pi)^n} \int e^{i\langle x,\eta\rangle} a\left(L^{-1}(x+h), L'\eta\right) \int e^{-i\langle z,\eta\rangle} u(z)\, dz\, d\eta$$

which proves (37.27).

Evidently $a_1(x,\eta)$ belongs to the same class as $a(x,\eta)$ and

$$\lim_{x\to\infty} a_1(x,\eta) = a_1(\infty,\eta) = a(\infty, L'\eta) .$$

The proof is complete. □

37.19 This result enables us to construct the symbolic calculus for pseudodifferential operators with affine shifts. Let $\{\alpha_g\}_{g\in G}$ be a certain group of affine mappings of $\mathbb{R}^n$, that is every α_g has the form

$$\alpha_g(x) = L_g x - h_g \tag{37.29}$$

where L_g is a non-degenerate matrix and $h_g \in \mathbb{R}^n$.

Recalling (37.24) and (37.25) we introduce the operators U_g and T_g via the formulae

$$\left(U_g u\right)(x) = u\left(L_g^{-1}\left(x+h_g\right)\right) \tag{37.30}$$

$$\left(T_g u\right)(x) = \left|L_g\right|^{-\frac{1}{2}} u\left(L_g^{-1}\left(x+h_g\right)\right) \tag{37.31}$$

and consider the operator

$$\mathcal{b} = \sum_{k=1}^{m} \mathfrak{a}_k T_{g_k} \tag{37.32}$$

where $\mathfrak{a}_k = a_k(x,D) \in \mathcal{A}$ (see (37.1)).

We shall call the subalgebra $\mathcal{B} = C^*\left(\mathcal{A}, T_g\right) \subset L\left(L^2\left(\mathbb{R}^n\right)\right)$ generated by operators of the form (37.32) the *algebra of pseudodifferential operators with affine shifts.*

If the group $\{\alpha_g\}_{g\in G}$ is fixed then we shall call the elements $\mathcal{b}$ of $\mathcal{B}$ *G*-pseudodifferential operators (*G*-PDOs).

37.20 In the space $L^2(Q)$ $\left(Q=\overline{\mathbf{R}}^n \times S^{n-1}\right)$ consider the unitary operator V_g given by

$$\left(V_g f\right)(x,\xi)=\left|\beta_g^{-1'}(x,\xi)\right|^{\frac{1}{2}} f\left(\beta_g^{-1}(x,\xi)\right) \tag{37.33}$$

where

$$\beta_g(x,\xi)=\begin{cases}\left(\alpha_g(x), \dfrac{L_g^{-1'}\xi}{\left\|L_g^{-1'}\xi\right\|}\right), & x\neq\infty \\ \left(\infty, \dfrac{L_g^{-1'}\xi}{\left\|L_g^{-1'}\xi\right\|}\right), & x=\infty ,\end{cases} \tag{37.34}$$

$\left|\beta_g^{-1'}\right|$ is the modulus of the Jacobian of the mapping $\beta_g^{-1}: Q\to Q$, and $L_g^{-1'}$ is the transpose of L_g^{-1}.

In view of the explicit form of α_g (37.29) we can write out β_g^{-1} in explicit form as well:

$$\beta_g^{-1}(x,\xi)=\begin{cases}\left(L_g^{-1}\left(x+h_g\right), \dfrac{L_g'\xi}{\left\|L_g'\xi\right\|}\right), & x\neq\infty \\ \left(\infty, \dfrac{L_g'\xi}{\left\|L_g'\xi\right\|}\right), & x=\infty .\end{cases} \tag{37.35}$$

Remark Clearly since $L^2(Q)\cong L^2\left(\mathbf{R}^n\times S^{n-1}\right)$ we can consider the operator V_g as acting in $L^2\left(\mathbf{R}^n\times S^{n-1}\right)$ and given by the same formula (37.33).

37.21 We define the *symbol* of the G-PDO $\mathcal{b}$ of the form (37.32) to be the operator $\sigma(\mathcal{b})$ acting in $L^2(Q)$ and defined by

$$\sigma(\mathcal{b})=\sum_{k=1}^{m} a_k(x,\xi)V_{g_k} \tag{37.36}$$

where $a_k(x,\xi)$ are the operators of multiplication by the symbols $a_k(x,\xi)=\sigma(\mathcal{a}_k)$ and V_{g_k} are defined by (37.33).

The following theorem shows that we have really obtained the symbolic calculus for these objects (cf. Theorem 37.4).

Theorem 37.22 *Let $\{\alpha_g\}_{g\in G}$ be a certain group of affine mappings of $\mathbb{R}^n$ (see (37.29)) with G being a discrete amenable group. Let $\mathcal{B} = C^*(\mathcal{A}, T_g)$ be the C*-algebra of pseudodifferential operators with affine shifts described in 37.19 and $B = C^*(A, V_g) \subset L(L^2(Q))$ be the C*-algebra generated by the operators of multiplication by the symbols $\sigma(a) \in A$ of the elements of $\mathcal{A}$ and the operators V_g defined by (37.33).*

Then the mapping

$$\sigma : b \to \sigma(b)$$

*given by (37.36) can be extended up to the *-epimorphism*

$$\sigma : \mathcal{B} \to B$$

the kernel of which is the ideal K of compact operators.

In particular the operator b is Fredholm iff its symbol $\sigma(b)$ is invertible.

Proof The scheme of the proof repeats that of Theorem 37.4. We have to establish the isomorphism

$$[\mathcal{B}] \overset{\sigma}{\cong} B \, .$$

Lemma 37.18 means that

$$\sigma\left(T_g a T_g^*\right) = V_g \sigma(a) V_g^* \, .$$

Now to complete the proof (in view of the isomorphism theorem (Corollary 12.17)) it is enough to check that the action $\{\beta_g\}_{g\in G}$ on Q is topologically free.

If we fix α_g $(g \neq e)$ then the set of fixed points of the mapping

$$\alpha_g(x) = L_g x - h_g, \; x \in \mathbb{R}^n$$

is a linear manifold of dimension less than n (or empty), and for the mapping β_g (37.34) a point (x_0, ξ_0) could be fixed only if $\alpha_g(x_0) = x_0$ or $x_0 = \infty$ which, along with the foregoing observation, shows that the group $\{\beta_g\}_{g \in G}$ acts topologically freely. Thus the theorem is proved. □

Remarks 37.23 (1) In general a group G generated by affine (or even linear) transformations of an argument is not necessarily amenable. In the case of $\mathbb{R}^n$ $(n = 1,2)$ the group of motions (that is, generated by rotations and shifts by vectors of $\mathbb{R}^2$) is amenable (in fact even solvable). Thus if the matrices L_g in (37.9) are orthogonal (and $n = 2$) then the group G generated by these mappings is amenable and we can apply Theorem 37.22 for any group of this type.

On the other hand, in $\mathbb{R}^3$ the orthogonal group is not amenable; in particular it contains a free subgroup with two generators (see Greenleaf [1], 1.2.8) (this difference between the orthogonal groups in $\mathbb{R}^n$, $n = 1,2$, and $\mathbb{R}^n$, $n \geq 3$, leads to the Hausdorff-Banach-Tarski paradox (see Hausdorff [1], Banach-Tarski [1])). So if $n = 3$ then we cannot apply Theorem 37.22 automatically even presupposing that all L_g are orthogonal.

The group of all affine transformations of $\mathbb{R}^2$ unfortunately is not amenable either (see Greenleaf [1], the reasoning in 1.2.8), so even in the case of $\mathbb{R}^2$ the theorem cannot be used for arbitrary matrices L_g defining α_g.

It is worth mentioning that for discrete finitely generated groups the criterion for their amenability can be written out in terms of the growth rate of the number of different words containing k letters ($k \to \infty$) and representing the identity of the group (this criterion was found by Grigorchuk [1] (see also Cohen [1]).

(2) In connection with the previous remark we mention that the amenability condition on G is linked with the application of general isomorphism theorems (like Corollary 12.17) which are not true without this assumption (see, for example, Pedersen [1], 7.3.9). But at the moment we do not know whether the statement about the symbolic calculus of pseudodifferential operators with shifts (which is in fact the statement about *concrete* objects related to *concrete* group of automorphisms) is true without the amenability condition.

(3) Observe that in the pseudodifferential difference operator case the argument ξ (of $(x,\xi) \in Q$) is not shifted under the action of the generated homeomorphism of Q and thus the symbols of the operators $\mathcal{b} \in \mathcal{B}$ can be decomposed into the operators $(b(\xi))$ acting in $L^2(\mathbb{R}^n)$.

The corresponding picture for the pseudodifferential operators with affine shifts is in fact much more complicated since here the argument ξ is shifted along with the argument x. Thus the passage to the consideration of the space $L^2(Q) = L^2\left(\overline{\mathbb{R}}^n \times S^{n-1}\right)$ is essential in this situation and this derivation made it possible to build the symbols with 'good' properties.

Further development of this approach will bring us to the general construction of the symbol for G-PDOs (see Sections 40-42).

38. The measures on the sphere ergodic under the induced linear actions

When considering the examples in subsequent sections we shall need to describe the ergodic measures for the mappings of the bundle S^*M of unit cotangent to a manifold M vectors. As an auxiliary problem, we consider the description of the ergodic measures on the sphere with respect to the mappings induced by linear transformations, that is having the form

$$\alpha(x) = \frac{Lx}{\|Lx\|}, \quad x \in S^{n-1}$$

where L is a non-degenerate $n \times n$ matrix. In this section we present the solution of this problem.

We distinguish the complex and real situations (that is, the sphere in $\mathbb{C}^n$ and $\mathbb{R}^n$) though in the following sections for direct applications we shall use only the real case.

38.1 To begin with we recall some facts about measures on the p-dimensional torus invariant with respect to the standard shifts. The main results of this section (Theorems 38.5 and 38.9) will be formulated in terms of such measures.

Let

$$\mathbb{T}^p = \left\{z = \left(z_1, \dots, z_p\right): z_i \in \mathbb{C}, \ |z_i| = 1\right\}$$

be the p-dimensional torus. For any fixed element $\gamma = \left(\gamma_1, \dots, \gamma_p\right) \in \mathbb{T}^p$, we shall call the mapping $\alpha_\gamma: \mathbb{T}^p \to \mathbb{T}^p$ defined by

$$\alpha_\gamma(z) = \gamma z = \left(\gamma_1 z_1, \dots, \gamma_p z_p\right)$$

the *standard shift of the torus* $\mathbb{T}^p$ (generated by γ).

38.2 The ergodic measures (with respect to the standard shift of the torus) were considered in Examples 5.19, 5.20. Since in this section these measures play an essential role we shall investigate them more thoroughly.

These measures depend on the closed subgroup Γ of $\mathbb{T}^p$ generated by γ. The form of Γ depends on the interrelations between the numbers $\gamma_1, \ldots, \gamma_p$. Let

$$\gamma_j = \mathrm{e}^{\mathrm{i}2\pi h_j},\ j = \overline{1,p}\ ,$$

and let m be the maximal number of the rationally independent numbers among $h_1, \ldots, h_p$. (Here, we say that the numbers $\tau_1, \ldots, \tau_s$ are *rationally independent* if the equality

$$q_1\tau_1 + \ldots + q_s\tau_s = q_0\ ,$$

where $q_i \in Q$ are rational, can be satisfied only when all the q_i are zero. This definition differs from 24.9. It will coincide with 24.9 if one considers the system $\{1, \tau_1, \ldots, \tau_s\}$).

In the situation we are considering, the subgroup Γ is a submanifold of $\mathbb{T}^p$ of dimension m and is isomorphic either to $\mathbb{T}^m$ or to $\mathbb{T}^m \times \mathbb{Z}_N$ where N is an integer depending on the type of rational correlations between the h_j.

Example. Let $p = 3$.

(1) If h_1, h_2, h_3 are rationally independent then $\Gamma = \mathbb{T}^3$.

(2) If h_1, h_2 are rationally independent and $h_3 = \dfrac{M}{N}$ (where $\dfrac{M}{N}$ is an irreducible fraction) then $\Gamma = \mathbb{T}^2 \times \mathbb{Z}_N$.

(3) If h_1 is irrational and, for example, $h_2 = 2h_1 + \frac{1}{5}$ and $h_3 = 3h_1$ then $\Gamma = \mathbb{T}^1 \times \mathbb{Z}_5$.

The general rule of how to calculate N is the following (see Weyl [1]). One has to find all relations with integral coefficients $n_1, \ldots, n_p$ of the form

$$n_1h_1 + \ldots + n_ph_p = q$$

where $q \in Q$, and for the rational q so obtained, find the least common denominator. It is equal to N.

On every torus $\mathbb{T}^m$ forming the group Γ the Haar measure is equal to $\frac{1}{N}\,dx$ where dx is the normalized Lebesque measure on $\mathbb{T}^m$.

The conjugacy classes of $\mathbb{T}^p$ with respect to the group Γ are arranged analogous to Γ and on each such class there is defined an analogous measure. We call these measures the *standard measures* on $\mathbb{T}^p$ (related to γ).

We can summarize all these facts in the next

Proposition 38.3 *The ergodic measures (with respect the standard shift of the torus $\mathbb{T}^p$) are the standard measures and only them.*

Now we pass to the main object of the section.

We begin with

The complex case

38.4 Thus we are considering a linear non-degenerate mapping $L: \mathbb{C}^m \to \mathbb{C}^m$ (which we shall identify with its matrix) and the mapping α of the unit sphere $S = \{x: |x| = 1\} \subset \mathbb{C}^m$ given by

$$\alpha(x) = \frac{Lx}{\|Lx\|}, \; x \in S \; . \tag{38.1}$$

Our goal is to describe the ergodic measures with respect to α on S.

Remark In fact the mapping α naturally acts on the space of (real) rays. Any particular norm in $\mathbb{C}^m$ makes it possible to establish a one to one correspondence between the set of rays and the unit sphere (with respect to the norm chosen) and thus to define the mapping α in a more visual way. But the ergodic measures (on the space of rays) do not depend on the norm and thus the choice of any norm will not influence the picture.

We start with the introduction of certain objects and notations.

Take a basis in $\mathbb{C}^m$ with respect to which the matrix L has a normal Jordan form. Let q be the number of different absolute values of the eigenvalues of L. Let us enumerate these absolute values in decreasing order:

$$r_1 > r_2 > \ldots > r_q > 0 \; .$$

We shall denote by $q(k)$ the number of Jordan boxes for which the absolute value of its eigenvalue is equal to r_k.

For every fixed k we enumerate these boxes in decreasing order of the dimension. Thus we obtain a system of Jordan boxes J_{kj}, $1 \le k \le q$, $1 \le j \le q(k)$. We denote the dimension of the box J_{kj} by $\nu(k,j)$, and the corresponding eigenvalue by $\lambda(k,j)$.

For the vectors in the basis in which L has the Jordan form we now obtain an enumeration by three indexes:

$$e(k,j,l),\ 1 \le k \le q,\ 1 \le j \le q(k),\ 1 \le l \le \nu(k,j)\ .$$

We shall denote the coordinates of a vector $x \in \mathbb{C}^m$ in this basis by $x(k,j,l)$.

Bearing in mind these remarks we recall that the ergodic measures do not depend on the particular choice of the norm in $\mathbb{C}^m$. Having this remark in mind we henceforth *assume that the norm of* $x \in \mathbb{C}^m$ *is given by*

$$\|x\| = \sqrt{\sum_{k,j,l} |x(k,j,l)|^2}\ . \tag{38.2}$$

With this norm the vectors $e(k,j,l)$ clearly form an orthonormal basis.

Let M_k be an invariant vector subspace of $\mathbb{C}^m$, with respect to L, generated by the eigenvectors of the matrix L corresponding to the eigenvalues $\lambda(k,j)$, $1 \le j \le q(k)$. Thus M_k is generated by the vectors $e(k,j,1)$, $1 \le j \le q(k)$.

The set

$$S_k = S \cap M_k \tag{38.3}$$

is the unit sphere in the subspace M_k. Its (real) dimension is equal to $2q(k)-1$.

The sphere S_k can be stratified into submanifolds invariant with respect to the mapping α in the following way.

Let us define the mapping $\pi_k : S_k \to \mathbb{R}^{q(k)}$ via the formula

$$\pi_k(x) = (|x(k,1,1)|, |x(k,2,1)|, \ldots, |x(k,q(k),1)|)\ .$$

The image of S_k under the mapping π_k is a part of the $(q(k)-1)$-dimensional sphere, namely the set

$$B_k = \{\xi = \{\xi_i\} \in \mathbb{R}^{q(k)} : \xi_i \geq 0,\ \|\xi\| = 1\}$$

which is the intersection of the unit sphere with the positive cone.

For every $\xi \in B_k$ its inverse image $\pi_k^{-1}(\xi)$ has the natural structure of the torus, the dimension of which is equal to the number of non-zero coordinates of ξ.

On the set S_k the action of α is written as

$$\alpha(x) = (v(k,1)x(k,1,1), v(k,2)x(k,2,1), \ldots, v(k,q(k))x(k,q(k),1)) \tag{38.4}$$

where

$$v(k,j) = \frac{\lambda(k,j)}{|\lambda(k,j)|}, \tag{38.5}$$

which means that on the torus $\mathbb{T}_{k,\xi} = \pi_k^{-1}(\xi)$ the action of α coincides with the standard shift (see 38.1) generated by the element $\gamma_{k,\xi}$ composed from those numbers $v(k,j)$ for which the corresponding coordinates of the vector ξ are non-zero.

We denote by $\Gamma_{k,\xi}$ the subgroup of $\mathbb{T}_{k,\xi}$ generated by $\gamma_{k,\xi}$.

Having all these assumptions and notations in mind we can now state the first principal result of this section.

Theorem 38.5 *A measure μ on the sphere S is ergodic with respect to the mapping $\alpha(x) = \dfrac{Lx}{\|Lx\|}$ iff its support belongs to one of the invariant tori $\mathbb{T}_{k,\xi}$ and coincides with a certain conjugacy class of $\mathbb{T}_{k,\xi}$ with respect to the subgroup $\Gamma_{k,\xi}$; and the measure μ coincides with the standard measure on this class (see 38.2).*

Proof To prove the theorem we need certain information about the behaviour of the sequence of vectors $L^n x$, $n \to \infty$. This information is useful in its own right (for example, for the subtle analysis of the convergence of iterative methods).

Let us introduce a number of auxiliary objects. For a vector $x \in \mathbb{C}^m$, $x \neq 0$, let us define the following integral-valued indexes.

$$k(x) = \min\{k : x(k,j,l) \neq 0\} \tag{38.6}$$

$$v_j(x) = \max\{l : x(k(x),j,l) \neq 0\} \tag{38.7}$$

$$\nu(x) = \max\left\{\nu_1(x), \nu_2(x), \dots, \nu_{q(k(x))}(x)\right\} . \tag{38.8}$$

With a vector x we shall associate three new vectors y, z, u, that is we shall define the three mappings

$$\varphi(x) = y,\ \psi(x) = z,\ \eta(x) = u$$

where

$$y(k,j,l) = \begin{cases} v(k,j)^{1-\nu(x)} x(k,j,\nu(x)), & \text{if } k = k(x), l = 1 \\ 0, & \text{if } k \neq k(x) \text{ or } l \neq 1 , \end{cases} \tag{38.9}$$

(where $v(k,j)$ is given by (38.5));

$$z(k,j,l) = \begin{cases} x(k,j,l), & k = k(x) \\ 0, & k \neq k(x) ; \end{cases} \tag{38.10}$$

$$u = x - z . \tag{38.11}$$

To describe the behaviour of $L^n x$ we shall observe individually the behaviour of the norm $\|L^n x\|$ and the direction $\dfrac{L^n x}{\|L^n x\|} = \alpha^n(x)$.

The point $\tilde{y} = \dfrac{\varphi(x)}{\|\varphi(x)\|}$ belongs to the sphere $S_{k(x)}$ (see (38.3)). The trajectory of this point $\left\{\alpha^n(\tilde{y})\right\}$ belongs to the invariant torus $\mathbb{T}_{k(x),\xi}$ where $\xi = \pi_{k(x)}(\tilde{y})$. The closure of this trajectory coincides with a certain conjugacy class of $\mathbb{T}_{k(x),\xi}$ with respect to the subgroup $\Gamma_{k(x),\xi}$. Let us denote this class by $\theta(x)$.

For number sequences $\{a_n\}$ and $\{b_n\}$ we shall write

$$a_n \sim b_n \text{ if } a_n = O(b_n) \text{ and } b_n = O(a_n) , \tag{38.12}$$

and we shall write

$$a_n \approx b_n \text{ if } \lim_{n\to\infty} \frac{a_n}{b_n} = 1 . \tag{38.13}$$

As the main technical step of the proof of the theorem we shall prove first the next

Theorem 38.6 *The dynamics of the mappings $x \to L^n x$, $n \to \infty$, is described via the following statements.*

I. $$\|L^n x\| \approx \frac{r_{k(x)}^{-\nu(x)+1}\|y\|}{(\nu(x)-1)!} \cdot r_{k(x)}^n \cdot n^{\nu(x)-1} \tag{38.14}$$

(r_k is defined in 38.4).

In particular for a general position vector we have

$$\|L^n x\| \sim r_1^n n^{\nu_1}. \tag{38.15}$$

II. *The half trajectory $\alpha^n(x)$, $n = 1, 2, \ldots$, of a point x tends to the half trajectory of the point $\tilde{y} = \dfrac{\varphi(x)}{\|\varphi(x)\|}$ (φ is given by (38.9)) and the limit set $\Omega(x)$ coincides with the closure of the half trajectory of the point $\tilde{y}$, that is with the conjugacy class $\theta(x)$ on the torus $\mathbb{T}_{k(x),\xi}$, where $\xi = \pi_{k(x)}(\tilde{y})$.*

The degree of convergence is estimated as

$$\|\alpha^n(x) - \alpha^n(\tilde{y})\| \sim \begin{cases} 0, & x = y = z \\ \left[\dfrac{r_{k(u)}}{r_{k(x)}}\right]^n \cdot n^{\nu(u)-\nu(x)}, & x \neq z, y = z \\ \dfrac{1}{n}, & y \neq z. \end{cases} \tag{38.16}$$

Proof As is well known (see for example Gantmaher [1], Chapter V, §3) the nth degree of the Jordan box J of dimension ν with the eigenvalue λ has the form (for $n > \nu$)

$$J^n = \begin{bmatrix} \lambda^n & n\lambda^{n-1} & C_n^2 \lambda^{n-2} & \ldots & C_n^{\nu-1}\lambda^{n-\nu+1} \\ 0 & \lambda^n & n\lambda^{n-1} & \ldots & C_n^{\nu-2}\lambda^{n-\nu} \\ \ldots & \ldots & \ldots & \ldots & \ldots \\ 0 & 0 & 0 & \ldots & \lambda^n \end{bmatrix}.$$

Thus the coordinates of the vector $x^n = L^n x$ are expressed by the formula

$$x^n(k,j,l) = \sum_{i=l}^{\nu(k,j)} \lambda(k,j)^{n-i+l} C_n^{i-l} x(k,j,i) \,. \tag{38.17}$$

Since

$$C_n^{i-l} \approx \frac{n^{i-l}}{(i-l)!} \tag{38.18}$$

it follows that the dominating term in the sum (38.17) for the fixed (k,j,l) (that is, having the maximal growth degree (or the minimal decrease degree)) is the non-zero term having the maximal index i.

Therefore

$$\left|x^n(k,j,l)\right| \approx \frac{r_k^{n-p+l} \cdot n^{p-l} \cdot |x(k,j,p)|}{(p-l)!} \tag{38.19}$$

where

$$p = p(k,j) = \max\{i : x(k,j,i) \neq 0\} \,.$$

Consequently the dominating coordinate among the coordinates corresponding to the Jordan box $J_{k,j}$ is the coordinate $x^n(k,j,1)$; and among all the coordinates of the vector x^n the dominating coordinates are those such that $k = k(x)$, $l = 1$ and j is determined by $\nu_j(x) = \nu(x)$.

The dominating coordinates determine the behaviour of $\left\|L^n x\right\|$ so we conclude that

$$\begin{aligned}\left\|L^n x\right\|^2 &\approx \sum_j \left|x^n(k(x),j,1)\right|^2 \\ &\approx \left(r_{k(x)}^{n-\nu(x)+1} C_n^{\nu(x)-1}\right)^2 \sum_j |x(k(x),j,\nu(x))|^2 \\ &= \left(r_{k(x)}^{n-\nu(x)+1} C_n^{\nu(x)-1}\right)^2 \|y\|^2 \,,\end{aligned}$$

which, along with (38.18), implies (38.14).

Analyzing the form of the remaining terms in (38.17) we obtain

$$\left\|L^n x\right\| = r_{k(x)}^{n-\nu(x)+1} C_n^{\nu(x)-1} \|y\| (1+\delta_n) \tag{38.20}$$

where $\delta_n = O\left(\frac{1}{n}\right)$. Equation (38.20) in turn implies

$$\left\|L^n x\right\|^{-1} = \left(r_{k(x)}^{n-\nu(x)+1} C_n^{\nu(x)-1} \|y\|\right)^{-1} (1+\varepsilon_n) \tag{38.21}$$

where $\varepsilon_n = O\left(\frac{1}{n}\right)$.

Now let us pass to the verification of II, namely to the checking of (38.16).

Observe that the vectors $L^n z$ and $L^n u$ (see (38.10), (38.11) belong to mutually complementing subspaces and therefore

$$\left\| \frac{L^n x}{\|L^n x\|} - \frac{L^n y}{\|L^n y\|} \right\| \sim \left\| \frac{L^n z}{\|L^n x\|} - \frac{L^n y}{\|L^n y\|} \right\| + \frac{\|L^n u\|}{\|L^n x\|} .$$

In view of the estimate (38.14) we have

$$\frac{\|L^n u\|}{\|L^n x\|} \sim \left[\frac{r_{k(u)}}{r_{k(x)}} \right]^n \frac{n^{\nu(u)}}{n^{\nu(x)}} . \tag{38.22}$$

Now let us look at the norm of the vector

$$s^n = \frac{L^n z}{\|L^n x\|} - \frac{L^n y}{\|L^n y\|} .$$

The picture here depends on the interrelation between x, y and z.

Let $y = z$. Noticing that $\|L^n y\| = r_{k(x)}^n \|y\|$ we obtain

$$\|s^n\| = \|L^n y\| \left(\frac{1}{\|L^n x\|} - \frac{1}{\|L^n y\|} \right) = \frac{\|L^n y\| - \|L^n x\|}{\|L^n x\|} \sim \frac{\|L^n u\|}{\|L^n x\|} .$$

Thus in view of (38.22) the estimate (38.16) for the case $y = z$ is true.

Now let $y \neq z$, that is $\nu(x) > 1$. Since, for the vector $y^n = L^n y$ we have $y^n(k(x), j, 2) = 0$, it follows (by using (38.14) and (38.17)) that

$$s^n(k(x), j, 2) \sim \frac{1}{n} \text{ for } j \text{ satisfying } \nu_j(x) = \nu(x) .$$

Thus the degree of convergence to zero of the norm $\|s^n\|$ cannot be higher than $\frac{1}{n}$.

For the indexes (k,j,l) for which the corresponding coordinate $y(k,j,l)$ is zero the upper estimate for $|s^n(k,j,l)|$ via $\frac{1}{n}$ comes out directly from (38.17).

Thus to finish the proof it is enough to show that for the remaining coordinates (that is, for the coordinates with the indexes (k,j,l) satisfying $k = k(x)$, $l = 1$ and j is such that $\nu_j(x) = \nu(x)$) the upper estimate via $\frac{1}{n}$ is also true.

Observe that (38.17) implies

$$|s^n(k(x), j, 1)| \leq \sum_{i=1}^{\nu(x)-1} \left|\lambda(k(x), j)^{n-i+1} C_n^{i-1} x(k(x), j, i)\right|$$

$$+ \left| \frac{\lambda(k(x), j)^{n-\nu(x)+1} C_n^{\nu(x)-1} x(k(x), j, \nu(x))}{\|L^n x\|} - \frac{\lambda(k(x), j)^n \nu(k(x), j)^{-\nu(x)+1} x(k(x), j, \nu(x))}{\|L^n y\|} \right| \tag{38.23}$$

(where $\nu(k,j)$ is given by (38.5)). In view of (38.14) for the first term in (38.23) we obtain the upper estimate via $\frac{1}{n}$.

Now applying (38.21) to the second term in (38.23) we conclude that

$$\left| \frac{\lambda(k(x), j)^{n-\nu(x)+1} C_n^{\nu(x)-1} x(k(x), j, \nu(x))}{\|L^n x\|} - \frac{\lambda(k(x), j)^n \nu(k(x), j)^{\nu(x)-1} x(k(x), j, \nu(x))}{\|L^n y\|} \right|$$

$$= \varepsilon_n \frac{|x(k(x), j, \nu(x))|}{\|y\|} = O\left(\frac{1}{n}\right),$$

and the proof is complete. □

Remark 38.7 The theorem just proved in particular contains the conditions under which the sequence of directions $\frac{L^n x}{\|L^n x\|}$ converges (that is $\Omega(x)$ consists of a single point). This convergence takes place iff the point $\tilde{y}$ is a fixed point of α and the last property is valid iff all the eigenvalues $\lambda(k(x), j)$ corresponding to the indexes j satisfying $\nu_j(x) = \nu(x)$ are positive.

Completion of the proof of Theorem 38.5 In view of Theorem 38.6, Theorem 38.5 can be proved by a general type of argument, namely statement II of Theorem 38.6 and Lemma 5.2 and Corollary 5.5 means that the support of every ergodic measure belongs to some torus $\mathbb{T}_{k(x),\xi}$ and as mentioned in 38.4 the action α on $\mathbb{T}_{k(x),\xi}$ coincides with the standard shift; and the ergodic measures for the standard shift are written out in Proposition 38.3.

The proof is complete. □

Now we proceed to

The real case.

38.8 Thus here we are considering a linear non-degenerate mapping $L: \mathbb{R}^m \to \mathbb{R}^m$ (which we are identifying with its matrix) and the induced mapping of the unit sphere $S^{m-1} \subset \mathbb{R}^m$ given by

$$\beta(x) = \frac{Lx}{\|Lx\|}, \; x \in S^{m-1}. \tag{38.24}$$

The description of the ergodic measures with respect to α here differs in certain details from the description of such measures in the complex case.

One can obviously identify $\mathbb{R}^m$ with a subspace of $\mathbb{C}^m$ and define the operator L in $\mathbb{C}^m$ by the same matrix. With this identification we have $S^{m-1} \subset S$ (S is the unit sphere in $\mathbb{C}^m$) and β is the restriction of the mapping α (defined by (38.1)) onto S^{m-1}. This means in particular that the ergodic measures for β coincide with those for α having supports in S^{m-1}. The description of these measures can be presented in the following way.

Let us enumerate, as in the complex case, the absolute values of the eigenvalues of L and the (real) Jordan boxes of L. Let us choose the basis in $\mathbb{R}^m$ in such a way that the operator L is written in this basis as a block matrix in which to the real eigenvalues there correspond the ordinary Jordan boxes and to the pairs of complex conjugate eigenvalues the blocks of the form

$$\begin{bmatrix} U_{r,\varphi} & E & 0 & \dots & 0 \\ 0 & U_{r,\varphi} & E & \dots & 0 \\ \dots & \dots & \dots & \dots & \dots \\ 0 & 0 & 0 & \dots & U_{r,\varphi} \end{bmatrix} \tag{38.25}$$

where

$$U_{r,\varphi} = \begin{bmatrix} r\cos\varphi & -r\sin\varphi \\ r\sin\varphi & r\cos\varphi \end{bmatrix}; \ E = \begin{bmatrix} 1 & 0 \\ 0 & 1 \end{bmatrix}.$$

Let $\chi(k)$ be the number of Jordan boxes corresponding to the real eigenvalues such that $|\lambda| = r_k$, and $e_1, \dots, e_{\chi(k)}$ be the corresponding eigenvectors.

We denote by $\zeta(k)$ the number of blocks of the form (38.25) corresponding to pairs of complex conjugate eigenvalues the absolute values of which are equal to r_k.

We denote

$$e^+_{\chi(k)+1}, e^+_{\chi(k)+2}, \dots, e^+_{\chi(k)+\zeta(k)}$$

the first basis vectors in each such block and by

$$e^-_{\chi(k)+1}, e^-_{\chi(k)+2}, \dots, e^-_{\chi(k)+\zeta(k)}$$

the second basis vectors (thus obtaining the basis corresponding to the block).

We denote by M'_k the subspace of $\mathbb{R}^m$ generated by the vectors

$$e_1, \dots, e_{\chi(k)}, e^\pm_{\chi(k)+1}, \dots, e^\pm_{\chi(k)+\zeta(k)}.$$

Note that the real dimension of this subspace is equal to $q(k)$ ($q(k)$ was defined in 38.4).

We denote by S'_k the set

$$S'_k = S^{m-1} \cap M'_k .$$

Now let us define the mapping

$$\pi'_k : S'_k \to \mathbb{R}^{\nu} ,$$

where $\nu = \chi(k) + \zeta(k)$, by the formula

$$\pi'_k(x) = \left\{ |x(1)|, \ldots, |x(\chi(k))|, \left| x^{+}(\chi(k)+1) + \mathrm{i}\, x^{-}(\chi(k)+1) \right|, \ldots, \left| x^{+}(\nu) + \mathrm{i}\, x^{-}(\nu) \right| \right\}$$

(cf. the definition of the mapping π_k in 38.4).

The inverse image of a point ξ under this mapping has the form of the product

$$\left(\Pi S^0\right) \cdot \left(\Pi S^1\right)$$

(S^0 is the 0-dimensional sphere (two points), S^1 is the 1-dimensional sphere) where the number of factors S^0 is equal to the number of non-zero coordinates of ξ with the numbers less or equal to $\chi(k)$, and the number of factors S^1 is equal to the number of non-zero coordinates among the remaining coordinates of ξ.

If we consider S^0 as a group ($= \{+1, -1\}$) then the product

$$G_{pq} = \left(S^0\right)^q \left(S^1\right)^p$$

is a group (isomorphic to $\mathbb{T}^p \times \left\{\text{the group of order } 2^{\chi(k)}\right\}$).

In the description of the ergodic measures in the real case the groups G_{pq} play the same role as the torus $\mathbb{T}^p$ in the complex case; namely, any element γ of G_{pq} generates a certain set of *standard ergodic measures* on G_{pq} (cf. Proposition 38.3).

All of the foregoing discussion and Theorem 38.5 allow us to write out the desired result for the real case.

Theorem 38.9 *A measure* μ *on the sphere* $S^{m-1} \subset \mathbb{R}^m$ *is ergodic with respect to the mapping* $\beta(x) = \dfrac{Lx}{\|Lx\|}$ *(*L *is a non-degenerate linear mapping in* $\mathbb{R}^m$*) iff its support belongs to a certain submanifold having the structure of the group* G_{pq} *(defined above), and the measure* μ *coincides with the standard measure on this group.*

Example 38.10 As an illustration of Theorem 38.9 we present here an example where the ergodic measures are written out in a very simple way. Let L be a diagonal matrix:

$$L = \begin{bmatrix} \lambda_1 & & 0 \\ & \ddots & \\ 0 & & \lambda_m \end{bmatrix}$$

and $|\lambda_i| \neq |\lambda_j|$, $i \neq j$. In this case the set S'_k consists of two points

$$e^{\pm}(k) = (0,\ldots,\pm 1, 0,\ldots,0)$$

where ± 1 is situated at the kth position.

If $\lambda_k > 0$ then there exist two ergodic measures $\mu_k^+ = \delta(e^+(k))$, $\mu_k^- = \delta(e^-(k))$.

If $\lambda_k < 0$ there exists one ergodic measure $\mu_k(e^+(k)) = \mu_k(e^-(k)) = \frac{1}{2}$.

All the ergodic measures can be written out in this way.

39. *G*-PDOs on $\mathbb{R}^n$ with 'linear' shifts. Examples.

In this section we consider a number of examples of the calculation of the Fredholmity conditions for concrete *G*-PDOs on $\mathbb{R}^n$ with shifts being generated by linear and affine transformations of an argument. The results of the preceding sections and Theorem 17.3 enable us to obtain the explicit form of the Fredholmity conditions.

We begin with a consideration of the operators of the form

$$\mathcal{b} = a_0(x,D) + a_1(x,D)U \tag{39.1}$$

where U is the shift operator generated by the affine (invertible) mapping

$$\alpha(x) = Lx - h$$

that is

$$(Uf)(x) = f(\alpha^{-1}(x)) = f(L^{-1}(x+h)) . \tag{39.2}$$

Remark We mention some 'ideological' differences between the operator U and the operator T (37.31) which, if one uses the notation α for the mapping, has the form

$$(Tf)(x) = \left|\alpha^{-1}(x)\right|^{\frac{1}{2}} f\left(\alpha^{-1}(x)\right).$$

The factor $\left|\alpha^{-1\prime}(x)\right|^{\frac{1}{2}}$ (with differing U and T) makes the operator T unitary (in $L^2(\mathbb{R}^n)$) which is quite reasonable (bearing in mind the ideas developed in the first part of the monograph) if one fixes the functional space (here $L^2(\mathbb{R}^n)$). On the other hand, if we consider the formula (39.2) and the corresponding operator U as being in different functional spaces (for example in $L^2(\mathbb{R}^n, \mu)$ with different measures μ) then the factor depends on the choice of measure μ. Moreover, if we consider the operator U as being in a Sobolev space and the mapping α not necessarily being affine then in general there is no factor transforming this operator into a unitary operator.

Therefore in the general situation, if we wish to investigate an operator in different spaces it is more natural to write the shift operator in the form (39.2) without the factor $\left|\alpha^{-1\prime}\right|^{\frac{1}{2}}$ and the singling out of this factor is related to the method of investigation and depends on the functional space being considered.

Returning to the consideration of the operator (39.1) in $L^2(\mathbb{R}^n)$ we recall (see Theorem 37.22) that its symbol is an operator in $L^2\left(\overline{\mathbb{R}}^n \times S^{n-1}\right)$ of the form

$$\sigma(\mathcal{b}) = b = a_0(x,\xi) + a_1(x,\xi)W \tag{39.3}$$

where

$$(W\varphi)(x,\xi) = |L|^{\frac{1}{2}} \left|\beta^{-1\prime}(x,\xi)\right|^{\frac{1}{2}} \varphi\left(\beta^{-1}(x,\xi)\right) \tag{39.4}$$

and where β and β^{-1} are described in (37.34), (37.35) and consequently

$$W = |L|^{\frac{1}{2}} V$$

where V is given by (37.33) (with $\beta = \beta_g$).

We start the presentation of the examples (bearing in mind that our method of investigation is related to the dynamical properties of the shifts) with the description of the set $\Omega(\alpha)$ of non-wandering points of the mapping $\alpha(x) = Lx - h$.

39.1 The form of the set $\Omega(\alpha)$ of non-wandering points of the mapping $\alpha(x): Lx-h$ depends in particular on the properties of the fixed points of the mapping α, that is on the solutions of the equation

$$Lx-h=x,\ x\in\mathbb{R}^n. \tag{39.5}$$

The possible situations here are the following.

I. $\det(L-I)\neq 0$, that is all the eigenvalues of L are not equal to 1. In this case there exists a unique fixed point

$$x_0=(L-I)^{-1}h$$

and under the transformation of an argument

$$x'=x-x_0 \tag{39.6}$$

the mapping α is transformed into the mapping

$$\tilde{\alpha}(x')=Lx'. \tag{39.7}$$

Indeed, denoting

$$y'=y-x_0,\ \ x'=x-x_0$$

we rewrite the equation

$$y=\alpha(x)$$

as

$$y'+x_0=L(x'+x_0)-h,$$

which means (recalling (39.5)) that

$$y'=Lx'+[(L-I)x_0-h]=Lx'$$

and consequently (39.7) is true.

Thus the case $\det(L-I) \neq 0$ in fact can be reduced to the case $h=0$.

So let us consider the picture of the non-wandering points under this assumption. That is we look at the mapping (39.7) and give, in terms of it, the description of the properties of

$$\alpha(x) = Lx - h,\ x \in \overline{\mathbb{R}}^n \tag{39.8}$$

with the singled-out fixed point

$$x_0 = (L-I)^{-1}h\ . \tag{39.9}$$

The simple 'basic' pictures are drawn in I(a)-(c), and the general case in I(d).

I(a). Let all the absolute values of the eigenvalues of L be not equal to 1. In this case in view of Theorem 38.6 the half-trajectory of any point x tends to x_0 or to ∞ and the set $\Omega(\alpha)$ consists of two fixed points x_0 and ∞.

I(b). Let there exist only one eigenvalue λ with $|\lambda|=1$, namely $\lambda=-1$, and to this eigenvalue let there correspond k linearly independent eigenvectors.

Passing to the origin x_0 (see (39.6), (39.7) we denote by N_{-1} the linear subspace generated by these eigenvectors.

All the points of the set $N_{-1} \setminus \{x_0\}$ are periodic of period 2 and the half trajectories of the remaining points tend to one of the fixed points x_0 or ∞. Thus in this situation we have

$$\Omega(\alpha) = N_{-1} \cup \{\infty\}\ .$$

I(c). Let L have a number of eigenvalues λ such that $|\lambda|=1$ and suppose that all of them are not equal to -1. The dynamics of the mapping α generated by this L were in fact described in 38.8.

Let us choose a basis in which the matrix L is written in block-diagonal form with blocks of the form (38.25). Corresponding to $r=|\lambda|=1$ take the two first basis vectors corresponding to the block.

Passing to the origin x_0 via $N_{|1|}$ we denote the subspace by these two basis vectors. If we set the inner product in $N_{|1|}$ such that the chosen basis will consist of orthonormal vectors then the spheres (with respect to the induced norm) will be invariant under α. These spheres in turn will be stratified into tori of various dimensions (recall 38.8) and moreover on each such torus the action of α coincides with the standard shift (see 38.1) and every such torus belongs to $\Omega(\alpha)$.

For the points $x \overline{\in} N_{|1|}$ we have $\alpha^n(x) \to x_0$ or $\alpha^n(x) \to \infty$. Thus in this case

$$\Omega(\alpha) = N_{|1|} \cup \{\infty\} .$$

I(d). In general when the operator L possesses non-trivial invariant subspaces N_{-1} and $N_{|1|}$ (see I(b) and I(c)) generated by the corresponding eigenvectors, the argument just presented, along with Theorem 38.6 and the argument used in 38.8, shows that

$$\Omega(\alpha) = \left[N_{-1} + N_{|1|}\right] \cup \{\infty\}$$

and here the set $\left[N_{-1} + N_{|1|}\right] \setminus N_{-1}$ is stratified into pairs of tori mapped into each other under α and on each of these tori the action of α^2 coincides with the standard shift.

II. Now let $\det(L - I) = 0$.

II(a). Suppose that $h \in \mathrm{Im}(L - I)$. In this case there exists a point x_0 such that

$$Lx_0 - h = x_0 ,$$

and as in the situation considered in I the transformation of the argument

$$x' = x - x_0$$

reduces the problem to the investigation of the linear mapping

$$y' = Lx' .$$

Analogously to case I, here the set $\Omega(\alpha)$ coincides with the union of $\{\infty\}$, $\{x_0\}$ and the subspace generated by the eigenvectors corresponding to the eigenvalues having modulus equal to 1. As in I(d), $\Omega(\alpha)$ is stratified into tori and invariant pairs of tori where the action of α or α^2 coincides with the standard shift.

II(b). Now suppose that $h \overline{\in} \mathrm{Im}(L - I)$. In this situation there exists a linear functions φ such that

$$\varphi(\mathrm{Im}(L - I)) = 0 \quad \text{and} \quad \varphi(h) \neq 0 .$$

For any fixed x set

$$x^{(1)} = \alpha(x) = (L-I)x + (x-h)$$

and

$$x^{(n)} = \alpha\left(x^{(n-1)}\right) = (L-I)x^{(n-1)} + \left(x^{(n-1)} - h\right).$$

Now for this φ we have

$$\varphi\left(x^{(n)}\right) = \varphi\left[(L-I)x^{(n-1)}\right] + \varphi\left(x^{(n-1)} - h\right)$$

$$= \varphi\left(x^{(n-1)} - h\right) = \varphi(x) - \varphi(h) \xrightarrow[n\to\infty]{} \infty ,$$

which means that $x^{(n)} \to \infty$ and thus in this situation

$$\Omega(\alpha) = \{\infty\} .$$

39.2 By applying these results on the properties of the action of $\alpha: \overline{\mathbb{R}}^n \to \overline{\mathbb{R}}^n$, along with the description of the ergodic measures on tori and the results of the previous Section 38, it is not difficult to obtain a total description of all the ergodic measures on $\overline{\mathbb{R}}^n \times S^{n-1}$ with respect to the mappings β_g given by (37.34).

In view of the awkwardness of this description we shall confine ourselves to the consideration of a number of concrete examples, and we shall discover that even in the case $n=2$ $\left(n = \dim\left(\mathbb{R}^n\right)\right)$ and when the matrix L is diagonal we can come across a variety of qualitatively different situations (see Example 39.3).

Remark Note meanwhile that in the case $n=1$ the description of the set $\Omega(\alpha)$ for the mapping

$$\alpha(x) = \lambda x - h, \ x \in \overline{\mathbb{R}}$$

is elementary and is as follows:

(1) $|\lambda| \neq 1$.

$\Omega(\alpha)$ consists of two fixed points $\dfrac{h}{\lambda - 1}$ and ∞.

(2) $\lambda = 1,\ h = 0.$

All the points of $\overline{\mathbb{R}}$ are fixed.

$\Omega(\alpha) = \overline{\mathbb{R}}$.

(3) $\lambda = 1,\ h \neq 0.$

$\Omega(\alpha)$ coincides with only one fixed point ∞.

(4) $\lambda = -1.$

The fixed points here are $\left(-\frac{h}{2}\right)$ and ∞. All the other points are periodic of period 2.

$\Omega(\alpha) = \overline{\mathbb{R}}$.

Example 39.3 Now let us consider the operator $\mathcal{b}$ of the form (39.1) in the space $L^2\left(\mathbb{R}^2\right)$ with U given by (39.2) where L is the diagonal matrix

$$L = \begin{bmatrix} \lambda_1 & 0 \\ 0 & \lambda_2 \end{bmatrix}. \tag{39.10}$$

Since the symbol $\sigma(\mathcal{b})$ has the form (39.3) Theorem 37.22, along with the results of Section 38 and 39.1 (which gives us the possibility of writing the ergodic measures with respect to the mapping β, given by (37.34) and Theorem 17.3 lead to the following total description of the Fredholmity conditions of $\mathcal{b}$.

Let $\mathcal{b}$ be the above-mentioned operator with L given by (39.10) and $h = (h_1, h_2)$. The operator $\mathcal{b}$ is Fredholm iff the following conditions (depending on the interrelations between λ_k and h_k) are satisfied.

(1) $\lambda_1 = \lambda_2 = 1,\ h_1 = h_2 = 0$.

The condition:

$a_0(x,\xi) + a_1(x,\xi) \neq 0$ for any $(x,\xi) \in Q$

$\left(Q = \overline{\mathbb{R}}^2 \times S^1\right)$.

(2) $\lambda_1 = \lambda_2 = 1,\ h \neq (0,0)$.

The condition:

(2a) $a_0(x,\xi) \neq 0$, for any $(x,\xi) \in Q$

and

$|a_0(\infty,\xi)| > |a_1(\infty,\xi)|$, for every $\xi \in S^1$;

or

(2b) $a_1(x,\xi) \neq 0$, for any $(x,\xi) \in Q$

and

$|a_0(\infty,\xi)| < |a_1(\infty,\xi)|$, for every $\xi \in S^1$.

(3) $\lambda_1 = 1,\ \lambda_2 = -1,\ h_1 = 0$ or $\lambda_1 = \lambda_2 = -1$.

The condition:

$$a_0(x,\xi)a_0\left(\alpha(x), L^{-1}\xi\right) \neq a_1(x,\xi)a_1\left(\alpha(x), L^{-1}\xi\right), \text{ for any } (x,\xi) \in Q .$$

(4) $\lambda_1 = 1,\ \lambda_2 > 0$ and $\lambda_2 \neq 1;\ h_1 = 0$.

The condition:

(4a) $a_0(x,\xi) \neq 0$, for any $(x,\xi) \in Q$

and

$$\left|a_0\left(x, \pm e_k\right)\right| > \left|a_1\left(x, \pm e_k\right)\right| \lambda_2^{\frac{1}{2}},\ k = 1,2$$

for $x = \infty$ and $x = (x_1, 0)$, with $e_1 = (1,0)$ and $e_2 = (0,1)$;

or

(4b) $a_1(x,\xi) \neq 0$, for any $(x,\xi) \in Q$

and

$$\left|a_0\left(x, \pm e_k\right)\right| < \left|a_1\left(x, \pm e_k\right)\right| \lambda_2^{\frac{1}{2}},\ k = 1,2$$

for $x = \infty$ and $x = (x_1, 0)$.

(5) $\lambda_1 = 1,\ \lambda_2 < 0$ and $\lambda_2 \neq -1;\ h_1 = 0$

The condition:

(5a) $a_0(x,\xi) \neq 0$, for any $(x,\xi) \in Q$

and

$$\left|a_0(x,\pm e_1)\right| > \left|a_1(x,\pm e_1)\right| \left|\lambda_2\right|^{\frac{1}{2}},$$

$$\left|a_0(x,e_2)a_0(x,-e_2)\right| > \left|a_1(x,e_2)a_1(x,-e_2)\right| \left|\lambda_2\right|$$

for $x=\infty$ and $x=(x_1,0)$;

or

(5b) $a_1(x,\xi)\neq 0$, for any $(x,\xi)\in Q$

and

$$\left|a_0(x,\pm e_1)\right| < \left|a_1(x,\pm e_1)\right| \left|\lambda_2\right|^{\frac{1}{2}},$$

$$\left|a_0(x,e_2)a_0(x,-e_2)\right| < \left|a_1(x,e_2)a_1(x,-e_2)\right| \left|\lambda_2\right|$$

for $x=\infty$ and $x=(x_1,0)$.

(6) $\lambda_1=1,\ \lambda_2=-1,\ h_1\neq 0$.

The condition:

(6a) $a_0(x,\xi)\neq 0$, for any $(x,\xi)\in Q$

and

$$\left|a_0(\infty,\xi_1,\xi_2)a_0(\infty,\xi_1,-\xi_2)\right| > \left|a_1(\infty,\xi_1,\xi_2)a_1(\infty,\xi_1,-\xi_2)\right|,$$

or

(6b) $a_1(x,\xi)\neq 0$, for any $(x,\xi)\in Q$

and

$$\left|a_0(\infty,\xi_1,\xi_2)a_0(\infty,\xi_1,-\xi_2)\right| < \left|a_1(\infty,\xi_1,\xi_2)a_1(\infty,\xi_1,-\xi_2)\right|.$$

(7) $\lambda_1=1,\ \lambda_2>0,\ \lambda_2\neq 1,\ h_1\neq 0$.

The condition:

(7a) $a_0(x,\xi)\neq 0$, for any $(x,\xi)\in Q$

and

$$\left|a_0(\infty,\pm e_k)\right| > \left|a_1(\infty,\pm e_k)\right|\lambda_2^{\frac{1}{2}},\ k=1,2;$$

or

(7b) $a_1(x,\xi)\neq 0$, for any $(x,\xi)\in Q$

and

$$\left|a_0(\infty,\pm e_k)\right| < \left|a_1(\infty,\pm e_k)\right|\lambda_2^{\frac{1}{2}},\ k=1,2\ .$$

(8) $\lambda_1 = 1,\ \lambda_2 < 0,\ \lambda_2 \neq -1,\ h_1 \neq 0.$

The condition:

(8a) $a_0(x,\xi) \neq 0$, for any $(x,\xi) \in Q$

and

$$\left|a_0(\infty,\pm e_1)\right| > \left|a_1(\infty,\pm e_1)\right| |\lambda_2|^{\frac{1}{2}},$$

$$\left|a_0(x,e_2)a_0(x,-e_2)\right| > \left|a_1(x,e_2)a_1(x,-e_2)\right| |\lambda_2|\ ;$$

or

(8b) $a_1(x,\xi) \neq 0$, for any $(x,\xi) \in Q$

and

$$\left|a_0(\infty,\pm e_1)\right| < \left|a_1(\infty,\pm e_1)\right| |\lambda_2|^{\frac{1}{2}},$$

$$\left|a_0(x,e_2)a_0(x,-e_2)\right| < \left|a_1(x,e_2)a_1(x,-e_2)\right| |\lambda_2|\ .$$

(9) $\lambda_1 = 1,\ \lambda_2 > 0,\ \lambda_2 \neq 1$.

The condition:

(9a) $a_0(x,\xi) \neq 0$, for any $(x,\xi) \in Q$

and

$$\left|a_0(x,\pm e_2)a_0(\alpha(x),\pm e_2)\right| > \left|a_1(x,\pm e_2)a_1(\alpha(x),\pm e_2)\right|\lambda_2\ ,$$

$$\left|a_0(x,e_1)a_0(\alpha(x),-e_1)\right| > \left|a_1(x,e_1)a_1(\alpha(x),-e_1)\right|\lambda_2$$

for $x = \infty$ and $x = \left(x_1, \dfrac{h_2}{\lambda_2 - 1}\right)$, $x_1 \in \mathbb{R}$;

or

(9b) $a_1(x,\xi) \neq 0$, for any $(x,\xi) \in Q$

and

$$\left|a_0(x,\pm e_2)a_0(\alpha(x),\pm e_2)\right| < \left|a_1(x,\pm e_2)a_1(\alpha(x),\pm e_2)\right|\lambda_2$$

$$\left|a_0(x,e_1)a_0(\alpha(x),-e_1)\right| < \left|a_1(x,e_1)a_1(\alpha(x),-e_1)\right|\lambda_2$$

for $x = \infty$ and $x = \left(x_1, \dfrac{h_2}{\lambda_2 - 1}\right)$, $x_1 \in \mathbb{R}$.

(10) $\lambda_1 = -1,\ \lambda_2 = 0,\ \lambda_2 \neq -1.$

The condition:

(10a) $a_0(x,\xi) \neq 0,$ for any $(x,\xi) \in Q$

and

$$\left|a_0\left(x,e_k\right)a_0\left(\alpha(x),-e_k\right)\right| > \left|a_1\left(x,e_k\right)a_1\left(\alpha(x),-e_k\right)\right|\left|\lambda_2\right|,\ k=1,2$$

for $x = \infty$ and $x = \left(x_1, \dfrac{h_2}{\lambda_2 - 1}\right)$;

or

(10b) $a_1(x,\xi) \neq 0,$ for any $(x,\xi) \in Q$

and

$$\left|a_0\left(x,e_k\right)a_0\left(\alpha(x),-e_k\right)\right| > \left|a_1\left(x,e_k\right)a_1\left(\alpha(x),-e_k\right)\right|\left|\lambda_2\right|,\ k=1,2$$

for $x = \infty$ and $x = \left(x_1, \dfrac{h_2}{\lambda_2 - 1}\right)$.

(11) $\lambda_1 > 0,\ \lambda_2 > 0,\ \lambda_1 \neq 1,\ \lambda_2 \neq 1,\ \lambda_1 \neq \lambda_2.$

The condition:

(11a) $a_0(x,\xi) \neq 0,$ for any $(x,\xi) \in Q$

and

$$\left|a_0\left(x,\pm e_k\right)\right| > \left|a_1\left(x,\pm e_k\right)\right|\left(\lambda_1 \cdot \lambda_2\right)^{\frac{1}{2}},\ k=1,2$$

for $x = \infty$ and $x = \left(\dfrac{h_1}{\lambda_1 - 1}, \dfrac{h_2}{\lambda_2 - 1}\right)$;

or

(11b) $a_1(x,\xi) \neq 0,$ for any $(x,\xi) \in Q$

and

$$\left|a_0\left(x,\pm e_k\right)\right| < \left|a_1\left(x,\pm e_k\right)\right|\left(\lambda_1 \cdot \lambda_2\right)^{\frac{1}{2}},\ k=\overline{1,2}$$

for $x = \infty$ and $x = \left(\dfrac{h_1}{\lambda_1 - 1}, \dfrac{h_2}{\lambda_2 - 1}\right)$.

(12) $\lambda_1 = \lambda_2 > 0,\ \lambda_1 \neq 1.$

The condition:

(12a) $a_0(x,\xi) \neq 0$, for any $(x,\xi) \in Q$

and

$$|a_0(x,\xi)| > |a_1(x,\xi)|\lambda_1$$

for any ξ and for $x = \infty$ and $x = \left(\dfrac{h_1}{\lambda_1 - 1}, \dfrac{h_2}{\lambda_2 - 1}\right)$;

or

(12b) $a_1(x,\xi) \neq 0$, for any $(x,\xi) \in Q$

and

$$|a_0(x,\xi)| < |a_1(x,\xi)|\lambda_1$$

for any ξ and for $x = \infty$ and $x = \left(\dfrac{h_1}{\lambda_1 - 1}, \dfrac{h_2}{\lambda_2 - 1}\right)$.

(13) $\lambda_1 > 0,\ \lambda_2 < 0,\ |\lambda_1| \neq |\lambda_2|.$

The condition:

(13a) $a_0(x,\xi) \neq 0$, for any $(x,\xi) \in Q$

and

$$|a_0(x,\pm e_1)| > |a_1(x,\pm e_1)|\,|\lambda_1 \cdot \lambda_2|^{\frac{1}{2}},$$

$$|a_0(x,e_2)a_0(x,-e_2)| > |a_1(x,e_2)a_1(x,-e_2)|\,|\lambda_1 \cdot \lambda_2|$$

for $x = \infty$ and $x = \left(\dfrac{h_1}{\lambda_1 - 1}, \dfrac{h_2}{\lambda_2 - 1}\right)$;

or

(13b) $a_1(x,\xi) \neq 0$, for any $(x,\xi) \in Q$

and

$$|a_0(x,\pm e_1)| < |a_1(x,\pm e_1)|\,|\lambda_1 \cdot \lambda_2|^{\frac{1}{2}},$$

$$|a_0(x,e_2)a_0(x,-e_2)| < |a_1(x,e_2)a_1(x,-e_2)|\,|\lambda_1 \cdot \lambda_2|$$

for $x = \infty$ and $x = \left(\dfrac{h_1}{\lambda_1 - 1}, \dfrac{h_2}{\lambda_2 - 1}\right)$.

(14) $\lambda_1 > 0,\ \lambda_2 < 0,\ |\lambda_1| = |\lambda_2|$.

The condition:

(14a) $a_0(x,\xi) \neq 0$, for any $(x,\xi) \in Q$

and

$$\left|a_0(x,\xi_1,\xi_2)a_0(x,\xi_1,-\xi_2)\right| > \left|a_1(x,\xi_1,\xi_2)a_1(x,\xi_1,-\xi_2)\right|\left|\lambda_1\right|^2$$

for $x = \infty$ and $x = \left(\dfrac{h_1}{\lambda_1 - 1}, \dfrac{h_2}{\lambda_2 - 1}\right)$ and every $\xi = (\xi_1, \xi_2)$;

or

(14b) $a_1(x,\xi) \neq 0$, for any $(x,\xi) \in Q$

and

$$\left|a_0(x,\xi_1,\xi_2)a_0(x,\xi_1,-\xi_2)\right| < \left|a_1(x,\xi_1,\xi_2)a_1(x,\xi_1,-\xi_2)\right|\left|\lambda_1\right|^2$$

for $x = \infty$ and $x = \left(\dfrac{h_1}{\lambda_1 - 1}, \dfrac{h_2}{\lambda_2 - 1}\right)$ and every $\xi = (\xi_1, \xi_2)$.

(15) $\lambda_1 < 0,\ \lambda_2 < 0,\ \lambda_1 \neq \lambda_2,\ \lambda_1 \neq -1,\ \lambda_2 \neq -1$.

The condition:

(15a) $a_0(x,\xi) \neq 0$, for any $(x,\xi) \in Q$

and

$$\left|a_0(x,e_k)a_0(x,-e_k)\right| > \left|a_1(x,e_k)a_1(x,-e_k)\right|\left|\lambda_1 \cdot \lambda_2\right|,\ k = 1,2$$

for $x = \infty$ and $x = \left(\dfrac{h_1}{\lambda_1 - 1}, \dfrac{h_2}{\lambda_2 - 1}\right)$;

or

(15b) $a_1(x,\xi) \neq 0$, for any $(x,\xi) \in Q$

and

$$\left|a_0(x,e_k)a_0(x,-e_k)\right| < \left|a_1(x,e_k)a_1(x,-e_k)\right|\left|\lambda_1 \cdot \lambda_2\right|,\ k = 1,2$$

for $x = \infty$ and $x = \left(\dfrac{h_1}{\lambda_1 - 1}, \dfrac{h_2}{\lambda_2 - 1}\right)$.

(16) $\lambda_1 = \lambda_2 < 0,\ \lambda_1 \neq -1$

The condition:

(16a) $a_0(x,\xi) \neq 0$, for any $(x,\xi) \in Q$

and

$$\left|a_0(x,\xi)a_0(x,-\xi)\right| > \left|a_1(x,-\xi)a_1(x,\xi)\right| \left|\lambda_1 \cdot \lambda_2\right|$$

for $x = \infty$ and $x = \left(\dfrac{h_1}{\lambda_1 - 1}, \dfrac{h_2}{\lambda_2 - 1}\right)$ and every ξ;

or

(16b) $a_1(x,\xi) \neq 0$, for any $(x,\xi) \in Q$

and

$$\left|a_0(x,\xi)a_0(x,-\xi)\right| < \left|a_1(x,-\xi)a_1(x,\xi)\right| \left|\lambda_1 \cdot \lambda_2\right|$$

for $x = \infty$ and $x = \left(\dfrac{h_1}{\lambda_1 - 1}, \dfrac{h_2}{\lambda_2 - 1}\right)$ and every ξ.

Example 39.4 Let us consider one more example of the operator b of the form (39.1) in $L^2(\mathbb{R}^2)$ - the case when the shift is generated by a rotation, that is

$$h = 0 \text{ and } L = L_{\varphi_0} = \begin{bmatrix} \cos\varphi_0 & -\sin\varphi_0 \\ \sin\varphi_0 & \cos\varphi_0 \end{bmatrix}. \tag{39.11}$$

Thus

$$\alpha(x) = L_{\varphi_0}(x)\ . \tag{39.12}$$

In this situation the foregoing argument implies the following statement.

Let $\alpha(x)$ be defined by (39.12), (39.11). The operator

$$b = a_0(x,D) + a_1(x,D)U$$

is Fredholm in $L^2(\mathbb{R}^2)$ iff the following conditions (depending on φ_0) are satisfied.

(1) $\dfrac{\varphi_0}{2\pi} = \dfrac{m}{N}$ *($\dfrac{m}{N}$ is an irreducible fraction).*

The condition:

$$\prod_{k=0}^{N-1} a_0\left(L^k x, L^k \xi\right) \neq \prod_{k=0}^{N-1} a_1\left(L^k x, L^k \xi\right), \ \forall (x,\xi) \in Q .$$

(2) $\dfrac{\varphi_0}{2\pi}$ *is an irrational number.*

The condition:

(2a) $a_0(x,\xi) \neq 0, \ \forall (x,\xi) \in Q$

and

$$\int_0^{2\pi} \ln\left|a_0\left(L_\varphi x, L_\varphi \xi\right)\right| \mathrm{d}\varphi > \int_0^{2\pi} \ln\left|a_1\left(L_\varphi x, L_\varphi \xi\right)\right| \mathrm{d}\varphi, \ \forall (x,\xi) \in Q ;$$

or

(2b) $a_1(x,\xi) \neq 0, \ \forall (x,\xi) \in Q$

and

$$\int_0^{2\pi} \ln\left|a_0\left(L_\varphi x, L_\varphi \xi\right)\right| \mathrm{d}\varphi < \int_0^{2\pi} \ln\left|a_1\left(L_\varphi x, L_\varphi \xi\right)\right| \mathrm{d}\varphi, \ \forall (x,\xi) \in Q .$$

Example 39.5 (Pseudodifferential operators with the shifts generated by 'expansions') Let the group of the shifts in $\mathbb{R}^n$ be generated by the mappings

$$\alpha_\tau(x) = \tau x, \ \tau \in \mathbb{R} \setminus \{0\} .$$

Thus

$$L_\tau = \tau \cdot I .$$

This group is commutative and consequently amenable. Thus the symbolic calculus of pseudodifferential operators with these shifts is described in Theorem 37.22.

For a two-term operator

$$\mathcal{L} = a_0(x,D) + a_1(x,D) U_\tau$$

where

$$\left(U_\tau f\right)(x) = f\left(\tau^{-1} x\right) \tag{39.13}$$

the Fredholmity conditions can be written out in a form analogous to that described in Example 39.3, situation (12).

Note that in the case when the symbols $a_k(x,\xi)$ do not depend on x, that is

$$a_k(x,\xi) = a_k(\xi),$$

we can investigate the multiterm operators as well.

Let us consider this case in detail. The symbol of the operator

$$\mathcal{b} = \sum_{k=0}^{s} a_k(D) U_{\tau_k} \tag{39.14}$$

has the form

$$b = \sigma(\mathcal{b}) = \sum_{k=0}^{s} a_k(\xi) W_{\tau_k} \tag{39.15}$$

where

$$\left(W_{\tau_k} w\right)(x,\xi) = w\left(\tau_k^{-1} x, \xi\right) . \tag{39.16}$$

The argument analogous to that used when considering the pseudodifferential difference operators (see 37.5 and Lemma 37.6) shows that the invertibility of b is equivalent to the invertibility (in $L^2\left(\mathbb{R}^n\right)$) of every $b(\xi)$, $\xi \in S^{n-1}$, ($b(\xi)$ is given by the same formula (39.15) for a fixed ξ).

If $\tau_k > 0$, $k = \overline{0,s}$, then we can investigate these operators with the help of the Mellin transform (recall (31.43)), after application of which the operator $b(\xi)$ is transformed into the operator of multiplication by the function

$$\left[\tilde{b}(\xi)\right](t) = \sum a_k(\xi) \tau_k^{\frac{n}{2}} \tau_k^{-t}$$

and consequently $b(\xi)$ is invertible iff

$$\inf_t \left|[\tilde{b}(\xi)](t)\right| > 0 .$$

Analogous to the pseudodifferential difference operator case the invertibility conditions for operators of the form (39.15) can be written out in a more convenient form.

Let G be the subgroup of $\mathbb{R} \setminus \{0\}$ generated by the numbers τ_k, $k = \overline{0,m}$ (used in (39.15)). This group is a free commutative group isomorphic to $\mathbb{Z}^l$. Let $\lambda_1, \dots, \lambda_l$ be the generators of b (by the generators of $\mathbb{Z}^l$ we mean elements of the form $(0,\dots,0,1,0,\dots,0)$). The operator $b(\xi)$ can be written out as a finite sum

$$[b(\xi)u](x) = \sum_m a_m(\xi) u\left(\lambda^{-m} x\right) \tag{39.17}$$

where $m = (m_1, \dots, m_l) \in \mathbb{Z}^l$ and $\lambda = \lambda_1^{m_1} \cdot \dots \cdot \lambda_l^{m_l}$.

The operators

$$(V_m u)(x) = \left|\lambda^m\right|^{\frac{1}{2}} u\left(\lambda^{-m} x\right) \tag{39.18}$$

are unitary and the C^*-algebra B_1 generated by the operators $b(\xi)$ of the form (39.17) coincide with $C^*(\mathbb{C}, V_m)$.

Consider the C^*-algebra $B_2 = C\left(\mathbb{T}^l\right)$ where $T^l = \left\{z : z = (z_1, \dots, z_l), z_i \in \mathbb{C}, |z_i| = 1\right\}$ is the l-dimensional torus.

The correspondence

$$m \to z^m$$

$\left(z^m = z_1^{m_1} \cdot \dots \cdot z_l^{m_l}\right)$ defines the unitary representation of the group $\mathbb{Z}^l$ into B_2.

Observing that $B_2 = C^*\left(\mathbb{C}, Z^m\right)$ and noticing that both the algebras B_1 and B_2 possess the property (*) (see 12.2) we conclude (applying Corollary 12.9) that the mapping

$$V_m \to z^m$$

establishes the isomorphism $B_1 \cong B_2$.

Under this isomorphism the operator (39.17) is mapped into the function

$$[S(\xi)](z) = \sum_m a_m(\xi) \left|\lambda^m\right|^{-\frac{1}{2}} z^m .$$

Thus the invertibility of $b(\xi)$ is equivalent to the condition

$$[S(\xi)](z) \neq 0,\ z \in \mathbb{T}^l .$$

Remark In fact in this situation (as well as in the corresponding case for the pseudodifferential difference operators) Corollary 12.9 (or the isomorphism theorem) allows us to construct explicitly the Gelfand transform of the commutative C^*-algebra generated by operators of the form (39.17).

39.6 One more situation where we can effectively use the ideas exploited in Section 37 is the case when the shifts are generated by diagonal matrices. The nature of the pseudodifferential operators with such shifts is closely related to that of pseudodifferential operators with a convex rationally independent system of shifts.

We shall call the system of vectors $\{\lambda^1,\dots,\lambda^p\}$ where $\lambda^k = \{\lambda_1^k,\dots,\lambda_h^k\} \in \mathbb{R}_+^n$, $k=\overline{1,p}$, *logarithmically rationally independent* if the system of vectors $\{h^1,\dots,h^p\}$ where $h^k = \{\ln\lambda_1^k,\dots,\ln\lambda_n^k\}$ is rationally independent; and we shall call the system $\{\lambda^1,\dots,\lambda^p\}$ *logarithmically convex* if the system $\{h^1,\dots,h^p\}$ is convex (recall Example 37.10).

The next theorem is the 'logarithmic analogue' of the results of Example 37.10.

Theorem 39.7 *Let $a_k(x) = L_k x$, $k=\overline{1,p}$, where*

$$L_k = \begin{bmatrix} \lambda_1^k & & 0 \\ & \ddots & \\ 0 & & \lambda_n^k \end{bmatrix},\ \lambda_j^k > 0 .$$

If the system of vectors $\{\lambda^1,\dots,\lambda^p\}$ is logarithmically convex and logarithmically rationally independent and $a_k(0,\xi) = a_k(\infty,\xi) = a_k$ (thus not depending on ξ at 0 and ∞) then the operator $\mathcal{B}$ of the form

$$\mathcal{B} = \sum_{k=1}^{p} a_k(x,D)U_k \quad \left((U_k f)(x) = f\left(L_k^{-1}x\right)\right) \tag{39.19}$$

is Fredholm in $L^2(\mathbb{R}^n)$ iff there exists a number $k_0 \in \overline{1,p}$ such that:

(1) $a_{k_0}(x,\xi) \neq 0$ *for every* $(x,\xi) \in Q$; (39.20)

(2) $$\left|a_{k_0}(0,\xi)\right|\cdot\left|\lambda_1^{k_0}\cdot\ldots\cdot\lambda_n^{k_0}\right|^{\frac{1}{2}} > \sum_{k\neq k_0}\left|a_k(0,\xi)\right|\cdot\left|\lambda_1^{k}\cdot\ldots\cdot\lambda_n^{k}\right|^{\frac{1}{2}}. \qquad (39.21)$$

Proof By Theorem 37.22 the symbol $\sigma(\mathscr{b})$ is the operator acting in $L^2\left(\overline{\mathbb{R}}^n\times S^{n-1}\right)\cong L^2\left(\mathbb{R}^n\times S^{n-1}\right)$.

The decomposition

$$\mathbb{R}^n=\bigcup_{\varepsilon}\mathbb{R}^n_{\varepsilon}$$

where $\varepsilon=(\varepsilon_1,\ldots,\varepsilon_n)$, $\varepsilon_i=\pm 1$, $\mathbb{R}^n_{\varepsilon}=\left\{x\in\mathbb{R}^n:\text{sign } x_i=\text{sign } \varepsilon_i\right\}$, generates the decomposition of $L^2\left(\mathbb{R}^n\times S^{n-1}\right)$ into the direct sum of 2^n subspaces - the restrictions on $\mathbb{R}^n_{\varepsilon}\times S^{n-1}$. These subspaces are invariant with respect to the action of $\sigma(\mathscr{b})$ and therefore the operator $\sigma(\mathscr{b})$ can be decomposed into the direct sum of 2^n operators $\sigma^{\varepsilon}(\mathscr{b})$ defined by the same formula on the corresponding subspaces. Let us fix some ε and consider the restriction $\sigma^{\varepsilon}(\mathscr{b})$ of $\sigma(\mathscr{b})$ onto $L^2\left(\mathbb{R}^n_{\varepsilon}\times S^{n-1}\right)$.

We denote by B^{ε} the C^*-algebra of operators in $L^2\left(\mathbb{R}^n_{\varepsilon}\times S^{n-1}\right)$ generated by the corresponding restriction of the algebra A of the symbols $\sigma(\mathscr{a})$ of pseudodifferential operators $\mathscr{a}\in\mathscr{A}$ and the operators W_k, $k=\overline{1,p}$,

$$\left(W_k w\right)(x,\xi)=w\left(L_k^{-1}x,\ \frac{L_k'\xi}{\left\|L_k'\xi\right\|}\right) \qquad (39.22)$$

(recall (37.35)). Clearly $\sigma^{\varepsilon}(\mathscr{b})\in B^{\varepsilon}$.

Consider the group G generated by the matrices L_k, $k=\overline{1,p}$. This group is isomorphic to $\mathbb{Z}^p$ and the generators L_k in G could be identified with the elements (generators) (0,...,1,...,0) of $\mathbb{Z}^p$. Moreover, the trajectory of every point in $\mathbb{R}^n_+$ under the action of G is discrete (the limit point 0 does not belong to $\mathbb{R}^n_{\varepsilon}$). Applying the method of discretization (that is, passing to the operators $\sigma^{\varepsilon}(\mathscr{b})(\xi)_x$ (recall 37.12)) we conclude that $\sigma^{\varepsilon}(\mathscr{b})$ is invertible iff all the corresponding discrete operators are invertible. The discrete operators here are the multidimensional discrete convolution operators with the coefficients having limits at infinity (here we use the condition $a_k(0,\xi)=a_k(\infty,\xi)=a_k$); and it follows (in view of the argument of 37.12-37.15) that the conditions (1) and (2) of the theorem are necessary and sufficient for all $\sigma^{\varepsilon}(\mathscr{b})$ (and thus for $\sigma(\mathscr{b})$) to be invertible.

The proof is complete. □

40. The symbolic calculus of G-PDOs on a manifold (L^2-case)

This and the two subsequent sections are the main theoretical sections in this chapter (concerning the symbolic calculus of G-PDOs). We would like to emphasize at once that the construction of the symbol of G-PDOs is suggested in a natural way by the isomorphism theorem (Corollary 12.17). We begin the building of the symbolic calculus from consideration of the 'simplest' L^2-situation where in fact the main ideas can be presented in a visual form. After this is done we proceed to the general case of the symbolic calculus in Sobolev spaces on vector bundles and the general theorem on the symbolic calculus will be obtained in Section 42 (Theorem 42.4).

Here we consider only smooth compact manifolds. Other types of manifold could be investigated with the *localization* ideology (see Chapter 4) which will be presented (applied to G-PDOs with discontinuous coefficients) in the next chapter.

40.1 Let M be a C^∞-smooth compact manifold with a fixed Riemann metric which induces a C^∞-smooth measure μ on M thus generating the norm in $L^2(M)$.

Let $g \to \alpha_g$, $g \in G$, be a representation of a certain discrete amenable group G via C^∞-diffeomorphisms of M.

These diffeomorphisms α_g generate the operators U_g and T_g in $L^2(M)$ defined by the formulae

$$(U_g f)(x) = f\left(\alpha_g^{-1}(x)\right) \tag{40.1}$$

and

$$(T_g f)(x) = \left|\alpha_g^{-1\prime}(x)\right|^{\frac{1}{2}} f\left(\alpha_g^{-1}(x)\right) \tag{40.2}$$

where $\left|\alpha_g^{-1\prime}(x)\right|$ is the modulus of the Jacobian of α_g^{-1} at x. The operators $\{T_g\}_{g \in G}$ form a unitary representation of the group G.

It is more convenient to deal with the operators T_g (bearing in mind U_g as the initial generating objects related to the group $\{\alpha_g\}_{g \in G}$ (recall 37.16)).

The goal of this section is to construct the symbolic calculus for operators of the form

$$\mathcal{b} = \sum a_g U_g$$

where a_g are pseudodifferential operators in $L^2(M)$.

The commutation relations between pseudodifferential operators and the operators W_g are linked with Theorem 2.48 which suggests that we consider, along with the group of diffeomorphisms α_g of M, the group of diffeomorphisms β_g of S^*M (S^*M is the bundle of unit vectors of the cotangent bundle T^*M (see 2.35)) given by

$$\beta_g(\eta)=\frac{\partial\alpha_g(\eta)}{\left\|\partial\alpha_g(\eta)\right\|},\ \eta\in S^*M \tag{40.3}$$

(here $\partial\alpha_g:T^*M\to T^*M$ is the codifferential of α_g (see 2.38)).

Recall that if (x,ξ) are local coordinates on T^*M over a neighbourhood $w\subset M$ and (y,ζ) are the local coordinates over the neighbourhood $v=\alpha_g(w)$ and $y=\alpha_g(x)$ then the mapping $\partial\alpha_g$ acts via the formula

$$\partial\alpha_g(x,\xi)=\left(\alpha_g(x),L(x)\xi\right) \tag{40.4}$$

where $L(x)$ is the inverse of the transpose of the Jacobian matrix of the mapping $y=\alpha_g(x)$. Note that the mapping $\partial\alpha_g$ is a linear extension of α_g (onto T^*M).

The diffeomorphisms β_g generate the operators W_g and V_g in $L^2(S^*M)$ defined (analogously to U_g and T_g) via the formulae

$$\left(W_g\psi\right)(\eta)=\psi\left(\beta_g^{-1}(\eta)\right) \tag{40.5}$$

and

$$\left(V_g\psi\right)(\eta)=\left|\beta_g^{-1\prime}(\eta)\right|^{\frac{1}{2}}\psi\left(\beta_g^{-1}(\eta)\right). \tag{40.6}$$

The operators $\left\{V_g\right\}_{g\in G}$ form a unitary representation of G.

40.2 Now we come to the main object of the section. Let $\mathcal{A}^0$ be the algebra of pseudodifferential operators of zero order in $L^2(M)$ (see 2.45 and Theorem 2.46), and let $g\to\alpha_g$, $g\in G$, be a certain representation of a discrete group G via diffeomorphisms α_g of M.

We shall call the operator $\mathcal{b}$ of the form

$$\mathcal{b}=\sum_{k=1}^{m}a_{g_k}U_{g_k} \tag{40.7}$$

(where $a_{g_k} \in \mathcal{A}^0$ and U_{g_k} are defined by (40.1)) a G-pseudodifferential operator (G-PDOs) of order 0 acting $L^2(M)$ (associated with the group of diffeomorphisms $\{\alpha_g\}_{g\in G}$).

40.3 We define the symbol of the G-PDO or order 0 defined by (40.7) to be the operator $\sigma(\mathcal{b})$ (acting in $L^2(S^*M)$) given by the formula

$$\sigma(\mathcal{b}) = \sum_{k=1}^{m} \sigma\left(a_{g_k}\right)\cdot\left|\alpha_{g_k}^{-1'}(x)\right|^{-\frac{1}{2}} V_{g_k} \tag{40.8}$$

where we consider $\sigma\left(a_{g_k}\right)$ as an operator of multiplication by the function $\sigma\left(a_{g_k}\right)(x,\xi)$ and V_{g_k} is defined by (40.6).

Remark 40.4 In fact it is reasonable to write the G-PDOs in the form

$$\mathcal{b} = \sum a_{g_k} T_{g_k} \tag{40.9}$$

with T_g defined by (40.2). (Observe that the operators of the form (40.9) can be rewritten in the form (40.7) by inserting the multipliers $\left|\alpha_{g_k}^{-1'}(x)\right|^{\frac{1}{2}}\cdot\left|\alpha_{g_k}^{-1'}(x)\right|^{-\frac{1}{2}}$.)

The passage to the symbol for operators of the form (40.9) is more visual than the corresponding procedure for operators of the form (40.7). Indeed if $\mathcal{b}$ is an operator of the form (40.9) then (according to (40.8)) its symbol $\sigma(\mathcal{b})$ has the form

$$\sigma(\mathcal{b}) = \sum \sigma\left(a_{g_k}\right) V_{g_k} \tag{40.10}$$

which means that in this case the symbol of a PDO a is defined in the standard way and the symbol of the unitary shift operator T_g is the unitary shift operator V_g.

40.5 Bearing in mind the foregoing remark we consider the C^*-algebra

$$\mathcal{B} = C^*\left(\mathcal{A}^0, U_g\right) = C^*\left(\mathcal{A}^0, T_g\right) \subset L\left(L^2(M)\right)$$

generated by $\mathcal{A}^0$ and $\{T_g\}_{g\in G}$ (or $\{U_g\}_{g\in G}$) which it is reasonable to call the *algebra of G-pseudodifferential operators* of order 0 (associated with the group of diffeomorphisms $\{\alpha_g\}_{g\in G}$).

40.6 Let

$$A=\sigma(\mathcal{A}^0)$$

be the algebra of symbols of pseudodifferential operators of order zero (belonging to $\mathcal{A}^0$) considered as operators of multiplication in $L^2(S^*X)$.

We denote by

$$B=C^*(A,V_g)\subset L(L^2(S^*M))$$

the C^*-algebra generated by A and the operators V_g defined via (40.6).

The next theorem (in accordance with the idea of the symbol construction presented in Section 36) shows that B is the algebra of symbols of elements of $\mathcal{B}$.

Theorem 40.7 *Let an amenable group* $\{\alpha_g\}_{g\in G}$ *of diffeomorphisms act topologically freely on M. Then the mapping*

$$\sigma:\mathcal{b}\to\sigma(\mathcal{b})$$

*(see (40.8) or (40.10)) can be extended up to the *-epimorphism*

$$\sigma:\mathcal{B}\to B$$

($\mathcal{B}$ and B are the algebras defined in 40.5 and 40.6) the kernel of which is the idea K of compact operators in $L^2(M)$.

In particular the symbol of a G-PDO *possesses properties analogous to the properties of the symbol of a PDO:*

(1) $\sigma(\mathcal{b}_1+\mathcal{b}_2)=\sigma(\mathcal{b}_1+\mathcal{b}_2)$;

(2) $\sigma(\mathcal{b}_1\cdot\mathcal{b}_2)=\sigma(\mathcal{b}_1)\cdot\sigma(\mathcal{b}_2)$;

(3) $\sigma(\mathcal{b}^*)=\sigma(\mathcal{b})^*$;

(4) $\sigma(\mathcal{b})=0$ *iff* $\mathcal{b}$ *is compact;*

(5) *an operator* $\mathcal{b}$ *is Fredholm iff* $\sigma(\mathcal{b})$ *is invertible;*

(6) *for any element b of the algebra B of symbols there exists an operator* $\mathcal{b} \in \mathcal{B}$ *such that* $\sigma(\mathcal{b}) = b$.

Proof The proof is absolutely analogous to that of Theorem 37.4. We have to check that the mapping σ defined via (40.10) establishes the isomorphism

$$\mathcal{B}/K = [\mathcal{B}] \overset{\sigma}{\cong} B\,. \tag{40.11}$$

Recall that for the algebra $\mathcal{A}^0$ of pseudodifferential operators we have (Theorem 2.46) that

$$\mathcal{A}^0/K = \left[\mathcal{A}^0\right] \overset{\sigma}{\cong} A = \sigma(\mathcal{A})\,,$$

and Theorem 2.48 means that

$$\sigma\left(T_g \mathcal{a} T_g^*\right) = V_g \sigma(\mathcal{a}) V_g^*\,. \tag{40.12}$$

Observe that on the space

$$\operatorname{Prim} A = S^* M \tag{40.13}$$

the group of homeomorphisms $\{\beta_g\}$ generated by the automorphisms $V_g(\cdot)V_g^*$ acts topologically freely (since the action of the group $\{\alpha_g\}$ is topologically free on *M*).

In view of this and (40.12) we conclude (by the isomorphism theorem (Corollary 12.17)) that

$$[\mathcal{B}] = C^*\left(\left[\mathcal{A}^0\right], \left[T_g\right]\right) \overset{\sigma}{\cong} C^*\left(A, V_g\right) = B$$

which completes the proof. □

Remark 40.8 In the previous theorem we confined ourselves to the class of topologically freely acting amenable groups of diffeomorphisms. As already mentioned (see Remarks 37.23) we don't know whether the amenability condition can be discarded.

On the other hand, the assumption on the group $\{\alpha_g\}$ of diffeomorphisms of M to act topologically freely could certainly be discarded, and we shall obtain a result of this sort in Theorem 40.12. The derivation of these results is based not on the isomorphism theorem (Corollary 12.17) but on the ideas and methods developed in Section 16 and in particular on Theorem 16.21 which generalizes the isomorphism theorem up to the case of certain 'normally' acting groups of homeomorphisms.

Before discussing these objects we would like to emphasize that when the group $\{\alpha_g\}$ does not act topologically freely (thus among the diffeomorphisms α_g, $g \neq e$, there are some for which the set of fixed points

$$M_g = \{x \in M: \alpha_g(x) = x\}$$

has a non-empty interior) then the coefficients a_{g_k} in the expression of b in formulae (40.7) or (40.9) are not determined uniquely and thus even the correctness of the definition of $\sigma(b)$ via formula (40.8) is not obvious.

40.9 For convenience we recall now the terminology which was exploited in Theorem 16.21 and which we shall use while stating a generalization of Theorem 40.7.

Let us fix a group $\{\alpha_g\}_{g\in G}$ of homeomorphisms of M. For any $g \in G$, we define the set

$$M_g = \{x \in M: \alpha_g(x) = x\}. \tag{40.14}$$

Let $\mathring{M}_g$ be the interior of M_g and

$$M_\infty = M \setminus \left[\bigcup_{g\neq e} \mathring{M}_g \right]. \tag{40.15}$$

For every $x \in M$ we denote by G_x the subgroup of G defined by

$$G_x = \left\{ g \in G: x \in \mathring{M}_g \right\}. \tag{40.16}$$

Remark Clearly if $x \in M_\infty$ then $G_x = \{e\}$ and in particular the group $\{\alpha_g\}$ acts topologically freely on M iff $G_x = \{e\}$ for every $x \in M$.

If G_x is normal then along with this group we consider the factor group

$$G^x = G/G_x \,. \tag{40.17}$$

40.10 We shall say that G^x *acts topologically freely at* x if for any finite set $g_1,\dots,g_n$ such that $[g_i] \neq [e]$ (that is $g_i \,\overline{\in}\, G_x$, $i=\overline{1,n}$) the set $\left[\bigcup_{i=1}^{n} M_{g_i}\right]$ does not contain any neighbourhood of x.

40.11 We shall say that the group $\{\alpha_g\}$ acts normally on M if it satisfies the following two conditions:

(1) G_x is a normal group for every $x \in M$;

(2) G^x acts topologically freely at every $x \in M$.

Condition (1) was investigated in Lemma 16.17 and Corollary 16.18.

Remark (i) If G is commutative then condition (1) is clearly satisfied for every x while if M is an arbitrary topological space then condition (2) is not necessarily satisfied even in the case of finite commutative groups (see Remark 16.24).
(ii) If $\{\alpha_g\}$ acts topologically freely then as we have already noticed $G_x = \{e\}$ for every $x \in M$ and in this case both conditions (1) and (2) are satisfied. Thus the class of topologically freely acting groups is a subclass of normally acting groups.
(iii) We would like to make an observation which will be important for our purposes; namely, that

if α *is a diffeomorphism of a manifold M then the group* $\{\alpha^z\}_{z\in\mathbb{Z}}$ *acts normally on M.*

Indeed condition (1) here is satisfied so it is enough to verify (2).

Let us fix any point $x \in M$ and let $G^x = \mathbb{Z}_N$. This means that choosing a sufficiently small neighbourhood U of x we obtain a representation of $\mathbb{Z}_N$ on $\tilde{M} = \left[\bigcup_{n=0}^{N-1} \alpha^n(u)\right]$ by diffeomorphisms α^n and each of the sets

$$\tilde{M}_n = \{x \in \tilde{M} : \alpha^n(x) = x\},\ n = \overline{1, N-1}$$

does not cover $\tilde{M}$.

But since each of $\tilde{M}_n$, $n=\overline{1,N-1}$, consists of a finite set of submanifolds of $\tilde{M}$ (see Conner, Floyd [1], §22) the latter property means that

$$\dim \tilde{M}_n = m_n < m, \; n=\overline{1,N-1}$$

($m = \dim \tilde{M}$) and thus

$$\left[\bigcup_{n=1}^{N-1} \tilde{M}_n\right] \not\supset \tilde{M}$$

which proves the topological freedom of the action of G^x at x. □

Theorem 40.12 *Let a countable amenable group* $\{\alpha_g\}_{g\in G}$ *of diffeomorphisms act normally on M. Then the mapping*

$$\sigma : \mathcal{b} \to \sigma(\mathcal{b})$$

*can be extended up to the *-epimorphism*

$$\sigma : \mathcal{B} \to B$$

the kernel of which is the ideal K of compact operators $L^2(M)$.

Proof For the same reasons as those used in the proof of Theorem 40.7, (40.12) and (40.13) are true in this case.

In view of the interrelation between the groups of diffeomorphisms $\{\alpha_g\}$ and $\{\beta_g\}$ (recall (40.3) and (40.4)) we conclude that the normal action of $\{\alpha_g\}$ on M implies the normal action of $\{\beta_g\}$ on $S^*M = \text{Prim}\, A$ (in particular

$$G_\eta = G_{\pi(\eta)}$$

where $\pi : S^*M \to M$ is the canonical projection). Thus the C^*-algebra $[\mathcal{B}] = C^*\left([\mathcal{A}^0],[T_g]\right)$ satisfies conditions (1) and (2) of Theorem 16.21.

Observe that the elements $[T_g]$ satisfy condition (3) of Theorem 16.21.

Indeed, the ideal I_g of $\left[\mathcal{A}^0\right]$ is generated by the pseudodifferential operators a satisfying the condition

$$\sigma(a)(\eta)=0 \text{ for every } \eta \overline{\in} \pi^{-1}\left(\overset{\circ}{M}_g\right). \tag{40.18}$$

For every fixed operator a satisfying (40.18) one can find a sequence of operators of multiplication by the functions $\varphi_n \in C(M)$ with

$$\varphi_n(x)=0 \text{ for every } x \overline{\in} \overset{\circ}{M}_g \tag{40.19}$$

and such that

$$\left[\varphi_n a\right] \to [a]. \tag{40.20}$$

Note that for any φ_n satisfying (40.19) we have

$$T_g \varphi_n = \varphi_n. \tag{40.21}$$

Now (40.20) and (40.21) imply

$$\left[T_g\right][a]=\left[T_g a\right]=\lim_{n\to\infty}\left[T_g \varphi_n a\right]=\lim_{n\to\infty}\left[\varphi_n a\right]=[a]$$

which means that $\left[T_g\right]$ satisfies condition (3) of Theorem 16.21.

The algebra $B=C^*\left(A,V_g\right)$ also satisfies all the conditions of Theorem 16.21. In fact, conditions (1) and (2) here coincide with those already verified for $C^*\left(\left[\mathcal{A}^0\right],\left[T_g\right]\right)$ and condition (3) for $C^*\left(A,V_g\right)$ is fulfilled in view of the explicit form of V_g.

All these facts, along with Theorem 16.21 (observing that the right-hand side of (16.37) does not depend on the explicit form of A and T_g) mean that the mapping σ establishes the isomorphism

$$[\mathcal{B}]=C^*\left(\left[\mathcal{A}^0\right],\left[T_g\right]\right)\overset{\sigma}{\cong}C^*\left(A,V_g\right)=B.$$

The proof is complete. □

Remark 40.13 Observe that in view of Remark (iii) in 40.11 the theorem just proved establishes the symbolic calculus for G-PDOs in the case of an *arbitrary* action of group $\mathbb{Z}$.

41. The symbolic calculus of G-PDOs in Sobolev spaces.

In this section we consider G-PDOs of order r acting from the Sobolev space $H^s(M)$ into $H^{s-r}(M)$ (we work here with scalar functions; the vector bundle situation will be presented in the next section).

It is known (see Theorem 2.44) that a PDO of order r is Fredholm (as an operator acting from $H^s(M)$ into $H^{s-r}(M)$) iff it is elliptic and this condition does not depend on the index s of the Sobolev space. In the G-PDO case we shall find that the situation is more complicated and the symbol and the Fredholmity conditions depend on s.

We present two variants of the construction of the symbol, each of which has its own merits.

41.1 As in the PDO case the investigation is based on the L^2-theory of G-PDOs developed (in Section 40) and on the isomorphism between the spaces $H^s(M)$. It is known (see Seeley [1], Lemma 14.1) that there exists a family of positively defined PDOs Λ^t, $t \in \mathbb{R}$, such that:

(1) $\sigma(\Lambda^t) = \|\xi\|^t$ (in the fixed Riemann metric);

(2) Λ^t maps $H^s(M)$ onto $H^{s-t}(M)$ bijectively;

(3) $\Lambda^{t_1+t_2} = \Lambda^{t_1} \cdot \Lambda^{t_2}$.

Henceforth in this section we shall assume such a family of operator to be given and fixed.

41.2 The objects we shall deal with in this section are the generalization of the G-PDOs of zero order acting in $L^2(M)$ considered in Section 40. To define them we recall (see 40.1) that we assume a certain group $\{\alpha_g\}_{g\in G}$ of diffeomorphisms of M to be fixed (with G being discrete and amenable) and we denote by U_g and T_g the operators in $H^s(M)$ defined via formulae (40.1) and (40.2) respectively. Note that both these operators are bounded in every $H^s(M)$, $s \in \mathbb{R}$.

41.3 We now come to the description of the G-PDOs in Sobolev spaces.

Let $\mathcal{A}_{sr}$, $s, r \in \mathbb{R}$, be the Banach space consisting of PDOs of order r and acting from the Sobolev space $H^s(M)$ into $H^{s-r}(M)$ (that is the closure in the norm $L(H^s(M), H^{s-r}(M))$ of the set of PDOs of order r (see 2.42)).

We shall call the set

$$\mathcal{A} = \bigcup_{s,r} \mathcal{A}_{sr} \tag{41.1}$$

the *bigraded pseudoalgebra of PDOs.*

The set $\mathcal{A}$ is not an algebra in the ordinary sense since the sum of two elements here is defined if both of them lie in a certain $\mathcal{A}_{sr}$ and the multiplication $a_1 \cdot a_2$ $(a_1 \in \mathcal{A}_{sr}, a_2 \in \mathcal{A}_{tq})$ is defined iff $s = t - q$ and in this case

$$a_1 \cdot a_2 \in \mathcal{A}_{t,r+q} .$$

We shall denote by $\mathcal{B}_{s,r}$ the subspace of $L(H^s(M), H^{s-r}(M))$ generated by operators of the form

$$\sum a_g T_g, \; a_g \in \mathcal{A}_{sr} . \tag{41.2}$$

The operators (41.2) indeed belong to $L(H^s(M), H^{s-r}(M))$ since, as was noted, T_g is bounded in every $H^s(M)$ (see 41.2).

Clearly the same class $\mathcal{B}_{sr}$ of operators will be obtained if we use U_g instead of T_g in (41.2).

We shall call the set

$$\mathcal{B} = \bigcup_{s,r} \mathcal{B}_{s,r} \tag{41.3}$$

the *bigraded pseudoalgebra of G-PDOs* and its elements will be called G-PDOs (acting in the scale of Sobolev spaces).

Remark The G-PDOs of order 0 acting in $L^2(M)$ considered in the previous section correspond here to the elements of $\mathcal{B}_{00}$.

41.4 For the operators acting in a scale of Hilbert spaces the conjunction can be introduced in a number of ways depending on the method of representation of the dual space (that is, the space of all the linear bounded functionals).

The Banach space F is called the *antidual* of the Banach space E if there exists a sesquilinear mapping

$$\begin{aligned} &E \times F \to \mathbb{C} \\ &(e,f) \to \langle e,f \rangle,\ e \in E, f \in F \end{aligned} \tag{41.4}$$

such that

(i) for any $f \in F$ the formula

$$f(e) = \langle e,f \rangle$$

defines a linear bounded functional on E and the standard norm $\|f(\cdot)\|$ of a linear functional is equivalent to $\|f\|$ in F;

(ii) for any linear bounded functional $f(\cdot)$ on E there exists $f \in F$ such that

$$f(e) = \langle e,f \rangle,\ e \in E.$$

If $a \in L(E_1, E_2)$ and F_i, $i = \overline{1,2}$, are the antidual spaces for E_i in the sense described above then the operator $a' \in L(F_2, F_1)$ satisfying

$$\langle ae_1, f_2 \rangle = \langle e_1, a'f_2 \rangle,\ e_1 \in E_1, f_2 \in F_2$$

is called the adjoint to a.

Remark The definition of the adjoint operator here clearly depends on the method of representation of linear functionals in the form (41.4), that is on a particular choice of antidual spaces F_1 and F_2.

41.5 For the Sobolev space $H^s(M)$ there are at least two reasonable ways to choose the antidual space.

(1) We can take $H^{-s}(M)$ as the antidual to $H^s(M)$. Indeed, for every pair of smooth functions $(e,f)\in H^s(M)\times H^{-s}(M)$ set

$$\langle e,f\rangle=\langle e,f\rangle_0=\int_M e(x)\overline{f(x)}\,d\mu \tag{41.5}$$

where the right-hand side is the standard inner product in $H^0(M)=L^2(M)$. By continuity we can uniquely extend the mapping onto the whole of $H^s(M)\times H^{-s}(M)$ (recall in particular 2.39 and 2.40).

With this choice of the antidual space the adjoint to an operator $b\in L\big(H^s(M),H^{s-r}(M)\big)$ belongs to $L\big(H^{r-s}(M),H^{-s}(M)\big)$. We shall denote this adjoint to b as b^*.

In particular if $a\in\mathcal{A}_{sr}$ then $a^*\in\mathcal{A}_{r-s,r}$ and if $b\in\mathcal{B}_{sr}$ then $b^*\in\mathcal{B}_{r-s,r}$.

(2) As $H^s(M)$ is a Hilbert space one can take itself as the antidual (in this case we have to use the inner product $\langle e,f\rangle_s$ in $H^s(M)$).

The adjoint to an operator $b\in L\big(H^s(M),H^{s-r}(M)\big)$ here will belong to $L\big(H^{s-r}(M),H^s(M)\big)$. We shall denote it by $b^\#$.

Note also that if $a\in\mathcal{A}_{sr}$ then $a^\#\in\mathcal{A}_{s-r,-r}$ and if $b\in\mathcal{B}_{sr}$ then $b^\#\in\mathcal{B}_{s-r,-r}$.

Depending on the situation, we shall use both these procedures * and # for obtaining the adjoint in $\mathcal{B}$.

Remark The interrelation between the operators b^* and $b^\#$ can be established easily under a special choice of the inner product in $H^s(M)$. Indeed, consider the operator Λ described in 41.1. The formula

$$\langle u,v\rangle_s=\langle\Lambda^s u,\Lambda^s v\rangle_0,\quad u,v\in H^s(M) \tag{41.6}$$

defines the inner product which is equivalent to the initial inner product (any of those defined via the construction of the norm described in 2.40).

By the definition of * and # for any pair of smooth functions u,v we have

$$\langle bu,v\rangle_0=\langle u,b^*v\rangle_0 \tag{41.7}$$

and

$$\langle bu,v\rangle_{s-r}=\langle u,b^\#v\rangle_s\,. \tag{41.8}$$

Note that (41.6) along with (41.8) implies

$$\left\langle \Lambda^{s-r} bu, \Lambda^{s-r} v \right\rangle_0 = \left\langle \Lambda^s u, \Lambda^s b^\# v \right\rangle_0$$

and consequently

$$\left\langle bu, \Lambda^{2(s-r)} v \right\rangle_0 = \left\langle u, \Lambda^{2s} b^\# \Lambda^{-2(s-r)} \cdot \Lambda^{2(s-r)} v \right\rangle_0 . \tag{41.9}$$

Since the elements of the form $\Lambda^{2(s-r)} v$, $v \in H^{+\infty}(M)$, are dense in $H^0(M)$, formulae (41.7) and (41.9) mean that

$$b^* = \Lambda^{2s} b^\# \Lambda^{-2(s-r)} \tag{41.10}$$

and this is the interrelation we were looking for.

Now we come to the first approach to the symbolic calculus in $\mathscr{B}$.

The reduction to $\mathscr{B}_{00}$

41.6 Let us show that the investigation of the elements of $\mathscr{B}_{sr}$ can be reduced by means of the operator Λ to the investigation of the elements of $\mathscr{B}_{00}$ (the symbolic calculus for this last algebra was obtained in the previous section).

Consider the mapping $R_{sr}: \mathscr{B}_{sr} \to L(H^0(M))$ given by the formula

$$R_{sr}(\ell) = \ell_0 = \Lambda^{s-r} \ell \Lambda^{-s}, \; \ell = \mathscr{B}_{sr} . \tag{41.11}$$

The system of mappings R_{sr}, $s, r \in \mathbb{R}$, defines the mapping

$$R: \mathscr{B} \to L(H^0(M))$$

which we shall call the *reduction to* $\mathscr{B}_{00}$.

The next lemma shows that this mapping indeed reduces the consideration of the elements of $\mathscr{B}$ to that of the elements of $\mathscr{B}_{00}$.

Lemma 41.7 *The reduction R possesses the following properties.*

(1) *For any $\ell \in \mathscr{B}_{sr}$ the operator $R(\ell) = \ell_0$ belongs to $\mathscr{B}_{00}$.*

(2) *R is a 'homeomorphism' of the pseudoalgebra $\mathcal{B}$ into $\mathcal{B}_{00}$, that is*

(i) $R(\ell_1+\ell_2)=R(\ell_1)+R(\ell_2)$
(ii) $R(\ell_1\cdot\ell_2)=R(\ell_1)\cdot R(\ell_2)$

and Ker $R=\{0\}$.

(3) *The procedure of passing to the adjoint $b\to b^*$ is transformed under the mapping R into the standard procedure of passing to the adjoint in $H^0(M)$.*

(4) *If the inner product in $H^s(M)$ is given by (41.6) then the procedure of passing to the adjoint $b\to b^{\#}$ is transformed under the mapping R into the standard procedure of passing to the adjoint in $H^0(M)$.*

(5) *$\ell\in\mathcal{B}$ is compact (invertible, Fredholm) iff $R(\ell)$ is compact (invertible, Fredholm) respectively.*
If ℓ is Fredholm then

$$\operatorname{ind}\ell=\operatorname{ind}R(\ell)\ .$$

Proof (1) Let the operator $\ell\in\mathcal{B}_{sr}$ have the form

$$\ell=\sum a_g T_g,\ a_g\in\mathcal{A}_{sr}$$

where the sum is finite. Then (41.11) implies

$$\ell_0=\sum\Lambda^{s-r}a_gT_g\Lambda^{-s}T_g^{-1}T_g\ . \tag{41.12}$$

In view of Theorem 2.48 the operator

$$T_g\Lambda^{-s}T_g^{-1}$$

is a PDO of order $-s$ having the symbol

$$\left\|\partial\alpha_g^{-1}(\eta)\right\|^{-s},\ \eta\in T_0^*M \tag{41.13}$$

and consequently the operator

$$\Lambda^{s-r}a_gT_g\Lambda^{-s}T_g^{-1}$$

belongs to $\mathcal{A}_{00}$ and has the symbol

$$\sigma(a_g)\left\|\partial\alpha_g^{-1}(\eta)\right\|^{-s},\ \eta \in S^*M\ . \tag{41.14}$$

As Λ is an isometry, it follows also that

$$\|\mathcal{b}\|_{H^s \to H^{s-r}} = \|\mathcal{b}_0\|_{H^0} \tag{41.15}$$

which means in particular that R establishes the isometrical bijection between $\mathcal{B}_{sr}$ and $\mathcal{B}_{00}$.

The proof of (1) is complete.

(2) The 'linearity' (i) of R is obvious. So let us pass to the verification of (ii).
Let $\mathcal{b}_1 \in \mathcal{B}_{s-r,q}$ and $\mathcal{b}_2 = \mathcal{B}_{s,r}$. Then

$$\begin{aligned} &R(\mathcal{b}_1 \cdot \mathcal{b}_2) \\ &= R_{s,r+q}(\mathcal{b}_1 \cdot \mathcal{b}_2) = \Lambda^{s-r-q}\mathcal{b}_1 \cdot \mathcal{b}_2\Lambda^{-s} \\ &= \left[\Lambda^{s-r-q}\mathcal{b}_1\Lambda^{r-s}\right]\left[\Lambda^{s-r}\mathcal{b}_2\Lambda^{-s}\right] \\ &= R(\mathcal{b}_1)\cdot R(\mathcal{b}_2)\ . \end{aligned}$$

The property

$$\ker R = \{0\}$$

has in fact already been established while proving (1).

Remark Observe that $\mathcal{B}$ is *not* a vector space and though $\ker R = \{0\}$, R is not injective. For any non-zero element $\mathcal{b} \in \mathcal{B}_{00}$, $R^{-1}(\mathcal{b})$ consists of a countable number of elements but all of them belong to different spaces $\mathcal{B}_{sr}$.

(3) Let $\mathcal{b} \in \mathcal{B}_{sr}$. Then (recall 41.5 (1)) $\mathcal{b}^* \in \mathcal{B}_{r-s,r}$ and we have

$$R(\mathcal{b})^* = \Lambda^{-s}\mathcal{b}^*\Lambda^{s-r} = \left[\left(\Lambda^{s-r}\right)^*\mathcal{b}\left(\Lambda^{-s}\right)^*\right]^* = R(\mathcal{b})^*$$

(4) In view of (40.10) we obtain

$$R(\mathcal{L}^*) = R(\Lambda^{2s}\mathcal{L}^{\#}\Lambda^{-2(s-r)}) . \tag{41.16}$$

Recalling that $\mathcal{L}^{\#} \in \mathcal{B}_{s-r,-r}$ (see 41.5 (2)) we deduce from (41.16) that

$$R(\mathcal{L}^*) = \Lambda^{-s}\Lambda^{2s}\mathcal{L}^{\#}\Lambda^{-2(s-r)}\Lambda^{s-r} = \Lambda^{s}\mathcal{L}^{\#}\Lambda^{r-s} = R(\mathcal{L}^{\#})$$

and thus (4) follows from (3).

(5) This is evident in view of the invertibility of Λ^{-s} and Λ^{s-r}.

The proof is finished. □

41.8 Now since the symbolic calculus in $\mathcal{B}_{00}$ has already been constructed in Section 40 the reduction to $\mathcal{B}_{00}$ suggests the corresponding symbolic calculus in $\mathcal{B}$. Thus, let $\mathcal{L} \in \mathcal{B}_{sr}$ be an operator of the form

$$\mathcal{L} = \sum a_g T_g,\ a_g \in \mathcal{A}_{sr} \tag{41.17}$$

where the sum is finite.

We define the *s-symbol* of the operator $\mathcal{L}$ to be the operator $\sigma_s(\mathcal{L})$ acting in $L^2(S^*M)$ via the formula

$$\sigma_s(\mathcal{L}) = \sum d_g V_g \tag{41.18}$$

where V_g is defined by (40.6) and d_g is the operator of multiplication by the function

$$d_g(\eta) = \sigma(a_g) \cdot \left\| \partial \alpha_g^{-1}(\eta) \right\|^{-s},\ \eta \in S^*M . \tag{41.19}$$

Remarks (1) The symbol of G-PDOs of order 0 in $L^2(M)$ introduced in 40.3 corresponds here to the 0-symbol for the elements of $\mathcal{B}_{00}$.
(2) In view of the interrelation between U_g and T_g the s-symbol for the operator of the form (41.17) with U_g instead of T_g can be written out in the obvious way.

The next theorem shows that s-symbols indeed solve the problem of the symbolic calculus in $\mathcal{B}$.

Theorem 41.9 *(s-symbolic calculus of G-PDOs) Let an amenable group G act topologically freely on M via diffeomorphisms α_g (or let G be countable and act normally on M). Then the mapping*

$$\sigma:\mathcal{B} \to L\left(L^2\left(S^*M\right)\right)$$

which coincides on every $\mathcal{B}_{sr}$ with the mapping of taking the s-symbol (41.18) possesses the following properties:

(1) $\sigma(b_1 + b_2) = \sigma(b_1) + \sigma(b_2)$, $b_1, b_2 \in \mathcal{B}_{sr}$,

(2) $\sigma(b_1 \cdot b_2) = \sigma(b_1) \cdot \sigma(b_2)$ *if the product $b_1 \cdot b_2$ is defined; that is, if $b \in \mathcal{B}_{s-r,q}$, $b_2 \in \mathcal{B}_{sr}$ then*

$$\sigma_s(b_1 \cdot b_2) = \sigma_{s-r}(b_1) \cdot \sigma_s(b_2)\,;$$

(3) $\sigma(I) = 1$;

(4) *an operator $b \in \mathcal{B}_{sr}$ is Fredholm iff its s-symbol $\sigma_s(b)$ is invertible*;

(5) *an operator $b \in \mathcal{B}_{sr}$ is compact iff its s-symbol $\sigma_s(b)$ is zero*;

(6) *if b^* is defined as in 41.5 (1) then*

$$\sigma(b^*) = \sigma(b)^*\,;$$

if the inner product in $H^s(M)$ is given by (41.6) and $b^{\#}$ is defined as in 41.5 (2) then

$$\sigma(b^{\#}) = \sigma(b)^*\,.$$

Proof The 0-symbol on $\mathcal{B}_{00}$ possesses the properties described in Theorem 40.7.

Observe that via the definition of the s-symbol we have

$$\sigma_s(b) = \sigma_0(R(b))$$

(just compare (41.8) and (40.10) recalling (41.19) and (41.14)). Thus all the statements of the theorem follow from the corresponding properties of 0-symbols (Theorems 40.7 or 40.12) and the properties of the reduction R (Lemma 41.7). □

Remarks 41.10 (1) Observe that the statement of the theorem differs in some specified sense from analogous statements for PDOs, and this is related to the nature of the object being considered; namely, in the PDO case the symbol of an operator does not depend on the index s of the Sobolev space $H^s(M)$ from which the operator acts. The dependence of the s-symbol in the G-PDO case is essential and this will be demonstrated by the forthcoming examples.

Let us clarify this difference by considering zero-order G-PDOs, that is the elements of $\mathcal{B}_{s0}$. Let a be a pseudodifferential operator of zero order with smooth symbol. We shall consider a to be the element of $\mathcal{A}_{00}$. We denote by a_s the restriction of a onto $H^s(M)$, $s > 0$. The mapping

$$\rho_s : a \to a_s$$

generates the homomorphism of (non-complete) algebras $\tilde{\mathcal{A}}_{00} \to \tilde{\mathcal{A}}_{s0}$ generated by the PDOs with smooth symbols. This homomorphism in turn establishes the isomorphism

$$\mathcal{A}_{00}/\mathrm{K} \cong \mathcal{A}_{s0}/K$$

which in particular implies the simultaneous Fredholmity of a and a_s.

If we consider, in the analogous situation, a G-PDO $b \in \mathcal{B}_{00}$ and its restriction b_s onto $H^s(M)$ then the mapping

$$f_s : b \to b_s$$

establishes the isomorphism of (non-complete) subspaces of G-PDOs (having a *finite* sum form) in $H^s(M)$ and $H^0(M)$. However contrary to the ordinary PDO case this isomorphism cannot be extended onto the complete subspaces.

The theorem just proved shows that it is 'more correct' to establish the isomorphism between the subspaces of G-PDOs consisting of finite sums with the help of the reduction mapping R. While in the PDO case the mapping R^{-1} coincides (modulo compact operators) with the restriction of the operator onto the suitable space, in the G-PDO case we need a more complicated formula (41.11).

(2) The design of the s-symbol has a certain 'drawback', namely the s-symbol $\sigma_s(b)$ does not show how to reconstruct the initial operator b up to a compact summand. This is in view of the fact that $\sigma_s(b)$ does not preserve the information about the order of b and the index of the Sobolev space where b is defined.

So now we shall present a modification of this symbol built in a way which allows us to eliminate this 'drawback'.

The 'homogeneous' symbol of a G-PDO

41.11 Let us denote $\mathcal{H}_s\left(T_0^* M\right)$, $s \in \mathbb{R}$, (or simply by $\mathcal{H}_s$) the set of functions on $T_0^{\ *} M$ of the form

$$u(\eta) = \|\eta\|^{-s} w\left(\frac{\eta}{\|\eta\|}\right), \ \eta \in T_0^* M$$

where $w \in L^2\left(S^* M\right)$ and the norm of u is given by

$$\|u\|_s = \|w\|_{L^2(S^* M)} \cdot$$

Thus $\mathcal{H}_s$ is the set of functions on $T_0^{\ *} M$ which are obtained from the functions belonging to $L^2\left(S^* M\right)$ by extension via positive homogeneity of degree $-s$.

The spaces $\mathcal{H}_s$ and $\mathcal{H}_t$ are isomorphic and the isomorphism is established via the operator $\tilde{\Lambda}^{s-t}$ of multiplication by the function $\|\eta\|^{s-t}$.

41.12 Let $b \in \mathcal{B}_{sr}$ be an operator of the form

$$\sum_{g \in F} a_g T_g, \ a_g \in \mathcal{A}_{sr}, |F| < \infty . \tag{41.20}$$

We shall construct the symbol of this operator in the following way.

In the space $\mathcal{H}_s$ we define the operator $\overline{V}_g$ via the formula

$$\left(\overline{V}_g u\right)(\eta) = \left|\beta_g^{-1'}\left(\frac{\eta}{\|\eta\|}\right)\right|^{\frac{1}{2}} u\left(\partial \alpha_g^{-1}(\eta)\right) \tag{41.21}$$

where β_g is defined by (40.3).

Remark Note that $\overline{V}_g$ is not a unitary operator in $\mathcal{H}_s$ if $s \neq 0$.

The symbol of a PDO $\mathcal{a} \in \mathcal{A}_{sr}$ is a positive homogeneous function $\sigma(\mathcal{a})$ of degree r on $T_0^* M$ (see 2.43). Thus the multiplication by $\sigma(\mathcal{a})$ transforms a function $u \in \mathcal{H}_s$ into a function belonging to $\mathcal{H}_{s-r}$.

Bearing this in mind we can define the operator $\sigma_{sr}(\mathcal{b}) \in L(\mathcal{H}_s, \mathcal{H}_{s-r})$ via the formula

$$\sigma_{sr}(\mathcal{b}) = \sum_{g \in F} \sigma(\mathcal{a}_g)\overline{V}_g \tag{41.22}$$

where we consider $\sigma(\mathcal{a}_g)$ as the operator of multiplication by $\sigma(\mathcal{a}_g)(\eta)$.

We shall call the operator $\sigma_{sr}(\mathcal{b})$ the *'homogeneous' sr-symbol* of the operator $\mathcal{b} \in \mathcal{B}_{sr}$ of the form (41.20).

Remark We have used the word 'homogeneous' here to emphasize the fact that the structure of the operator $\sigma_{sr}(\mathcal{b})$ involves the homogeneous coefficients $\mathcal{a}_g$ and the operator itself acts in the spaces of homogeneous functions.

We now denote by $\Sigma_{s,r}$ the closure in $L(\mathcal{H}_s, \mathcal{H}_{s-r})$ of the set of operators $\sigma_{sr}(\mathcal{b})$ of the form (41.22). We denote by Σ the *bigraded pseudoalgebra*

$$\Sigma = \bigcup_{s,r} \Sigma_{s,r}$$

of the 'homogeneous' symbols of G-PDOs.

Let $[\mathcal{B}_{s,r}]$ be the factor space of $\mathcal{B}_{s,r}$ modulo the subspace of compact operators (in $L(H^s(M), H^{s-r}(M))$), and let $[\mathcal{B}]$ be the bigraded algebra

$$[\mathcal{B}] = \bigcup_{s,r} [\mathcal{B}_{s,r}]$$

with the natural operations of addition and multiplication.

41.13 As in the Sobolev space case (see 41.5) there are two reasonable ways to choose the antidual space for $\mathcal{H}_s$.

(1) We can take $\mathcal{H}_{-s}$ as the antidual to $\mathcal{H}_s$ setting, for every pair $(e, f) \in \mathcal{H}_s \times \mathcal{H}_{-s}$,

$$\langle e, f \rangle = \langle e, f \rangle_0 = \int_{S^*M} e \cdot \bar{f} \, d\mu \ . \tag{41.23}$$

With this choice of the antidual space the adjoint to an operator $\sigma \in L(\mathcal{H}_s, \mathcal{H}_{s-r})$ belongs to $L(\mathcal{H}_{s-r}, \mathcal{H}_{-s})$ and we shall denote it by σ^*.

In particular if $\sigma \in \Sigma_{sr}$ then $\sigma^* \in \Sigma_{r-s,r}$.

(2) As in 41.5 (2) we can take $\mathcal{H}_s$ as the antidual to itself and set

$$\langle e, f\rangle_s = \left\langle \tilde{\Lambda}^s e, \tilde{\Lambda}^s f\right\rangle_0 = \int_{S^*M} e \cdot \bar{f}\, \mathrm{d}\mu \tag{41.24}$$

where $\tilde{\Lambda}^s$ is the operator of multiplication by the function $\|\eta\|^s$.

In this case the adjoint operator to $\sigma \in L(\mathcal{H}_s, \mathcal{H}_{s-r})$ belongs to $L(\mathcal{H}_{s-r}, \mathcal{H}_s)$ and we shall denote it by $\sigma^\#$.

The next result shows that the 'homogeneous' symbol indeed establishes the symbolic calculus for $\mathcal{B}$ with the properties we are looking for.

Theorem 41.14 *('Homogeneous' sr-symbolic calculus of G-PDOs) Let G act topologically freely on M via diffeomorphisms σ_g (or let G be countable and act normally on M). Then the mapping*

$$\sigma : \mathcal{B} \to \Sigma$$

which coincides on every $\mathcal{B}_{sr}$ with the mapping of taking the 'homogeneous' sr-symbol (41.22) establishes the isomorphism of pseudoalgebras

$$[\mathcal{B}] \overset{\sigma}{\cong} \Sigma. \tag{41.25}$$

In particular an operator $\ell \in \mathcal{B}_{sr}$ is Fredholm iff $\sigma_{sr}(\ell)$ is invertible.

Moreover we have

$$\sigma_{sr}(\ell^*) = \sigma_{sr}(\ell)^* \tag{41.26}$$

and

$$\sigma_{sr}(\ell^\#) = \sigma_{sr}(\ell)^\#. \tag{41.27}$$

Proof Let us reduce this theorem to Theorem 41.9 on s-symbols. Let $\sigma_{sr}(\ell)$ have the form (41.22). Consider the operator

$$\sigma_{00} = \tilde{\Lambda}^{s-r} \sigma_{sr}(\ell) \tilde{\Lambda}^{-s} \tag{41.28}$$

acting in $\mathcal{H}_0$.

Identifying $\mathcal{H}_0$ with $L^2(S^*M)$ one can easily verify that

$$\sigma_{00} = \sigma_0(R(\mathcal{b})) = \sigma_s(\mathcal{b})$$

where R is the reduction to $\mathcal{B}_{00}$ and $\sigma_s(\mathcal{b})$ is the s-symbol.

Now the statement of the theorem follows from Theorem 41.9 and Lemma 41.7. □

42. The symbolic calculus of G-PDOs on vector bundles. The main theorem.

In this section we consider the general situation when G-PDOs act on vector bundles and obtain *the main theorem on the symbol of a G-PDO*. In fact the results of the preceding Sections 40, 41 provide us with the apparatus which makes it possible to carry out the corresponding investigation with minor difficulties.

We start with the consideration of G-PDOs of order 0 in $H^0(E)$, that is we shall construct the 0-symbol.

42.1 Let E be a complex m-dimensional vector bundle over a compact smooth n-dimensional manifold M (see 2.28-2.32). Let $G = \{g\}$ be a discrete group, $\{\alpha_g\}$ be a given action of G by diffeomorphisms of M and θ_g be a given action of G on E by smooth linear extensions θ_g of α_g which preserve the Hermitian metrics on E (recall that θ_g is called the linear extension of α_g if its restriction $\theta_{g,x}$ onto a fibre E_x maps this fibre linearly into $E_{\alpha_g(x)}$).

Remark We do not discuss here the possibility of choosing such a θ_g (for a given α_g) and the method of its definition.

In the Sobolev space $H^s(E)$ of cross-sections of E (see 2.40) we define the operators T_g, $g \in G$, via the formula

$$\left(T_g u\right)(x) = \left|\alpha_g^{-1'}(x)\right|^{\frac{1}{2}} \theta_g\left(u\left(\alpha_g^{-1}(x)\right)\right) \tag{42.1}$$

(cf. (40.2)).

The operators $\left\{T_g\right\}_{g \in G}$ form a unitary representation of G in $H^0(E)$.

Consider the operators $\mathcal{b}$ of the form

$$\mathcal{b} = \sum_{g \in F} a_g T_g,\ a_g \in \mathcal{A}^0, |F| < \infty \tag{42.2}$$

where $\mathcal{A}^0$ is the algebra of PDOs of zero-order acting in $H^0(E)$ (see 2.45).

We shall call these operators the *G-PDOs* of order 0 acting in $H^0(E)$ (associated with the group of linear extensions $\{\theta_g\}_{g \in G}$ of diffeomorphisms $\{\alpha_g\}_{g \in G}$).

Let

$$\mathcal{B} = C^*\left(\mathcal{A}^0, T_g\right) \subset L\left(H^0(E)\right)$$

be the C^*-algebra generated by the operators of the form (42.2). We shall call this algebra the *algebra of G-PDOs* (of order 0 associated with the group $\{\theta_g\}_{g \in G}$ of linear extensions of diffeomorphisms $\{\alpha_g\}_{g \in G}$).

42.2 Before considering the construction of the symbols of elements of $\mathcal{B}$ let us recall certain notions and introduce a number of objects.

The canonical projection $\pi: S^* M \to M$ defines the bundle E_2 over $S^* M$ given by

$$E_2 = \pi^* E\ . \tag{42.3}$$

Remark It is worth mentioning that if $x \in M$ is fixed then the fibres $E_{2\xi}$, $\xi \in \pi^{-1}(x)$, 'do not depend' on ξ and all of them are copies of E_x.

The principal symbol $\sigma(a)$ of an operator $a \in \mathcal{A}^0$ is an element of $\mathrm{HOM}(E_2)$ (see Theorem 2.46) and any element $\bar{a} \in \mathrm{HOM}(E_2)$ defines the operator a of 'multiplication' in $H^0(E_2) = L^2(E_2)$, that is acting via the formula

$$(au)(\eta) = \bar{a}(\eta) u(\eta)\ , \quad \eta \in S^* M\ .$$

The norm of a is equal to

$$\|a\| = \sup_{\xi \in S^* M} \|\bar{a}(\xi)\|$$

where $\|\bar{a}(\xi)\|$ is the norm of a linear operator $\bar{a}(\xi): E_{2\xi} \to E_{2\xi}$ in the given Hermitian metric.

According to Theorem 2.46 the algebra A generated by these operators a in $H^0(E_2)$ is isomorphic to the C^*-algebra of symbols of operators from $\mathcal{A}^0$. In particular, we have

$$\hat{A} = \text{Prim } A = \text{Prim HOM}(E_2) = S^* M \tag{42.4}$$

(HOM(E_2) is a *continuous field of elementary* C^**-algebras* HOM$(E_{2\eta})(\eta \in S^* M)$ isomorphic to the algebra of $m \times m$ matrices (see Dixmier [1], 10.4.4)).

The diffeomorphisms α_g of M generate the diffeomorphisms β_g of $S^* M$ defined by (40.3). The linear extensions θ_g generate in a natural way the linear extensions $\overset{\approx}{\theta}_g$ of β_g; namely, if $e_\eta \in E_{2\eta}$ is such that

$$e_\eta = \pi^* e_x \cap E_{2\eta}$$

for a certain $e_x \in E_x$ with $\pi(\eta) = x$ (that is e_η is the inverse image of e_x in $E_{2\eta}$ for the mapping $\pi \cdot p$) then we define $\overset{\approx}{\theta}_g e_\eta$ by

$$\overset{\approx}{\theta}_g e_\eta = \pi^* \left[\theta_g e_x\right] \cap \left[\pi^* E\right]_{\beta_g(\eta)} \tag{42.5}$$

(that is, $\overset{\approx}{\theta}_g e_\eta$ is the inverse image of $\theta_g e_x$ in $E_{2\beta_g(\eta)} = \left[\pi^* E\right]_{\beta_g(\eta)}$ for the mapping $\pi \cdot p$).

In particular, in local coordinates (x, ξ, z), $x \in X$, $\xi \in S^* M_x$, $z \in E_x$, $\overset{\approx}{\theta}_g$ has the form

$$\overset{\approx}{\theta}_g(x, \xi, z) = \left(\beta_g(x, \xi), \theta(z)\right) \tag{42.6}$$

(cf. the corresponding object in 2.47).

Since, by assumption, θ_g preserves the Hermitian metric on E, $\overset{\approx}{\theta}_g$ also preserves the induced Hermitia metric on E_2. In view of this we can define the operators V_g in $H^0(E_2)$ by a formula analogous to that for T_g (42.1), namely

$$\left(V_g \psi\right)(\eta) = \left|\beta_g^{-1\prime}(\eta)\right|^{\frac{1}{2}} \overset{\approx}{\theta}_g\left(\psi\left(\beta_g^{-1}(\eta)\right)\right) \tag{42.7}$$

(cf. (40.6)).

The operators $\{V_g\}_{g\in G}$ form a unitary representation of G in $H^0(E_2)$.

42.3 Now let us define the symbols of G-PDOs. In view of the considerations of 42.2 we are identifying the symbols $\sigma(a)$, operators $a \in \mathcal{A}^0$ and the corresponding multiplication operators from $A \subset L(L^2(E_2))$.

We shall define the *symbol of a G-PDO* in $H^0(E)$ of order 0 defined via (42.2) to be the operator $\sigma(\ell) \in L(L^2(E_2))$ given by the formula

$$\sigma(\ell) = \sum_{g\in F} \sigma(a_g) V_g \tag{42.8}$$

where V_g is defined by (42.7) (cf. (40.10)).

We denote by

$$B = C^*(A, V_g) \subset L(L^2(E_2))$$

the C^*-algebra generated by $A \cong \sigma(\mathcal{A}^0)$ and V_g.

Now we present the vector-bundle generalization of Theorems 40.7 and 40.12.

Theorem 42.4 *(The main theorem on the symbol of a G-PDO) Let G be a discrete amenable group acting normally on M via diffeomorphisms α_g. Then the mapping*

$$\sigma : \ell \to \sigma(\ell)$$

*(see (42.8)) can be extended up to the *-epimorphism*

$$\sigma : \mathcal{B} \to B$$

($\mathcal{B}$ and B are the algebras defined in 42.1 and 42.3), the kernel of which is the ideal K of compact operators in $L^2(E)$.

In particular the symbol of a G-PDO possesses the properties (1)-(6) listed in the statement of Theorem 40.7.

Proof The argument is analogous to that used in the proofs of Theorems 40.7 and 40.12.

We have to establish the isomorphism

$$\mathcal{B}/K = [\mathcal{B}] \overset{\sigma}{\cong} B . \tag{42.9}$$

Consider first the case of a topologically freely acting group. The argument here is *no* different from that exploited in the proof of Theorem 40.7 when verifying (40.11) (note in particular that by (42.4)

$$\text{Prim } A = S^* M$$

which coincides with (40.13)).

Now let us pass to the case of a normally acting group. This situation can be treated in an analogous way to that presented in the proof of Theorem 40.12 using Corollary B.30 instead of Theorem 16.21.

Indeed the C^*-algebra $[\mathcal{B}] = C^*([\mathcal{A}^0], [T_g])$ satisfies the condition (i) of B.26 for the same reasons as those mentioned in the proof of Theorem 40.12 (concerning the conditions (1) and (2) of Theorem 16.21).

The ideal I_g here is also given by (40.18) and the property (ii) of B.26 can be verified by reasoning analogous to that used for the verification of property (3) of Theorem 16.21 in the proof of Theorem 40.12 (by exploiting the functions φ_n (40.19), (40.20)).

After this is done the proof can be completed in the same way as the proof of Theorem 40.12 by applying Corollary B.30 instead of Theorem 16.21. □

42.5 Now let us consider the symbolic calculus in Sobolev spaces. In view of the derivation of the principal instrument - Theorem 42.4 - this job can be done easily following the ideas of Section 41.

Thus, for the space $H^s(E)$ consisting of cross-sections of a vector bundle E, just as in the scalar case (see 41.1), there exists a family of PDOs Λ^t, $t \in \mathbb{R}$ such that Λ^t establishes the isomorphism between $H^s(E)$ and $H^{s-t}(E)$ and

$$\sigma(\Lambda^t)(\eta) = \|\eta\|^t, \; \eta \in T_0^* M$$

Thus we can reduce word for word as in the scalar case, the consideration of G-PDOs of order r acting from $H^s(E)$ into $H^{s-r}(E)$ to the consideration of G-PDOs of order 0 acting in $L^2(E)$ and obtain the corresponding analogue of Lemma 41.7 and the construction of the *s-symbol* (see 41.8) and the *s*-symbolic calculus (Theorem 41.9).

We can also construct, in a natural way, the analogue of the '*homogeneous' sr-symbol* belonging to the space $L(\mathcal{H}_s(E_2), \mathcal{H}_{s-r}(E_2))$ (where $\mathcal{H}_t(E_2)$ is the space of positive homogeneous cross-sections of degree $(-t)$ of the bundle E_2 (recall (42.3))); and the argument of 41.11-41.14 brings us to the derivation of the *'homogeneous' sr-symbolic calculus* for the G-PDOs acting on a vector bundle (the vector bundle analogue of Theorem 41.14). We leave the details to the reader.

43. *G*-PDOs with a finite group of shifts.

The results obtained in Sections 40-42 are evidently true in the situation when a group of diffeomorphisms generating the shift operators is finite. Nevertheless this particular case is worthy of individual consideration for a number of reasons.

Firstly in this case the conditions for the invertibility of the symbol of a G-PDO have a simple algebraic character (cf. Section 15).

Secondly here it is possible to reduce the investigation of a G-PDO to that of a PDO of higher dimension (*without shifts*) and to apply the known results on elliptic PDOs. For this reason the finite group situation for G-PDOs had been studied before than the construction of the general C^*-theory presented in Volume I was developed.

Thirdly when calculating the index of G-PDOs with a finite group of shifts we discover a natural interrelation between this and equivariant K-theory and in terms of the latter theory we manage to develop the index of G-PDO.

One more peculiarity of the finite group case (related in fact to compactness) is that in this case the Fredholmity conditions for G-PDOs do not depend on the index of the Sobolev space in which they are considered (cf. Section 41).

All these and certain other features give us the possibility of investigating G-PDOs with a finite group of shifts in more detail, and to consider this particular situation not simply as an example of the application of the general theory.

A number of ideas and results presented in this section in fact is not related *directly* to the nature of G-PDOs but has a general C^*-algebraic character. This part of the theory is carried out in Appendix A.

43.1 Let M be a compact smooth manifold and $\alpha_g : M \to M$, $g \in G$, be a given finite group of diffeomorphisms of M. It is known in the case of a finite (or simply compact) group of diffeomorphisms that one could choose a Riemann metric on M to be invariant with respect to the action of this group. For example, such metrics could be obtained from an arbitrary Riemann metric by means of averaging over the action of the group.

In view of the invariance of the Riemann metric the differential $\alpha_g^{-1\prime}$ is defined (on every fibre of the tangent bundle) by an orthogonal matrix and thus in all the formulae of Sections 40-42 where we used the factors $\left|\alpha_g^{-1\prime}(x)\right|$, $x \in M$ and $\left\|\partial\alpha_g^{-1}(\eta)\right\|$, $\eta \in S^*M$, these factors are equal to 1. Thus we can omit them. In particular it follows that the s-symbol (see 41.8) of a G-PDO with a finite group does not depend on s.

43.2 Henceforth we assume that there is fixed on M a certain invariant Riemann metric with respect to $\{\alpha_g\}_{g\in G}$.

If we consider the operators acting in cross-sections of a vector-bundle E over M then we could also choose a Hermite metric on E to be invariant with respect to the action of G (by linear extensions θ_g of α_g), and we shall presume this to be fulfilled as well.

Under these assumptions the operators

$$(T_g u)(x) = \theta_g\left(u\left(\alpha_g^{-1}(x)\right)\right) \tag{43.1}$$

(cf. (42.1)) are unitary in $L^2(E)$.

Consider a G-PDO $\mathcal{b}: H^s(E) \to H^{s-r}(E)$ of the form

$$\mathcal{b} = \sum_{g\in G} a_g T_g \tag{43.2}$$

where a_g are PDOs of order r.

The operator V_g (42.7) acting in $H^0(E_2)$ (E_2 is defined in (42.3)) in this situation is also simplified and has the form

$$(V_g \psi)(\eta) = \tilde{\tilde{\theta}}_g\left(\psi\left(\left(\partial\alpha_g\right)^{-1}(\eta)\right)\right) \tag{43.3}$$

and the *s-symbol* of $\mathcal{b}$ (43.2) is the operator $\sigma_s(\mathcal{b}): H^0(E_2) \to H^0(E_2)$ given by

$$\sigma_s(\mathcal{b}) = \sigma_0(\mathcal{b}) = \sum \sigma\left(a_g\right) V_g \tag{43.4}$$

(cf. (41.18)).

Remark 43.3 All these simplifications in fact hold not only for the finite group case but also for an infinite group in the situations when a Riemann metric on M is invariant with respect to the group action (in particular we could always build such metrics if the action of the group is included in an action of a certain compact group).

43.4 Since G is finite one could write out the condition on the group to act topologically freely on M in a simpler way.

The point is that any diffeomorphism α_g belonging to a finite group is periodic and thus the set M_g (40.14) of the fixed points of α_g consists of a finite number of submanifolds of M (see Conner, Floyd [1], §22), and consequently if M is connected and every α_g, $g \neq e$, is non-trivial then the set $\bigcup_{g \neq e} M_g$ is a nowhere dense subset of M which means that in this situation G acts topologically freely.

If M is not connected then the submanifold M_g for a non-trivial diffeomorphism α_g may contain a connected component of M (if α_g acts trivially on this component). Thus in this case M_g may consist only of certain connected components of M and a finite number of submanifolds of smaller dimensions on the rest of the connected components. Thus a finite group G of diffeomorphisms of M acts topologically freely on M iff *any diffeomorphism* α_g, $g \neq e$, *is non-trivial on any connected component of* M.

If the action of G possesses this property we say that G acts *effectively* on M.

Remarks 43.5 (1) We have already pointed out that the basic property which here makes the topologically free action coincide with the effective action is that every α_g, $g \in G$, is periodic. If we consider non-periodic actions then this coincidence is no longer true and the sets M_g do not necessarily consist of a finite number of submanifolds of M (cf. Section 47).

(2) Since (as has already been observed) in the case when M is connected the action of a finite group G is topologically free iff every α_g, $g \neq e$, is non-trivial, it is quite natural that, in the study of boundary value problems with a shift for analytical functions and of singular integral equations with a shift on a curve, the topologically free action condition did not appear while only the case of periodic mappings was under consideration.

Now Theorem 41.9 (reinterpreted for the vector bundle situation (see Section 42)) implies

Theorem 43.6 *Let a finite group G act effectively via diffeomorphisms $\{\alpha_g\}_{g \in G}$ of M. Then the mapping*

$$\mathcal{b} \to \sigma_0(\mathcal{b})$$

(here $\mathcal{b}$ is given by (43.2) and $\sigma_0(\mathcal{b})$ by (43.4)) possesses the properties mentioned in the statement of Theorem 41.9 and in particular the operator $\mathcal{b}$ is Fredholm (as acting from $H^s(E)$ into $H^{s-r}(E)$) iff its symbol $\sigma_0(\mathcal{b})$ is invertible.

43.7 Applying the procedure described in 41.6 (reinterpreted for the vector bundle case) we associate with the operator $\mathcal{b}$ (43.2) the operator $R(\mathcal{b})$ acting in $H^0(E)$ and having the same symbol (43.4) (recall the proof of Lemma 41.7). It follows from Lemma 41.7 that:

(i) *$\mathcal{b}$ is Fredholm iff $R(\mathcal{b})$ is Fredholm;*

(ii) *if $\mathcal{b}$ is Fredholm then*

$$\operatorname{ind} \mathcal{b} = \operatorname{ind} R(\mathcal{b}) \,. \tag{43.5}$$

Set

$$\tilde{H} = \left[H^0(E)\right]^{|G|} = \underset{g\in G}{\oplus} H_g \quad \left(H_g \approx H^0(E)\right). \tag{43.6}$$

Clearly

$$\tilde{H} \approx H^0(\tilde{E}) \tag{43.7}$$

where

$$\tilde{E} = \underset{g\in G}{\oplus} E_g \left(E_g \approx E\right). \tag{43.8}$$

From the operator $R(\mathcal{b})$ by means of the construction A.3 one could build the operator

$$\tilde{\mathcal{b}} \overset{\text{def}}{=} \widetilde{R(\mathcal{b})} \tag{43.9}$$

acting in $\tilde{H}$.

Clearly $\tilde{\mathcal{b}}$ is a PDO of order zero and its symbol (see 2.45) $\sigma\left(\tilde{\mathcal{b}}\right) \in \mathrm{HOM}\left(\tilde{E}_2\right)$ is given in the block-matrix form

$$\sigma\left(\tilde{\mathcal{b}}\right) = \left\{b_{h,g}\right\}_{h,g\in G} \tag{43.10}$$

where

$$b_{h,g} = \hat{V}_h\left(\sigma\left(a_{h^{-1}g}\right)\right) \tag{43.11}$$

(here V_h and $\sigma(a_g)$) are those used in (43.4)).

Now the foregoing Theorem 43.6 along with Lemma A.4 and Theorem 2.46 imply

Theorem 43.8 *Let a finite group G act effectively on M. Then the operator* $\mathcal{L}$ *(43.2) is Fredholm iff the operator* $\tilde{\mathcal{L}}$ *(43.9) is Fredholm and thus* $\mathcal{L}$ *is Fredholm iff* $\sigma(\tilde{\mathcal{L}})$ *(43.10) is invertible.*

In fact the results of Appendix A give us also a method to calculate $\operatorname{ind}\mathcal{L}$ in terms of $\tilde{\mathcal{L}}$; namely, Theorem A.17 combined with the previous theorem gives us

Theorem 43.9 *Let* $\mathcal{L}$ *be the operator (43.2) and* $\tilde{\mathcal{L}}$ *be the operator (43.9) and let G act effectively on M. Then* $\tilde{\mathcal{L}}$ *is the direct sum of the operators* $\tilde{\mathcal{L}}^i$, $i=\overline{i,p}$, *acting in the subspaces* F_i (= Im P_i, P_i *are given by (A.22)). Moreover* $\tilde{\mathcal{L}}^1$ *is conjugate to the initial operator* $\mathcal{L}$ *and if* $\mathcal{L}$ *(or* $\tilde{\mathcal{L}}$ *) is Fredholm then*

$$\operatorname{ind}\mathcal{L} = \operatorname{ind}\tilde{\mathcal{L}}^1 .$$

Extrapolating the reasoning of Appendix A and applying Theorem A.23 to the situation being considered we obtain

Theorem 43.10 *Let the operator* $\tilde{\mathcal{L}}$ *in (43.9) be Fredholm. Then:*

(i) *the index of the operator* $\mathcal{L}$ *(43.2) is equal to the multiplicity of the trivial representation of the group G in the decomposition of* $G\text{-}\operatorname{ind}\tilde{\mathcal{L}}$ *(A.27); and*

(ii) *it can be written out via the formula*

$$\operatorname{ind}\mathcal{L} = \frac{1}{|G|}\sum_g L(g,\tilde{\mathcal{L}}) . \tag{43.12}$$

Let us consider a simple example demonstrating the objects arising in the foregoing theorems.

Example 43.11 Let G be the group consisting of two elements $G=\{e,g_1\}$ with $g_1^2=e$ and let the operator $\mathcal{b}$ acting in $H^0(E)$ have the form

$$\mathcal{b}=a_e+a_{g_1}T_{g_1}\ .$$

The space $\tilde{H}=H^0(\tilde{E})$ (43.7) consists of pairs of sections (u_0,u_1) and the operator $\check{T}_{g_1}$ (A.18) is given by

$$\check{T}_{g_1}(u_0,u_1)=(T_{g_1}u_1,T_{g_1}u_0)$$

(see (A.19)).

There are only two non-equivalent irreducible representations of G. Both are one-dimensional and defined by

$$\nu_1{:}\nu_1(g_1)x=x,\quad \chi_1(e)=\chi_1(g_1)=1,$$

$$\nu_2{:}\nu_2(g_1)x=-x,\ \chi_2(e)=1,\chi_2(g_1)=-1.$$

The subspaces F_1 and F_2 (see A.15) here have the form

$$F_1=\left\{\bar{u}\in\tilde{H}{:}\check{T}_{g_1}\bar{u}=\bar{u}\right\},$$

$$F_2=\left\{\bar{u}\in\tilde{H}{:}\check{T}_{g_1}\bar{u}=-\bar{u}\right\}.$$

The operator S_2 acting from F_2 into $H^0(E)$ via the formula

$$S_2(u_0,u_1)=u_0$$

with the inverse

$$S_2^{-1}u=(u,-T_{g_1}u)$$

establishes the conjugacy

$$\tilde{\mathcal{b}}^2 = S_2^{-1}\left(a_e - a_{g_1} T_{g_1}\right) S_2$$

(recall the reasoning in the proof of Theorem A.16 where the analogous operator was written out for F_1 and (A.22)).

Thus the decomposition

$$\tilde{\mathcal{b}} = \tilde{\mathcal{b}}^1 \oplus \tilde{\mathcal{b}}^2$$

mentioned in Theorem 43.9 has the form

$$\tilde{\mathcal{b}} = \left(a_e + a_{g_1} T_{g_1}\right) \oplus \left(a_e - a_{g_1} T_{g_1}\right)$$

and the formulae (A.32) and (43.12) here look like

$$\operatorname{ind}\left(a_e + a_{g_1} T_{g_1}\right) + \operatorname{ind}\left(a_e - a_{g_1} T_{g_1}\right) = \operatorname{ind} \tilde{\mathcal{b}} \ ,$$

$$\operatorname{ind}\left(a_e + a_{g_1} T_{g_1}\right) - \operatorname{ind}\left(a_e - a_{g_1} T_{g_1}\right) = L\left(g_1, \tilde{\mathcal{b}}\right),$$

$$\operatorname{ind}\left(a_e + a_{g_1} T_{g_1}\right) = \frac{1}{2}\left[\operatorname{ind} \tilde{\mathcal{b}} + L\left(g_1, \tilde{\mathcal{b}}\right)\right] . \tag{43.13}$$

Thus the problem of the calculation of the Lefschetz number $L\left(g_1, \tilde{\mathcal{b}}\right)$ is reduced in this situation to the calculation of the difference of the indices of two operators.

In a number of examples it could happen that the operator $a_e - a_{g_1} T_{g_1}$ is homotopic in the class of Fredholm operators to $a_e + a_{g_1} T_{g_1}$ (see, say, the proof of Theorem 47.7) and in such situations we have $L\left(g_1, \tilde{\mathcal{b}}\right) = 0$ and, the formula (43.13) is simplified to

$$\operatorname{ind} \mathcal{b} = \frac{1}{2} \operatorname{ind} \tilde{\mathcal{b}} \ .$$

Remarks (1) In general $L(g, \tilde{\mathcal{b}}) \neq 0$ (see Example 43.16).

(2) Following the reasoning of the above example one could consider the operators $\tilde{\mathcal{b}}^i$ arising in the decomposition of $\tilde{\mathcal{b}}$ into a direct sum in the case of an arbitrary finite group (by constructing the operators S_i related to the subspaces F_i mentioned in A.15). These

operators turn out to be conjugate to *G*-PDOs of dimensions $d_i \times \dim E$ $(d_i - \dim v_i)$ of the form

$$\sum_{g \in G} a_g^i T_g$$

where the PDOs a_g^i are built with the help of the coefficients a_g of $\mathcal{L}$ and the matrix elements of v_i.

43.12 In the papers by Atiyah and Singer [1] there is constructed a mapping t-ind (the topological index) by means of which, to every equivariant PDO (in fact to its symbol) there corresponds a certain element of $R(G)$ and it is proved that

$$G - \operatorname{ind} \tilde{\mathcal{L}} = \mathrm{t} - \operatorname{ind} \tilde{\mathcal{L}} .$$

Thus since, by Theorem 43.10, $\operatorname{ind} \tilde{\mathcal{L}}$ could be calculated in terms of $G - \operatorname{ind} \tilde{\mathcal{L}}$ the problem of the calculation of $\operatorname{ind} \mathcal{L}$ is solved in principle.

Atiyah, Segal [1] obtains an expression for the Lefschetz numbers $L(g, \tilde{\mathcal{L}})$ in terms of the restriction of the symbol of $\tilde{\mathcal{L}}$ onto the set of fixed points of the mapping α_g. In particular this expression implies that if the set of fixed points of α_g consists of only a finite number of points then $L(g, \tilde{\mathcal{L}}) = 0$. Thus Theorem 43.10 (ii), along with this observation, leads to

Theorem 43.13 *Let G be a finite group and* $\mathcal{L}$ *be the operator (43.2). If all the mappings* α_g, $g \in G$, $g \neq e$, *have only finite sets of fixed points then*

$$\operatorname{ind} \mathcal{L} = \frac{1}{|G|} \operatorname{ind} \tilde{\mathcal{L}} . \tag{43.14}$$

Remark 43.14 The formula (43.14) is true in particular in the one-dimensional situation (that is, for singular integral operators on a curve) since a non-trivial periodic mapping of a simple curve may have only a finite number of fixed points.

The direct approach to the derivation of the formula (43.14) for *G*-SIOs (not using *K*-theoretical reasoning) is presented in Theorem 47.7.

We would like to stress once more (see Example 43.16) that in general (on manifolds of dimensions $m \geq 2$) the simpler formula (43.14) is no longer true and one has to use (43.12).

43.15 The principal results of this section were obtained under the assumption that G acts effectively on M. Now we would like to observe that in the case of a finite group G there is a sense in which one could always confine himself to the effective action situation.

Thus, let a manifold M be not connected and let the action of a finite group G be not effective. Let M_0 be a certain connected component of M. Set

$$\tilde{M}_0 = \bigcup_{g \in G} \alpha_g(M_0) . \tag{43.15}$$

The submanifold $\tilde{M}_0$ consists of a finite number of connected components

$$\tilde{M}_0 = M_0 \cup M_1 \cup \ldots \cup M_p$$

where $M_k = \alpha_{g_k}(M_0)$ for some g_k such that $\alpha_{g_l}(M_0) \neq \alpha_{g_k}(M_0)$, $l \neq k$. If we consider the operator $\mathfrak{b}$ (43.2), say, in the space $L^2(E) = L^2(M,E)$ (all a_g in this case have zero order) then the spaces $L^2(\tilde{M}_0, E)$ and $L^2(M \setminus \tilde{M}_0, E)$ are both invariant with respect to $\mathfrak{b}$. Thus $\mathfrak{b}$ decomposes into the corresponding direct sum of operators and respectively the investigation of $\mathfrak{b}$ could be reduced to the investigation of the operator $\mathfrak{b}_0$ (having the same form) in $L^2(\tilde{M}_0, E)$.

Observe that the space $L^2(\tilde{M}_0, E)$ is isomorphic to $\left[L^2(M_0, E)\right]^p$ and the isomorphism is given by

$$L^2(\tilde{M}_0, E) \ni u \xrightarrow{J} \left(u(x), \theta_{g_1}^{-1}\left(u\left(\alpha_{g_1}(x)\right)\right), \ldots, \theta_{g_p}^{-1}\left(u\left(\alpha_{g_p}(x)\right)\right)\right) \in \left[L^2(M_0, E)\right]^p. \tag{43.16}$$

Under this isomorphism the pseudodifferential operators a_g are transformed into the pseudodifferential operators acting in the space

$$L^2(M_0, \tilde{E}) = L^2\left(M_0, \bigoplus_{i=1}^{p} E_i\right), \quad E_i \approx E|_{M_0} , \tag{43.17}$$

and the action of the operator T_g (43.1) in this space could be written out in the following way. For any vector

$$\xi \in L^2(M_0, E)$$

we have

$$T_{g_i}\xi = f_i \in L^2(M_i, E), \quad i = \overline{i, p}$$

and if

$$T_g f_i \in L^2\left(M_{k_i}, E\right) \text{ for some } 0 \le k_i \le p$$

then there exists a vector $v \in L^2(M_0, E)$ such that

$$T_{g_{k_i}} v = T_g f_i$$

and consequently

$$\left(T_{g_{k_i}}^{-1} T_g T_{g_i}\right)\xi = v \ ,$$

which means that under the isomorphism (43.16) the operator T_g is transformed into the operator acting from $L^2(M_0, E_i)$ into $L^2\left(M_0, E_{k_i}\right)$, $i = \overline{1, p}$ $\left(E_i = E_{k_i} = E|_{M_0}\right)$ by the rule

$$L^2(M_0, E) \xrightarrow{T_{g_{k_i}}^{-1} T_g T_{g_i} = T_{g_{k_i}^{-1} g g_i}} L^2(M_0, E) \ . \tag{43.18}$$

In particular the diffeomorphism of M_0 corresponding to this operator is $\alpha_{g_{k_i}^{-1} g g_i}$.

Considering all the elements $g \in G$ and for these elements all the operators arising in (43.18) we obtain a certain subgroup of the restrictions of the operators T_h, $h \in G$ (leaving $L^2(M_0, E)$ invariant) onto $L^2(M_0, E)$. These operators form a representation of a certain subgroup $G_0 \subset G$ in $L^2(M_0, E)$ via formula (43.1).

Let $G^0 \subset G_0$ be the subgroup of G_0 consisting of the elements τ such that $\alpha_\tau(x) = x$, $x \in M_0$. G^0 is a normal subgroup.

Set

$$\overline{G} = {}^{G_0}\!/_{G_0} \ . \tag{43.19}$$

By the definition of $\overline{G}$ we have that

if $g_1, g_2 \in G_0$ *belong to one and the same equivalence class of* $\overline{G}$ *then* $\alpha_{g_1} = \alpha_{g_2}$ *(which means that* $\alpha_{[g]}$, $[g] \in \overline{G}$, *is well defined) and the group* $\{\alpha_{[g]}\}_{[g]\in G}$ *of homeomorphisms of* M_0 *acts effectively.*

Choosing, for every $[g] \in \overline{G}$, an operator $T_{\delta_{[g]}}$, $\delta_{[g]} \in [g]$, in such a way that

$$\delta_{[g]} = \left(\delta_{[g]^{-1}}\right)^{-1}$$

and

$$\delta_{[e]} = e$$

(which can always be done) we see that the family $\left\{T_{\delta_{[g]}}\right\}_{[g]\in\overline{G}}$ forms a *quasirepresentation* of $\overline{G}$ (see B.4) with the cocycle

$$u: \overline{G} \times \overline{G} \to \mathrm{HOM}\left(E|_{M_0}\right).$$

Observing now that any element of $\mathrm{HOM}\left(E|_{M_0}\right)$ could be identified with the pseudodifferential operator induced by it in $L^2(M_0, E)$ we conclude that the foregoing reasoning implies the next statement.

Under the isomorphism (43.16) a G-PDO $\mathfrak{b}_0$ *is transformed into a G-PDO* $\bar{\mathfrak{b}}_0$ *having the form*

$$\sum_{h\in\overline{G}} \overline{a}_h \overline{T}_h$$

where $\overline{G}$ *is defined by (43.19) and the operators* $\overline{T}_h$, $h \in \overline{G}$, *form a quasirepresentation (B.4) of* $\overline{G}$ *with a cocycle* $\overline{u} \in \mathrm{HOM}(\tilde{E})$ *(*$\tilde{E}$ *is given by (43.17)) and* $\overline{G}$ *acts effectively on* M_0 *via homeomorphisms* $\{\alpha_h\}_{h\in\overline{G}}$.

Remarks (1) In particular if $E|_{M_0}$ is a trivial vector bundle (which is of course true when we consider *G*-PDOs in the spaces of *scalar* functions) then one can choose $\overline{u} \equiv I$

(the trivial homomorphism) and consequently in this case $\{\overline{T}_h\}_{h\in\overline{G}}$ will form a *representation* of $\overline{G}$.

(2) $\{\overline{T}_h\}_{h\in\overline{G}}$ will also form a representation of $\overline{G}$ even in the situation when $E|_{M_0}$ is not trivial if we presume that

$$\textit{if } \alpha_g(x)=x, x\in M_0 \textit{ then } T_g\big|_{L^2(M_0,E)}=I\ .$$

Example 43.16 As we have already remarked, in general the formula (43.14) is not true due to the fact that $L(g,\widetilde{\ell})$ is not necessarily equal to zero. Here we present an example of a G-PDO of this type.

Let $\mathbb{T}^2$ be the torus realized as a factor space $\mathbb{R}^2/\mathbb{Z}^2$. We shall denote the coordinates on this space by $x=(x_1,x_2)$. As it is known that the cotangent bundle $T^*\mathbb{T}^2$ of $\mathbb{T}^2$ is trivial, $T^*\mathbb{T}^2=\mathbb{T}^2\times\mathbb{R}^2$ and the bundle $S^*\mathbb{T}^2$ of the normalized cotangent vectors has the form $\mathbb{T}^2\times S^1$. We shall denote the coordinates on $T^*\mathbb{T}^2$ by $(x,\xi)=(x_1,x_2,\xi_1,\xi_2)$ and therefore $\mathbb{T}^2\times S^1$ is given by $\{(x_1,x_2,\xi_1,\xi_2):\xi_1^2+\xi_2^2=1\}$.

Let $\alpha:\mathbb{T}^2\to\mathbb{T}^2$ be a diffeomorphism defined by the formula

$$\alpha(x_1,x_2)=(x_1,-x_2)\ .$$

Clearly $\alpha^2(x)=x$, that is $\alpha=\alpha^{-1}$, and thus (e,α) is a group consisting of two elements. We also have that

$$\partial\alpha(x_1,x_2,\xi_1,\xi_2)=(x_1,-x_2,\xi_1,-\xi_2)\ , \tag{43.20}$$

and the set of fixed points of α consists of two circles

$$\{(x_1,x_2):x_2=0\} \text{ and } \{(x_1,x_2):x_2=\tfrac{1}{2}\}\ .$$

In the space $L^2(\mathbb{T}^2)$ we consider the shift operator T given by

$$(Tf)(x_1,x_2)=(T_\alpha f)(x_1,x_2)=f(x_1,-x_2)\ .$$

We shall construct the operator

$$\mathfrak{b}=\mathfrak{a}_0+\mathfrak{a}_1 T \tag{43.21}$$

(where $\mathfrak{a}_i$, $i=0,1$ are pseudodifferential operators in $L^2(\mathbb{T}^2)$ of zero order) such that for the corresponding operator $\widetilde{\mathfrak{b}}$ in (43.9)

$$L(\alpha,\widetilde{\mathfrak{b}})=1\ . \tag{43.22}$$

To obtain (43.22) we shall build $\widetilde{\mathfrak{b}}$ in such a way that its symbol $\sigma(\widetilde{\mathfrak{b}})$ possesses certain special property (described below) which implies (43.22).

The desired property is the following.

Let $\sigma(\widetilde{\mathfrak{b}})$ be the symbol of $\widetilde{\mathfrak{b}}$. $\sigma(\widetilde{\mathfrak{b}})$ can be written out as a matrix function on $S^*\mathbb{T}$ of the form

$$d(x_1,x_2,\xi_1,\xi_2)=\begin{bmatrix} a_{11}(x_1,x_2,\xi_1,\xi_2) & a_{12}(x_1,x_2,\xi_1,\xi_2) \\ a_{21}(x_1,x_2,\xi_1,\xi_2) & a_{22}(x_1,x_2,\xi_1,\xi_2) \end{bmatrix} \tag{43.23}$$

where a_{11} and a_{12} are the symbols of the operators $\mathfrak{a}_0$ and $\mathfrak{a}_1$ in (43.21), respectively, and (in view of (43.20) and (43.10), (43.11))

$$\begin{aligned} a_{22}(x_1,x_2,\xi_1,\xi_2)&=a_{11}(x_1,-x_1,\xi_1,-\xi_2) \\ a_{21}(x_1,x_2,\xi_1,\xi_2)&=a_{12}(x_1,-x_1,\xi_1,-\xi_2) \end{aligned}\ . \tag{43.24}$$

Observe that

if for the matrix (43.23) we have

$$d(x_1,0,1,0)=\begin{bmatrix} \cos 2\pi x_1 & -i\sin 2\pi x_1 \\ -i\sin 2\pi x_1 & \cos 2\pi x_1 \end{bmatrix}, \tag{43.25}$$

$$d(x_1,0,-1,0)=\begin{bmatrix} 1 & 0 \\ 0 & 1 \end{bmatrix} \tag{43.26}$$

and

$$d\left(x_1,\tfrac{1}{2},\pm 1,0\right)=\begin{bmatrix} 1 & 0 \\ 0 & 1 \end{bmatrix} \tag{43.27}$$

then (43.22) is true.

Indeed, to calculate $L(\alpha,\widetilde{\mathcal{L}})$ we shall apply the localization formula from Atiyah, Segal [1]. This formula expresses the Lefshetz number of an equivariant PDO in terms of the Lefshetz numbers of a PDO on M_α (see 43.4 with $g=\alpha$) such that its symbol is a restriction of the symbol $\sigma(\widetilde{\mathcal{L}})$ onto S^*M_α. The formula has the simplest form in the case when the normal bundle of M_α is trivial (which is satisfied in the situation being considered), since here we have

$$M_\alpha = M_1 \cup M_2 \tag{43.28}$$

where

$$M_1 = \{(x_1,x_2): x_2 = 0\} \quad \text{and} \quad M_2 = \{(x_1,x_2): x_2 = \tfrac{1}{2}\}).$$

Namely in our case this form is the following (see Atiyah, Segal [1], Theorem 2.1 and Remark 2 after Theorem 3.3):

Let $\widetilde{\mathcal{L}}^i$ be elliptic operators in $L^2(M_i,\mathbb{C}^2)$ (where M_i, $i=1,2$ are those used in (43.28)) the symbols of which are restrictions of the symbol of $\widetilde{\mathcal{L}}$ onto S^*M_i. Then

$$L(\alpha,\widetilde{\mathcal{L}}) = \frac{1}{\det\left(1-\alpha|_{N_1}\right)} L\left(\alpha,\widetilde{\mathcal{L}}^1\right) + \frac{1}{\det\left(1-\alpha|_{N_2}\right)} L\left(\alpha,\widetilde{\mathcal{L}}^2\right) \tag{43.29}$$

where N_i, $i=1,2$ are the normals to M_i bundles (in the tangent space TM) and the $\alpha|_{N_i}$ are correspondingly induced by α linear mappings on N_i.

In view of (43.29), to calculate $L(\alpha,\widetilde{\mathcal{L}})$ we have to calculate $L\left(\alpha,\widetilde{\mathcal{L}}^i\right)$ $i=1,2$. Since by (43.27) $\widetilde{\mathcal{L}}^2$ is the identity operator, it follows that $L\left(\alpha,\widetilde{\mathcal{L}}^2\right)=0$ and therefore it is enough to calculate $L\left(\alpha,\widetilde{\mathcal{L}}^1\right)$. The formulae (43.25) and (43.26) imply that $\widetilde{\mathcal{L}}^1$ is a singular integral operator of the form

$$\widetilde{\mathcal{L}}^1 = \begin{bmatrix} \cos 2\pi x_1 & -i\sin 2\pi x_1 \\ -i\sin 2\pi x_1 & \cos 2\pi x_1 \end{bmatrix} P + Q \tag{43.30}$$

where P and Q are the projections (46.11).

The operator

$$\check{T}_\alpha : L^2(M_1) \oplus L^2(M_1) \to L^2(M_1) \oplus L^2(M_1)$$

(see (A.18), (A.19)) in the situation being considered has the form

$$\check{T}_\alpha = \begin{bmatrix} 0 & 1 \\ 1 & 0 \end{bmatrix}.$$

Following the argument of A.15 and A.16 we observe that the operator $\widetilde{\ell}^{\,1}$ is the direct sum of the operators $\widetilde{\ell}^{\,1,i}$, $i = 1,2$ acting in the subspaces

$$F_i \subset L^2(M_1) \oplus L^2(M_1)$$

which are defined by the conditions

$$(u_1, u_2) \in F_1 \Leftrightarrow (u_1, u_2) = (u_2, u_1) \Leftrightarrow u_1 = u_2 ,$$

$$(u_1, u_2) \in F_2 \Leftrightarrow (u_1, u_2) = (-u_2, -u_1) \Leftrightarrow u_1 = -u_2 .$$

The mappings

$$S_i : L^2(M_1) \to L^2(M_1) \oplus L^2(M_1) , \quad i = 1,2$$

given by

$$S_1 f = (f, f) ,$$

$$S_2 f = (f, -f)$$

define the isomorphisms between $L^2(M_1)$ and F_i. Under these isomorphisms the operators $\widetilde{\ell}^{\,1,i}$ are transformed into the singular integral operators $\bar{\ell}^{\,1,i}$ acting in $L^2(M_1)$ and having the respective forms

$$\bar{\ell}^{\,1,1} = (\cos 2\pi x_1 - i \sin 2\pi x_1) P + Q$$

and (43.31)

$$\bar{\ell}^{\,1,2} = (\cos 2\pi x_1 + i \sin 2\pi x_1) P + Q .$$

Note now that by the definition of $\widetilde{\ell}^{\,1}$ and $\bar{\ell}^{\,1,i}$ and formula (A.32) we have

$$L\left(\alpha,\tilde{\ell}^{1}\right)=\operatorname{ind}\bar{\ell}^{1,1}-\operatorname{ind}\bar{\ell}^{1,2}\ . \tag{43.32}$$

Applying Theorem 2.54 we obtain

$$\operatorname{ind}\bar{\ell}^{1,1}=1\ ,$$

$$\operatorname{ind}\bar{\ell}^{1,2}=-1$$

and therefore (43.32) implies

$$L\left(\alpha,\tilde{\ell}^{1}\right)=2\ . \tag{43.33}$$

Routine computation shows that in the situation considered

$$\det\left(1-\alpha|_{N_1}\right)=2$$

which along with the equality $L\left(\alpha,\tilde{\ell}^{2}\right)=0$ already observed, (43.33) and (43.29) proves that if (43.25)-(43.27) are satisfied then (43.22) is true as desired.

Now to obtain the example we are looking for it is enough to present an invertible continuous matrix function $d(x_1,x_2,\xi_1,\xi_2)$ (43.23) on $S^*\mathbb{T}^2$ satisfying (43.24) and (43.25)-(43.27). So we proceed with the construction of such a matrix.

In fact we shall start with the matrix function defined on S^*M_1 by (43.25) and (43.26), and extend this matrix function up to a matrix function on $S^*\mathbb{T}^2$ satisfying (43.24) and (43.27).

The set $\{(x_1,0,1,0)\}$ is a circle $\mathbb{T}^1$ embedded in the torus $\mathbb{T}^3=S^*\mathbb{T}^2$. Consider a neighbourhood W of $\mathbb{T}^1$ consisting of the points (x_1,x_2,ξ_1,ξ_2) satisfying the conditions

$$|x_2|\le\tfrac{3}{8}\ ,\quad |\xi_2|\le\tfrac{1}{4}\ .$$

W is homeomorphic to $\mathbb{T}^1\times D$ where D is a rectangle in $\mathbb{R}^2$

$$D=\left\{(x_2,\xi_2):\ |x_2|\le\tfrac{3}{8},\ |\xi_2|\le\tfrac{1}{4}\right\}\ .$$

As the local coordinates in W we take (x_1,x_2,ξ_2).

The matrix we shall build will be equal to the identity out of W. Thus it is enough to describe its form in W, and on this set (bearing in mind the local coordinates introduced above) we shall use the notation $d(x_1,x_2,\xi_2)$ instead of $d(x_1,x_2,\xi_1,\xi_2)$.

The description will be given in a number of steps.

Step I

On the set $\left\{(x_1,x_2,0)\colon\ |x_2|\le\frac{1}{8}\right\}$ we set

$$d(x_1,x_2,0)=\begin{bmatrix}(1-|\tau|)\cos 2\pi x_1+\tau i & -i(1-|\tau|)\sin 2\pi x_1\\ -i(1-|\tau|)\sin 2\pi x_1 & (1-|\tau|)\cos 2\pi x_1-\tau i\end{bmatrix} \tag{43.34}$$

where $\tau=8x_2$.

And on the set $\left\{(x_1,x_2,0)\colon\ \frac{1}{8}\le|x_2|\le\frac{1}{4}\right\}$ we set

$$d(x_1,x_2,0)=\begin{bmatrix}(1-\tau)i+\tau & 0\\ 0 & (\tau-1)i+\tau\end{bmatrix}, \tag{43.35}$$

where $\tau=8|x_2|-1$.

Step II

On the set $\left\{(x_1,x_2,\xi_2)\colon\ |x_2|\le\frac{1}{4},\ 0\le\xi_2\le\frac{1}{8}\right\}$ we set

$$d(x_1,x_2,\xi_2)=d(x_1,x_2,0)(1-8\xi_2)+8\xi_2\begin{bmatrix}0 & 1\\ -1 & 0\end{bmatrix}$$

where $d(x_1,x_2,0)$ is given by (43.34) or (43.35) respectively.

Step III

On the set $\left\{(x_1,x_2,\xi_2)\colon\ |x_2|\le\frac{1}{4},\ \frac{1}{8}\le\xi_2\le\frac{1}{4}\right\}$ we set

$$d(x_1,x_2,\xi_2)=(1-\tau)\begin{bmatrix}0 & 1\\ -1 & 0\end{bmatrix}+\tau\begin{bmatrix}1 & 0\\ 0 & 0\end{bmatrix}$$

where $\tau=8\left(\xi_2-\frac{1}{8}\right)$.

Step IV

On the set $\left\{(x_1,x_2,\xi_2):\ \frac{1}{4}\le x_2 \le \frac{3}{8},\ 0\le \xi_2 \le \frac{1}{4}\right\}$ we set

$$d(x_1,x_2,\xi_2) = (1-\tau)d\left(x_1,\tfrac{1}{4},\xi_2\right)+\tau\begin{bmatrix}1 & 0\\ 0 & 1\end{bmatrix}$$

where $\tau = 8\left(x_2-\frac{1}{4}\right)$.

Finally, on the set $\left\{(x_1,x_2,\xi_2):\ -\frac{1}{4}\ge x_2 \ge -\frac{3}{8},\ 0\le \xi_2 \le \frac{1}{4}\right\}$

we set

$$d(x_1,x_2,\xi_2) = (1-\tau)d\left(x_1,-\tfrac{1}{4},\xi_2\right)+\tau\begin{bmatrix}1 & 0\\ 0 & 1\end{bmatrix},$$

where $\tau = 8\left(-x_2-\frac{1}{4}\right)$.

Step V

Let U be the union of the sets mentioned in the Steps I-IV. Set

$$U' = \left\{(x_1,x_2,\xi_2):\ (x_1,x_2,-\xi_2)\in U\right\}$$

and define the matrix $d(x_1,x_2,\xi_2)$ on U' according to the formulae (43.24) (we recall that (x_1,x_2,ξ_2) are chosen to be the local coordinates on $U'\subset W$ and in particular ξ_1 is uniquely determined by ξ_2).

Now routine computation shows that all the matrixes described in Steps I-V are invertible and satisfy the relations (43.24). Since $U\cup U' = W$ the desired matrix has been constructed, and therefore (43.22) is proved.

Remark 43.17 In view of (43.13) and (43.22), for the example just presented, we have

$$\operatorname{ind}\mathcal{B} = \tfrac{1}{2}\left[\operatorname{ind}\widetilde{\mathcal{B}} + L(\alpha,\widetilde{\mathcal{B}})\right] = \tfrac{1}{2}\left[\operatorname{ind}\widetilde{\mathcal{B}}+1\right]$$

which implies

$$\operatorname{ind} \widetilde{\mathfrak{b}} \neq 0 .$$

One can compare this situation with the known fact that every scalar PDO on a two-dimensional manifold has zero index (thus the PDOs arising in the investigation of *G*-PDOs (even dealing in the spaces of scalar functions) have a qualitatively different nature).

44. *G*-PDOs on the torus. Examples.

In this section we present a number of examples of *G*-PDOs related to the shift operators on the *n*-dimensional torus

$$\mathbb{T}^n = \mathbb{R}^n \left(\operatorname{mod} \mathbb{Z}^n\right) .$$

44.1 The tangent and cotangent bundles of $\mathbb{T}^n$ are trivial. Thus we have

$$T^*\mathbb{T}^n = \mathbb{T}^n \times \mathbb{R}^n .$$

The natural Riemann metric on $T^*\mathbb{T}^n$ is invariant with respect to the shifts generated by elements of $\mathbb{T}^n$ and the smooth measure generated by this metric coincides locally with the Lebesgue measure.

Let $\alpha\colon \mathbb{T}^n \to \mathbb{T}^n$ be any diffeomorphism of the torus. It can be represented as a smooth mapping of $\mathbb{R}^n$ into itself satisfying the condition

$$\alpha(x+z) = \alpha(x) + z,\ z \in \mathbb{Z}^n .$$

Example I. Standard shifts.

44.2 We start with the consideration of the standard shifts on the torus, that is

$$\alpha_h(x) = x + h,\ h \in \mathbb{R}^n$$

(obviously we can always choose $h = (h_1, \ldots, h_n)$ satisfying the condition $0 \le h_j < 1,\ j = \overline{i,n}$).

The codifferrential of α_h has the form

$$\partial\alpha_h(x,\xi) = (x+h,\xi),\ (x,\xi) \in T^*\mathbb{T}^n \tag{44.1}$$

and consequently the mapping $\beta_h : S^*\mathbb{T}^n \to S^*\mathbb{T}^n$ (see (40.3)) can be written out as

$$\beta_h(x,\xi) = (x+h,\xi),\ \|\xi\| = 1\ . \tag{44.2}$$

Since the measure on $\mathbb{T}^n$ is invariant with respect to the standard shifts it follows that the operators U_h and T_h (40.1) and (40.2) coincide in this situation and are given (in $H^s(\mathbb{T}^n)$) by

$$(T_h f)(x) = f(x-h)\ . \tag{44.3}$$

For the same reasons, the operators W_h and V_h ((40.5) and (40.6)) coincide as well and are given by

$$(V_h\psi)(x,\xi) = \psi(x-h,\xi),\ (x,\xi) \in S^*\mathbb{T}^n\ . \tag{44.4}$$

Considering the operator $\mathcal{b} \in \mathcal{B}_{s,r}$ (see (41.17)) of the form

$$\mathcal{b} = \sum_{k=1}^{m} a_{h_k} T_{h_k} \tag{44.5}$$

and bearing in mind that in our case

$$\left\|\partial\alpha_h^{-1}(\eta)\right\| = 1,\ \eta \in S^*\mathbb{T}^n$$

we conclude that the s-symbol of $\mathcal{b}$ does not depend on s and the formula (41.18) reduces to

$$\sigma_s(\mathcal{b}) = \sum_{k=1}^{m} \sigma\left(a_{h_k}\right) V_{h_k}\ . \tag{44.6}$$

In view of this we shall omit the index s in the symbolic calculus in our reasoning up to 44.7.

44.3 Following the ideas presented in Section 37, for every $\xi \in S^{n-1}$ along with the operator $\mathcal{b}$ (44.5) we consider the operators

$$b(\xi)=\sigma(\mathcal{b})(\xi)$$

acting in $L^2(\mathbb{T}^n)$ and defined by

$$b(\xi)u(x)=\sum_{k=1}^{m}\sigma\left(a_{h_k}\right)(x,\xi)u(x-h_k)\ . \tag{44.7}$$

In terms of these operators we obtain the next statement on the conditions of Fredholmity of $\mathcal{b}$.

Lemma 44.4 *The operator $\mathcal{b}$ (44.5) is Fredholm iff $\sigma(\mathcal{b})$ is invertible and in turn $\sigma(\mathcal{b})$ is invertible iff all the operators $b(\xi), \xi\in S^{n-1}$, (44.7) are invertible.*

Proof The standard shifts α_h (see 44.2) act topologically freely. Thus the mapping $\mathcal{b}\to\sigma(\mathcal{b})$ given by (44.6) establishes the symbolic calculus of this class of operators (recall Theorem 41.9), which means that $\mathcal{b}$ is Fredholm iff $\sigma(\mathcal{b})$ is invertible. Now the statement of the theorem follows from the argument used in the proof of Lemma 37.6. □

Now we consider the situations when the results of the foregoing sections allow us to write out the explicit conditions of the invertibility of $\sigma(\mathcal{b})$ and thus of the Fredholmity of $\mathcal{b}$.

44.5 Rational shifts.

Let all the coordinates of the vectors h_k contained in (44.5) be rational and written as irreducible fractions. In this case the subgroup of G of $\mathbb{T}^n$ generated by these h_k is finite and $|G|$ does not exceed $N=\prod_{i=1}^{n}N_i$, where N_i is the least common denominator of all the coordinates $h_k^{(i)}$, $k=\overline{1,m}$. Indeed G is a subgroup of the group G_N consisting of all the 'vectors' $h\in\mathbb{T}^n$ having ith coordinate divisible by $\frac{1}{N_i}$, and $|G_N|=N$.

So in this situation, to obtain the invertibility conditions for $b=\sigma(\mathcal{b})$ we can use the ideas of Section 15 and in particular Corollary 15.6.

Observe that we can exploit here both the finite groups G and G_N. If we use G then we obtain an auxiliary matrix $\tilde{b}$ of smaller dimension but we have to describe beforehand the explicit form of G, while if we use G_N then the matrix $\tilde{b}$ turns out to be of higher dimension but the advantage here is that the group G_N is given explicitly at once.

Let us, for example, consider the second situation. We shall define the elements of the group $G_N \cong \bigotimes_{i=1}^{n} \mathbb{Z}_{N_i}$ by the multiindices $k = (k_1, \dots, k_n)$ where $k_j = 0, 1, \dots, N_j - 1$. Consider the operator $\mathcal{b}$ of the form

$$\mathcal{b} = \sum_{k \in G_N} \mathcal{a}_k T_{h_k} \tag{44.8}$$

where $h_k = \left(\frac{k_1}{N_1}, \dots, \frac{k_n}{N_n} \right)$. We have

$$\sigma(\mathcal{b}) = \sum_{k \in G_N} \sigma(\mathcal{a}_k) V_{h_k} \tag{44.9}$$

and $\widetilde{\sigma(\mathcal{b})}$ is defined by an operator matrix, the elements of which are enumerated by pairs of multiindices $k, l \in G_N$:

$$\widetilde{\sigma(\mathcal{b})}_{k,l}(x, \xi) = \sigma(\mathcal{a}_{l-k})(x - h_k, \xi), \quad (x, \xi) \in S^*\mathbb{T}^n . \tag{44.10}$$

In view of Corollary 15.6 we conclude that $\mathcal{b}$ is Fredholm iff $\det[\widetilde{\sigma(\mathcal{b})}](x, \xi) \neq 0$ for every $(x, \xi) \in S^*\mathbb{T}^n$.

Example Let $n = 2$, $N_1 = 2$, $N_2 = 3$ (thus $N = 6$) and let the operator $\mathcal{b}$ have the form

$$\mathcal{b} = a_0(x, D) + a_1(x, D) T_1 + a_2(x, D) T_2$$

where $T_1 = T_{h_{(1,0)}}$, $T_2 = T_{h_{(0,1)}}$, that is

$$(T_1 f)(x_1, x_2) = f\left(x_1 - \tfrac{1}{2}, x_2\right),$$

$$(T_2 f)(x_1, x_2) = f\left(x_1, x_2 - \tfrac{1}{3}\right).$$

Let us enumerate the elements of G_N in the following way:

$$h_{(0,0)} = (0,0) = g_0; \quad h_{(1,0)} = \left(\tfrac{1}{2},0\right) = g_1; \quad h_{(0,1)} = \left(0,\tfrac{1}{3}\right) = g_2;$$

$$h_{(0,2)} = \left(0,\tfrac{2}{3}\right) = g_3; \quad h_{(1,1)} = \left(\tfrac{1}{2},\tfrac{1}{3}\right) = g_4; \quad h_{(1,2)} = \left(\tfrac{1}{2},\tfrac{2}{3}\right) = g_5 .$$

With this enumeration the matrix $\widetilde{\sigma(\mathcal{A})}$ has the form

$$\begin{bmatrix} a_0(x,\xi) & a_1(x,\xi) & a_2(x,\xi) & 0 & 0 & 0 \\ a_1(x-h_{(1,0)},\xi) & a_0(x-h_{(1,0)},\xi) & 0 & 0 & a_2(x-h_{(1,0)},\xi) & 0 \\ 0 & 0 & a_0(x-h_{(0,1)},\xi) & a_2(x-h_{(0,1)},\xi) & a_1(x-h_{(0,1)},\xi) & 0 \\ a_2(x-h_{(0,2)},\xi) & 0 & 0 & a_0(x-h_{(0,2)},\xi) & 0 & a_1(x-h_{(0,2)},\xi) \\ 0 & 0 & a_1(x-h_{(1,1)},\xi) & 0 & a_0(x-h_{(1,1)},\xi) & a_2(x-h_{(1,1)},\xi) \\ 0 & a_2(x-h_{(1,2)},\xi) & 0 & a_1(x-h_{(1,2)},\xi) & 0 & a_0(x-h_{(1,2)},\xi) \end{bmatrix}.$$

44.6 Irrational shifts.
Now let us consider to the situation when among the coordinates of the vectors h_k (used in (44.5)) there are irrational numbers. In this case the subgroup $G \subset \mathbb{T}^n$ generated by these vectors is infinite. In general there is no hope of obtaining the effective invertibility conditions for $\sigma(\mathcal{A})$ in the case of arbitrary vectors h_k and arbitrary coefficients a_k.

Remark Recalling the results of Chapter 3 and in particular Sections 26, 27 we note that for any vectors h_k the invertibility conditions for $\sigma(\mathcal{A})$ can be formulated in terms of rational approximations of h_k, which means that one can exploit the results already obtained in 44.5.

In order to provide certain examples where the corresponding invertibility conditions could be written out explicitly we shall confine ourselves to the case of two-term operators that is the case when in (44.5) $m=1$, and thus we can rewrite $\mathcal{A}$ in the form (10.24).

The sufficient conditions for the Fredholmity of $\mathcal{A}$ where written out in Theorem 10.10, and Lemma 44.4 and the result of 18.7 show now that in the case when among the coordinates of h (in (10.24)) there are irrational numbers these conditions are also necessary.

In this situation one can single out the case when the *numbers* $1, h^{(1)}, \dots, h^{(n)}$ ($h^{(i)}$, $i = \overline{1,n}$, are the coordinates of h) are *rationally independent*, that is the equality

$$\sum_{i=1}^{n} s_i h^{(i)} = p \ ,$$

where p, s_i are integers, is possible only when $s_1 = \ldots = s_n = 0$.

Under this condition we have $\overline{G} = \overline{\mathbb{T}}^n$ (G is the group generated by h) and the mapping

$$\alpha_h(x) = x - h$$

admits only one ergodic measure - the Lebesque measure on $\mathbb{T}^n$ (see Kornfel'd, Sinaj, Fomin [1], Chapter 3, §1). This means that in this case

The operator

$$b = a_0 + a_1 T_h$$

is Fredholm iff one of the following two conditions is satisfied:

(i) $\sigma(a_0)(x,\xi) \neq 0$ *on* $S^*\mathbb{T}^n$

and

$$\int_{\mathbb{T}^n} \ln\left|\sigma(a_0)(x,\xi)\right| \mathrm{d}x > \int_{\mathbb{T}^n} \ln\left|\sigma(a_1)(x,\xi)\right| \mathrm{d}x \quad \textit{for every } \xi \in S^{n-1};$$

(ii) $\sigma(a_1)(x,\xi) \neq 0$ *on* $S^*\mathbb{T}^n$

and

$$\int_{\mathbb{T}^n} \ln\left|\sigma(a_1)(x,\xi)\right| \mathrm{d}x > \int_{\mathbb{T}^n} \ln\left|\sigma(a_0)(x,\xi)\right| \mathrm{d}x \quad \textit{for every } \xi \in S^{n-1}.$$

Example II.

44.7 Here we consider the example of a two-term operator on the two-dimensional torus $\mathbb{T}^2 = \mathbb{R}^2 / \mathbb{Z}^2$ where the shift operator is generated by the mapping

$$\alpha(x_1, x_2) = (x_1 - h(x_2), x_2) \tag{44.11}$$

with h being a smooth function (s times differentiable if the operator is acting from $H^s(\mathbb{T}^2)$).

The operator we shall investigate has the form

$$\mathcal{L} = a_0(x,D) + a_1(x,D)T \tag{44.12}$$

where $a_i(x,D)$, $i=0,1$ are PDOs of order r acting from $H^s(\mathbb{T}^2)$ into $H^{s-r}(\mathbb{T}^2)$ and the shift operator T is given by

$$(Tf)(x) = f\left(\alpha^{-1}(x)\right) . \tag{44.13}$$

Observe that

$$\alpha^{-1'}(x_1,x_2) = \begin{bmatrix} 1 & h'(x_2) \\ 0 & 1 \end{bmatrix} \tag{44.14}$$

and thus $\left|\det\left(\alpha^{-1'}\right)\right| = 1$; that is why we did not write this multiplier in (44.13) (cf. (40.2)).

The inverse transpose of the matrix (44.14) is

$$L(x_1,x_2) = \begin{bmatrix} 1 & 0 \\ -h'(x_2) & 1 \end{bmatrix}$$

and $|\det L| = 1$ as well.

Now in the local coordinates $(x,\xi) \in S^*\mathbb{T}^2 = \mathbb{T}^2 \times S^1$ the mapping β^{-1} (used in (40.6)) can be written as

$$\beta^{-1}(x,\xi) = \left(x_1 + h(x_2), x_2; \frac{\xi_1}{\delta(x,\xi)}, \frac{-h'(x_2)\xi_1 + \xi_2}{\delta(x,\xi)} \right) \tag{44.15}$$

where $x = (x_1,x_2)$, $\xi = (\xi_1,\xi_2)$, $\|\xi\| = 1$ and

$$\delta(x,\xi) = \|L(x)\xi\| = \left[|\xi_1|^2 + |-h'(x_2)\xi_1 + \xi_2|^2 \right]^{\frac{1}{2}} . \tag{44.16}$$

Changing to the coordinates

$$\xi_1 = \cos\varphi, \ \ \xi_2 = \sin\varphi$$

we rewrite (44.15) as

$$\beta^{-1}(x,\xi)=\left(x_1+h(x_2),x_2;\operatorname{arctg}\left(\operatorname{tg}\varphi-h'(x_2)\right)\right), \tag{44.17}$$

and in these coordinates the Jacobian matrix of β^{-1} has the form

$$\begin{bmatrix} 1 & h'(x_2) & 0 \\ 0 & 1 & 0 \\ 0 & \dfrac{-h''(x_2)}{1+\left(tg\varphi-h'(x_2)\right)^2} & \dfrac{1}{\left[1+\left(\operatorname{tg}\varphi-h'(x_2)\right)^2\right]\cos^2\varphi} \end{bmatrix}.$$

Thus the Jacobian is equal to

$$\gamma(x,\varphi)=\frac{1}{\left[1+\left(\operatorname{tg}\varphi-h'(x_2)\right)^2\right]\cos^2\varphi}=\frac{1}{\cos^2\varphi+\left(\sin\varphi-h'(x_2)\cos\varphi\right)^2}. \tag{44.18}$$

Now by (41.18) the s-symbol of the operator $\mathcal{L}$ (44.12) is equal to

$$\sigma_s(\mathcal{L})=a_0(x,\xi)+\delta^{-s}a_1(x,\xi)V \tag{44.19}$$

where δ is given by (44.16) and the operator V is defined by

$$(Vf)(\eta)=\gamma^{\frac{1}{2}}f\left(\beta^{-1}(\eta)\right),\ \eta\in S^*\mathbb{T}^2=\mathbb{T}^2\times S^1 \tag{44.20}$$

with γ given by (44.18) and β^{-1} by (44.15).

Considering the invertibility conditions of $\sigma_s(\mathcal{L})$ (that is, the Fredholmity conditions of $\mathcal{L}$) it is reasonable to single out two possible situations:

(A) α acts topologically freely on $\mathbb{T}^2$,

(B) α does not act topologically freely.

Situation (A) in terms of the function h can be described as

there are no non-trivial segments Δ such that $h(\Delta)$ is rational.

So, bearing in mind this description, we proceed to the investigation of the invertibility conditions of $\sigma_s(\mathcal{b})$ (44.19) in the case

(A) α acts topologically freely

44.8 The topological freedom of the action of α implies the topological freedom of the action of β^{-1} (44.15). In view of this the invertibility conditions of $\sigma_s(\mathcal{b})$ can be obtained by means of Theorem 17.3 and to write them out in explicit form we have to define all the ergodic measures with respect to β^{-1} on $\mathbb{T}^2 \times S^1$.

Let us do this. In view of Corollary 5.5 the support of every such measure is contained in the closure of the orbit of a certain point $(x,\xi) \in \mathbb{T}^2 \times S^1$. Since the coordinate x_2 of $x = (x_1, x_2)$ does not change under β^{-1} it follows that the support of an ergodic measure is contained in a set $x_2 = x_2^0$ which is the torus

$$\mathbb{T}^1 \times S^1 = \left\{\left\{x_1, x_2^0; \varphi\right\}\right\} \subset \mathbb{T}^2 \times S^1 . \tag{44.21}$$

Moreover the action β^{-1} on this torus is diagonal, that is

$$\beta^{-1}(x_1, \varphi) = (\beta_1(x_1), \beta_2(\varphi)) \tag{44.22}$$

where

$$\beta_1(x_1) = x_1 + h\left(x_2^0\right) ,$$

and $\beta_2(\varphi)$ is given by (44.17).

Thus we shall obtain the description of the desired ergodic measures from the description of the ergodic measures for β_1 and β_2.

Here one can arrive at the following possible cases:

I. $h\left(x_2^0\right)$ is irrational.

Under this condition the only ergodic measure with respect to β_1 on $\mathbb{T}^1$ is the normalized Lebesque measure. To investigate β_2 we have to look at the two feasible situations.

(i) $h'\left(x_2^0\right) = 0$ (thus $\beta_2(\varphi) = \varphi$).

The ergodic measures with respect to β_2 on S^1 are the Dirac point measures. Thus in this case the ergodic measures with respect to β^{-1} are concentrated on the circles

$$\mathbb{T}^1 = \{(x_1,\varphi) : \varphi = \varphi_0\} \subset \mathbb{T}^1 \times S^1$$

and coincide there with the Lebesque measure.

(ii) $h'(x_2^0) \neq 0$.

In this case the matrix $L(x_1, x_2^0) = L(x_2^0)$ defining the mapping β_2 (see (44.15), (44.16)) has only one eigenvalue 1 to which there corresponds only one linearly independent eigenvector $\xi = (0,1)$.

Thus the mapping β_2 of S^1 induced by this matrix has precisely two fixed points $\varphi_1 = \dfrac{\pi}{2}$ and $\varphi_2 = \dfrac{3\pi}{2}$ and in view of 5.11 (a) there exist precisely two ergodic measures with respect to β_2 - the Dirac point measures at the singled-out fixed points.

So in this situation the ergodic measures with respect to β^{-1} are concentrated on two circles

$$\mathbb{T}^1 \times \left\{\frac{\pi}{2}\right\} \text{ and } \mathbb{T}^1 \times \left\{\frac{3\pi}{2}\right\}$$

and coincide on this circles with the Lebesque measure.

II. $h(x_2^0) = \dfrac{M}{N}$ where $\dfrac{M}{N}$ is an irreducible fraction.

Bearing in mind the argument already presented in I and Example 5.11 (a) we conclude that in this case if

(i) $h'(x_2^0) = 0$.

then the trajectory of every point $(x_1, x_2^0; \xi) \in \mathbb{T}^1 \times S^1$ consists of N points and the ergodic measure with respect to β^{-1} concentrated on this trajectory is described in Example 5.11 (a);

while if

(ii) $h'(x_2^0) \neq 0$

then the corresponding ergodic measures have the same structure and are concentrated only on the trajectories of the points of the circles

$$\mathbb{T}^1 \times \left\{\frac{\pi}{2}\right\} \text{ and } \mathbb{T}^1 \times \left\{\frac{3\pi}{2}\right\}.$$

Thus we have considered all feasible situations.

44.9 Since to apply Theorem 17.3 we need only exploit the values of the coefficients on the supports of the ergodic measures let us evaluate the coefficient δ (see (44.19) and (44.16)) on the supports of the ergodic measures just described.

We have

(i) if $h'(x_2^0)=0$ then $\delta(x_1,x_2^0;\varphi)=1$;

(ii) if $h'(x_2^0)\neq 0$ then $\delta\left(x_1,x_2^0;\frac{\pi}{2}\right)=\delta\left(x_1,x_2^0;\frac{3\pi}{2}\right)=1$ (observe that in this case in general $\delta(x_1,x_2^0;\varphi)\neq 1$).

Now Theorems 41.9 and 17.3, along with the reasoning presented in 44.8, 44.9, imply the following

Theorem 44.10 *Let $\mathfrak{b}$ be an operator of the form (44.12) acting from $H^s(\mathbb{T}^2)$ into $H^{s-r}(\mathbb{T}^2)$. If there are no non-trivial segments Δ such that $h(\Delta)$ is rational then the operator $\mathfrak{b}$ is Fredholm iff one of the following conditions (1) or (2) holds.*

(1) $a_0(x,\xi)\neq 0$ *on* $\mathbb{T}^2\times S^1$ *and*

(i) *if $h(x_2)$ is irrational then*

$$\int_0^1 \ln\left|a_0(x_1,x_2;\xi)\right|\,dx_1 > \int_0^1 \ln\left|a_1(x_1,x_2;\xi)\right|\,dx_1$$

for every ξ in the case when $h'(x_2)=0$,

or

for every $\xi=(0,\pm 1)$ in the case when $h'(x_2)\neq 0$;

(ii) *if $h(x_2)=\dfrac{M(x_2)}{N(x_2)}$ is an irreducible fraction then*

$$\left|\prod_{k=0}^{N(x_2)-1} a_0(x_1+kh(x_2),x_2;\xi)\right| > \left|\prod_{k=0}^{N(x_2)-1} a_1(x_1+kh(x_2),x_2;\xi)\right|$$

for every ξ *in the case when* $h'(x_2) = 0$,

or

for $\xi = (0, \pm 1)$ *in the case when* $h'(x_2) \neq 0$.

(2) $a_1(x,\xi) \neq 0$ *on* $\mathbb{T}^2 \times S^1$ *and*

(i) *if* $h(x_2)$ *is irrational then*

$$\int_0^1 \left|\ln a_1(x_1,x_2;\xi)\right| \mathrm{d}x_1 > \int_0^1 \left|\ln a_0(x_1,x_2;\xi)\right| \mathrm{d}x_1$$

for every ξ *in the case when* $h'(x_2) = 0$,

or

for $\xi = (0, \pm 1)$ *in the case when* $h'(x_2) \neq 0$;

(ii) *if* $h(x_2) = \dfrac{M(x_2)}{N(x_2)}$ *is an irreducible fraction then*

$$\left|\prod_{k=0}^{N(x_2)-1} a_1(x_1 + kh(x_2), x_2;\xi)\right| > \left|\prod_{k=0}^{N(x_2)-1} a_0(x_1 + kh(x_2), x_2;\xi)\right|$$

for every ξ *in the case when* $h'(x_2) = 0$,

or

for $\xi = (0, \pm 1)$ *in the case when* $h'(x_2) \neq 0$.

The second possible situation is

(B) α does not act topologically freely

44.11 In this case there exists an interval Δ with $h(\Delta)$ being rational. Evidently the group $\mathbb{Z}$ acting by diffeomorphisms α^z (α is given by (44.11)) acts normally on $\mathbb{T}^2$ and in view of Theorem 41.9 the operator $\mathcal{L}$ (44.12) is Fredholm if its s-symbol $\sigma_s(\mathcal{L})$ (44.19) is invertible.

The explicit form of the mapping β^{-1} (44.17) shows that the action of the group $\mathbf{Z}$ on $\mathbb{T}^2 \times S^1$ by diffeomorphisms β^z is also normal and consequently the invertibility conditions for $\sigma_s(\mathcal{L})$ can be obtained by means of Theorem 21.3.

44.12 Observe that if Δ is an interval such that $h(\Delta)=\dfrac{M}{N}$ where $\dfrac{M}{N}$ is an irreducible fraction then the set $X_\Delta=\{(x_1,x_2;\xi):x_2\in\Delta\}\subset\mathbb{T}^2\times S^1$ is contained in the set X_N $(N\in\mathbb{Z})$ described in 16.1 for the algebra

$$A=\left[\mathcal{A}^0\right]=C\left(\mathbb{T}^2\times S^1\right)$$

of the symbols of zero-order PDOs in $H^0\left(\mathbb{T}^2\right)$ and the group of homeomorphisms $\{t_z\}_{z\in\mathbb{Z}}=\{\beta^z\}_{z\in\mathbb{Z}}$ of Prim $A=\hat{A}=\mathbb{T}^2\times S^1$.

For any $(x,\xi)\in X_\Delta$ we have

$$G^{(x,\xi)}=\mathbb{Z}_N$$

($G^{(x,\xi)}$ is defined in 16.19) and the corresponding representations 16.20 arise from the action of the group $\mathbb{Z}_N$ on $A|_{X_\Delta}$ by automorphisms $V^{[z]}(\cdot)V^{[z]*}$ (V is given by (44.20) and $[z]\in\mathbb{Z}_N$ is the equivalence class of z) which are totally defined by the homeomorphisms $\beta^{[z]}$ of X_Δ.

Note also that $h'(x_2)=0$ when $x_2\in\Delta$ which means that the function δ used in (44.19) satisfies the equality

$$\delta(x,\xi)=1,\ (x,\xi)\in X_\Delta\ .$$

Now for every natural number N we denote by $Y_N\subset\mathbb{T}^2\times S^1$ the union of all the sets X_Δ corresponding to the intervals Δ such that $h(\Delta)=\dfrac{M(\Delta)}{N}$ is an irreducible fraction.

All the foregoing reasoning, along with Corollary 15.6, means that

for every $(x,\xi)=(x_1,x_2;\xi)\in Y_N$ *the operator* $\pi_{(x,\xi)}(\sigma_s(\mathcal{L}))$ $\left(\pi_{(x,\xi)},(x,\xi)\in\hat{A}=\right.$ Prim $\left.A=T^2\times S^1\right)$ *is defined in 16.20) is invertible iff the* $N\times N$ *matrix*

$$\begin{bmatrix} a_0(x_1,x_2;\xi) & a_1(x_1,x_2;\xi) & 0 & \dots & 0\\ 0 & a_0(x_1+h(x_2),x_2;\xi) & a_1(x_1+h(x_2),x_2;\xi) & \dots & 0\\ \dots & \dots & \dots & \dots & \dots\\ a_1(x_1+(N-1)h(x_2),x_2;\xi) & 0 & 0 & \dots & a_0(x_1+(N-1)h(x_2),x_2;\xi)\end{bmatrix}$$

is invertible, that is

$$\prod_{k=0}^{N-1} a_0\left(x_1+kh(x_2),x_2;\xi\right) \neq \prod_{k=0}^{N-1} a_1\left(x_1+kh(x_2),x_2;\xi\right) \tag{44.23}$$

for every $(x_1,x_2;\xi)\in Y_N$.

44.13 Note that in this situation the set X_∞ defined in 16.1 is equal to

$$\left[\mathbb{T}^2\times S^1\right]\setminus\left[\bigcup_N Y_N\right] \tag{44.24}$$

where the Y_N are described in 44.12, and consequently for every point (x,ξ) belonging to the set (44.24) we can easily work out the invertibility conditions for $\pi_{(x,\xi)}(\sigma_s(\mathcal{b}))$ analogous to those presented in Theorem 44.10 (bearing in mind the description of the trajectories of these points, Theorem 17.3 and the form of the corresponding ergodic measures obtained in 44.8, 44.9).

All these arguments lead to the following result which is in fact a generalization of Theorem 44.10:

Theorem 44.14 *Let $\mathcal{b}$ be an operator of the form (44.12). Then $\mathcal{b}$ is Fredholm iff for every natural number N the condition (44.23) holds and for every connected component Ω of the set (44.24) one of the conditions (1) or (2) of Theorem 44.10 holds (with Ω instead of $\mathbb{T}^2\times S^1$).*

Example III.

44.15 Here we present another example of a *G*-PDO on the torus $\mathbb{T}^n$ where the Fredholmity conditions can be written out explicitly - the situation when the coefficients a_g of the operator

$$\mathcal{b}=\sum a_g T_g$$

commute with the shift operators T_g.

The idea behind the general approach to the investigation of operators of this sort is quite clear and can be demonstrated by the following simple example.

Let $\mathcal{b}$ be an operator of the form (44.5) where the coefficients a_{h_k} and vectors satisfy the following conditions (i) and (ii):

(i) all the vectors h_k have the form

$$h_k = (\gamma_k, 0, \ldots, 0) ,$$

thus

$$(T_{h_k} f)(x_1, \ldots, x_n) = f(x_1 - \gamma_k, x_2, \ldots, x_n) ;$$

(ii) the symbols $\sigma(a_{h_k}) = a_k(x,\xi)$, $(x,\xi) \in \mathbb{T}^n \times S^{n-1}$, do not depend on x_1.

The symbol $\sigma_s(\mathcal{L})$ (44.6) under conditions (i) and (ii) has the form

$$[\sigma_s(\mathcal{L})\psi](x,\xi) = \sum a_k(x_2, \ldots, x_n; \xi)\psi(x_1 - \gamma_k, x_2, \ldots, x_n; \xi) . \tag{44.25}$$

Expanding the function ψ in a Fourier series with respect to x_1,

$$\psi(x_1, \ldots, x_n; \xi) = \sum_j \psi_j(x_2, \ldots, x_n; \xi) e^{i2\pi j x_1} \tag{44.26}$$

one can rewrite (44.25) as

$$\sigma_s(\mathcal{L})\psi = \sum_j \left(\sum_k a_k(x_2, \ldots, x_n; \xi) e^{-i2\pi j \gamma_k} \right) \psi_j e^{i2\pi j x_1} \tag{44.27}$$

which means that the action $\sigma_s(\mathcal{L})$ is reduced to the multiplication of the coefficient ψ_j of ψ (44.26) by the functions

$$s_j(x_2, \ldots, x_n; \xi) = \sum_k a_k(x_2, \ldots, x_n; \xi) e^{-i2\pi j \gamma_k} . \tag{44.28}$$

In view of this

$\sigma_s(\mathcal{L})$ is invertible (and thus $\mathcal{L}$ is Fredholm) iff

$$\inf_{x,\xi,j} |s_j(x,\xi)| > 0 \tag{44.29}$$

where s_j are defined by (44.28).

44.16 We can also rewrite the conditions (44.29) of the invertibility of $\sigma_s(\mathcal{L})$ in a different form using the characters of the discrete group $G \subset \mathbb{T}^1$ generated by the numbers γ_k.

For example let G be a free group (that is it does not contain elements of finite order). This condition will be satisfied if there exists a system of numbers $\nu_1, \ldots, \nu_l$ such that

(1) the numbers $1, \nu_1, \ldots, \nu_l$ are rationally independent (see 44.7); and

(2) every γ_k can be represented in the form

$$\gamma_k = \sum_{j=1}^{l} c_{jk} \nu_j, \; c_{jk} \in \mathbb{Z} . \tag{44.30}$$

Remark Conditions (1) and (2) are evidently satisfied if $1, \gamma_1, \ldots, \gamma_m$ are rationally independent.

44.17 If conditions (1) and (2) hold then we shall associate with the operator $\sigma_s(\mathcal{L})$ (44.25) the function

$$W_{\mathcal{L}}(x,\xi,z) = \sum_{k=1}^{m} a_k(x_2,\ldots,x_n;\xi) \prod_{j=1}^{l} z_j^{c_{jk}} \tag{44.31}$$

on $\mathbb{T}^{n-1} \times S^{n-1} \times (S^1)^l$, where $(x_2,\ldots,x_n) \in \mathbb{T}^{n-1}$, $\xi \in S^{n-1}$, $\{z = (z_1,\ldots,z_l) : |z_i| = 1\} \in (S^1)^l$ and c_{jk} are those used in (44.30).

Lemma 44.18 *Let conditions (1) and (2) of 44.16 hold. Then the mapping*

$$\sigma_s(\mathcal{L}) \to W_{\mathcal{L}} \tag{44.32}$$

($\sigma_s(\mathcal{L})$ is given by (44.25), $W_{\mathcal{L}}$ by (44.31)) establishes an isomorphism between the C^*-algebra B_1 generated by the symbols $\sigma_s(\mathcal{L})$ of the operators satisfying conditions (i) and (ii) of 44.15 and the C^*-algebra B_2 generated by the functions $W_{\mathcal{L}}$ (44.31).

In particular $\mathcal{L}$ is Fredholm iff

$$W_{\mathcal{L}}(x,\xi,z) \neq 0 \text{ on } \mathbb{T}^{n-1} \times S^{n-1} \times (S^1)^l . \tag{44.33}$$

Proof The mapping (44.32) is clearly algebraic. Note that both the C^*-algebras B_1 and B_2 possesses the property (*) (for B_1 it follows from (44.27) and for B_2 from the definition of W_ξ (44.31)). Thus the desired result follows from Corollary 12.9. □

Remarks (1) In fact a result analogous to Lemma 44.18 can be obtained without assuming that conditions (1) and (2) of 44.16 hold.

In general we have to use in (44.31) not the characters of $\mathbb{Z}^l \left(\left(\hat{\mathbb{Z}^l} \right) = \left(S^1 \right)^l \right)$ which arise under the condition that the group generated by $\gamma_1, \ldots, \gamma_m$ is free, but simply the characters of the commutative discrete group generated by $\gamma_1, \ldots, \gamma_m$.

(2) Conditions (44.29) and (44.33) are equivalent (when the group generated by $\gamma_1, \ldots, \gamma_m$ is free) and though the function W_ξ is defined in a somewhat more complicated way (44.33) is more convenient since it does not contain the evaluation of an infinum.

Example IV. The hyperbolic shift.

44.19 It is known (see for example Kornfel'd, Sinaj, Fomin [1], Chapter 4, §3) that any automorphism of the torus $\mathbb{T}^m = \mathbb{R}^m / \mathbb{Z}^m$ has the form

$$\gamma(x) = Ex$$

where E is an $m \times m$ matrix with integer coefficients and such that $|\det E| = 1$.

The most interesting dynamical properties of these automorphisms (see the above-cited source) appear in the case when none of the eigenvalues of E lie on the unit circle, that is in the case when E defines a hyperbolic operator in $\mathbb{R}^m$ (recall 19.6). It is also reasonable to call the corresponding *automorphism hyperbolic.*

Now we shall consider the example of a G-PDO on the torus in the case when the shift operator is generated by a hyperbolic automorphism of the torus.

We confine ourselves to a concrete operator on the two-dimensional torus, though an analogous argument could be exploited in the general situation as well.

44.20 In the space $H^s\left(\mathbb{T}^2\right)$ consider the shift operator

$$(Tf)(x) = f(Lx) \tag{44.34}$$

where

$$L=\begin{bmatrix}2 & 1\\ 1 & 1\end{bmatrix}. \tag{44.35}$$

Since L defines the group automorphism of the torus it preserves the Haar measure.

Let $\mathcal{b}: H^s(\mathbb{T}^2) \to H^{s-r}(\mathbb{T}^2)$ be G-PDO of the form

$$\mathcal{b} = a_0(x,D) + a_1(x,D)T \tag{44.36}$$

where $a_i(x,D)$, $i=1,2$, are PDOs of order r.

The s-symbol of $\mathcal{b}$ is given by

$$[\sigma_s(\mathcal{b})\psi](x,\xi) = a_0(x,\xi)\psi(x,\xi) + a_1(x,\xi)\left\|L^{-1'}\xi\right\|^{-s}\gamma^{\frac{1}{2}}\psi\left(Lx,\frac{L^{-1'}\xi}{\|\xi\|}\right) \tag{44.37}$$

where $(x,\xi)\in S^*\mathbb{T}^2 = \mathbb{T}^2\times S^1$, γ is the modulus of the Jacobian of the mapping $\beta^{-1}:(x,\xi)\to\left(Lx,\frac{L^{-1'}\xi}{\|\xi\|}\right)$ and $L^{-1'}=\begin{bmatrix}1 & -1\\ -1 & 2\end{bmatrix}$ (observe that we do not need an explicit formula for γ since it is not used when calculating the invertibility conditions for $\sigma_s(\mathcal{b})$).

44.21 As the mapping

$$\beta^{-1}(x,\xi) = (\beta_1(x),\beta_2(\xi)) = \left(Lx,\frac{L^{-1'}\xi}{\|\xi\|}\right) \tag{44.38}$$

acts independently with respect to x and ξ the description of the ergodic measures for it can be obtained by consideration of its components β_1 and β_2.

The mapping $\beta_2: S^1\to S^1$ has four fixed points corresponding to the eigenvectors of the matrix $L^{-1'}$.

The eigenvalues of $L^{-1'}$ are

$$\lambda_{1,2} = \frac{3\pm\sqrt{5}}{2}$$

and the corresponding eigenvectors are

$$e_{1,2} = \frac{1}{\sqrt{\frac{5}{2} \mp \sqrt{5}}} \begin{pmatrix} \frac{1 \mp \sqrt{5}}{2} \\ 1 \end{pmatrix}.$$

Since $|\lambda_1| > 1$ and $0 < \lambda_2 < 1$ we conclude that the points $\pm e_1$ are attracting and $\pm e_2$ are repelling for β_2; and obviously the positive half trajectory (for β^{-1}) of every point (x, ξ), $\xi \neq \pm e_2$, is attracted to the positive half trajectory of one of the points $(x, \pm e_1)$.

If we look at the negative half trajectories then the picture is the same with the vectors e_1 and e_2 being interchanged.

Thus the limit set for every trajectory belongs to the union of the four sets $\mathbb{T}^2 \times \{\pm e_1\}, \mathbb{T}^2 \times \{\pm e_2\}$ (coinciding with the set of non-wandering points of β^{-1}). Each of these sets is invariant under β^{-1} and closed, which means that the support of every ergodic measure (for β^{-1}) is contained in one of these sets.

Observe now that the action β^{-1} on each of the singled-out sets coincides with the action of the automorphism of the torus given by the matrix (44.35), which means that the ergodic measures we are looking for are defined by this automorphism.

The ergodic measures for this automorphism are well known (see Kornfel'd, Sinaj, Fomin [1], Chapter 4, §3). A lot of effort has been devoted to the description of ergodic measures of automorphisms having a nature similar to the one we are considering - the so-called *group automorphisms* (see for example, Kornfel'd, Sinaj, Fomin [1], Chapter 4, §3). In our situation the set of periodic points of the automorphism is dense (periodic points are the points with rational coordinates) and on the trajectory of every periodic point there is concentrated a natural ergodic measure. The normalized Lebesque measure on the torus (the Haar measure) is also ergodic. In general group automorphisms may have many non-trivial continuous ergodic measures and a very useful observation for our purposes is that if the set of periodic points is dense then the spectral radius of the corresponding weighted shift operator (recall Theorem 4.4) could be written out in terms of these points.

Thus, we have in mind the next property:

if $\alpha: \mathbb{T}^m \to \mathbb{T}^m$ is a hyperbolic automorphism of the torus then for any continuous function $a \in C(\mathbb{T}^m)$ we have

$$\sup_{\mu \in \Lambda} \int_{\mathbb{T}^m} a(x)\, d\mu = \sup_{\nu \in \Lambda_p} \int_{\mathbb{T}^m} a(x)\, d\nu$$

where Λ is the set of all the ergodic measures with respect to α and Λ_p is the set of all the ergodic measures concentrated on the trajectories of periodic points.

Indeed, fix any $\mu \in \Lambda$. In view of the Birkhoff-Hinchin ergodic theorem (Theorem 2.19) there exists a point x_0 such that

$$\lim_{h\to\infty} \frac{1}{n}\sum_{k=1}^{n-1} a\left(\alpha^k(x_0)\right) = \int_{\mathbf{T}^m} a(x)\, d\mu .$$

Now approximating x_0 by periodic points we conclude (in view of the continuity of a) that

$$\sup_{\nu\in\Lambda_p} \int_{\mathbf{T}^m} a(x)\, d\nu \geq \int_{\mathbf{T}^m} a(x)\, d\mu$$

which proves the assertion.

Returning to our example, note that the set of non-periodic points for β^{-1} (44.38) is dense in $\mathbb{T}^2$ and thus β^{-1} acts topologically freely and we can exploit Theorem 17.3 to obtain the invertibility conditions for $\sigma_s(\mathcal{b})$.

Observe that to apply Theorem 17.3 we need to calculate the value of the multiplier $\left\|L^{-1'}\xi\right\|$ in (44.37) only on the supports of ergodic measures of β^{-1}. Since every such support is based on a set where $\xi = \pm e_1, \pm e_2$ it is enough to evaluate $\left\|L^{-1'}\xi\right\|$ for these vectors. We have

$$\left\|L^{-1'}(\pm e_1)\right\| = \lambda_1 = \frac{3+\sqrt{5}}{2}$$

$$\left\|L^{-1'}(\pm e_2)\right\| = \lambda_2 = \frac{3-\sqrt{5}}{2} .$$

Now the foregoing reasoning (and in particular the written-out assertion on the ergodic measures concentrated on the trajectories of periodic points) and Theorem 17.3 imply

Theorem 44.22 *The operator $\mathcal{b}: H^s\left(\mathbb{T}^2\right) \to H^{s-r}\left(\mathbb{T}^2\right)$ where $\mathcal{b}$ is given by (44.36) and T by (44.34) is Fredholm iff one of the following conditions (1) or (2) holds:*

(1) *$a_0(x,\xi) \neq 0$ for every $(x,\xi) \in S^*\mathbb{T}^2 = \mathbb{T}^2 \times S^1$ and there exists $1 > \lambda \geq 0$ such that for every x with rational coordinates*

$$\prod_{k=0}^{p(x)-1} \left| a_0\left(L_x^k, \pm e_1\right) \right| > \lambda^{p(x)} \left(\frac{3+\sqrt{5}}{2} \right)^{-s \cdot p(x)} \prod_{k=0}^{p(x)-1} \left| a_1\left(L_x^k, \pm e_1\right) \right| ,$$

$$\prod_{k=0}^{p(x)-1} \left| a_0\left(L_x^k, \pm e_2\right) \right| > \lambda^{p(x)} \left(\frac{3-\sqrt{5}}{2} \right)^{-s \cdot p(x)} \prod_{k=0}^{p(x)-1} \left| a_1\left(L_x^k, \pm e_2\right) \right|$$

where p(x) *is the period of* x;

(2) $a_1(x,\xi) \neq 0$ *for every* $(x,\xi) \in \mathbb{T}^2 \times S^1$ *and all the inequalities in (1) have the opposite sign.*

Remark 44.23 In this theorem for the first time we see that the dependence of the Fredholmity conditions on s can be essential. In particular, for example in the second inequality in (1) the right-hand part grows by s and when s is sufficiently large the inequality is violated.

Thus we find a qualitatively new effect - the concrete *G*-PDO is Fredholm in a certain scale range of the Sobolev space and outside this range it is not Fredholm.

This is the essential difference between the theory of *G*-PDOs and the theory of elliptic PDOs.

Now let us consider one more simpler example where the dependence of the Fredholmity conditions on s is essential.

Example V.

44.24 Let α_1, α_2 be certain orientation preserving diffeomorphisms of the circle S^1 both having finite sets of fixed points $t_1,\dots,t_{m_1}$ for α_1 and $\tau_1,\dots,\tau_{m_2}$ for α_2 and having no other periodic points.

Consider the diffeomorphism α^{-1} of the torus $\mathbb{T}^2$ given by

$$\alpha^{-1}(x_1,x_2) = (\alpha_1(x_1), \alpha_2(x_2)) . \tag{44.39}$$

The Jacobian matrix of this diffeomorphism is diagonal, thus the corresponding induced diffeomorphism β^{-1} of $S^*\mathbb{T}^2 = \mathbb{T}^2 \times S^1$ (see (40.3)) has the form (cf. (40.4))

$$\beta^{-1}(x;\xi)=\left(\alpha_1(x_1),\alpha_2(x_2);\frac{1}{\sqrt{\frac{\xi_1^2}{[\alpha_1'(x_1)]^2}+\frac{\xi_2^2}{[\alpha_2'(x_2)]^2}}}\left(\frac{\xi_1}{\alpha_1'(x_1)},\frac{\xi_2}{\alpha_2'(x_2)}\right)\right) \tag{44.40}$$

(where $\xi=(\xi_1,\xi_2)\in S^1\subset\mathbb{R}^2$).

In the example presented below we shall investigate the Fredholmity conditions of a two-term G-PDO with the shift operator generated by the diffeomorphism (44.39).

The invertibility conditions for its s-symbol can be obtained with the help of Theorem 17.3. In order to apply this theorem we have to describe all the ergodic measures for the mapping β^{-1} defined by (44.40). So let us do this.

44.25 In view of the definition of the diffeomorphism α^{-1} (44.39) and the properties of α_1 and α_2 the set of non-wandering points of α^{-1} consists of $m_1\times m_2$ fixed points

$$\Omega(\alpha^{-1})=\{(t_i,\tau_j)\}_{i=1,m_1;j=1,m_2}\,.$$

which means that there exist precisely $m_1\times m_2$ ergodic measures for α^{-1}, namely the Dirac measures at these fixed points.

Further the explicit form of β^{-1} (44.40) shows that the ergodic measures for β^{-1} are concentrated on the circles in $\mathbb{T}^2\times S^1$ corresponding to these points.

Let us consider a certain point (t_i,τ_j). The mapping $\overline{\beta}=\beta^{-1}_{(t_i,\tau_j)}:S^1\to S^1$ induced by the matrix

$$\begin{bmatrix} d_1 & 0 \\ 0 & d_2 \end{bmatrix}=\begin{bmatrix} \frac{1}{\alpha_1'(t_i)} & 0 \\ 0 & \frac{1}{\alpha_2'(\tau_j)} \end{bmatrix} \tag{44.41}$$

has a different character (and correspondingly different ergodic measures) depending on the interrelations between the numbers d_1 and d_2.

The possibilities here are the following:

I. $d_1 > 0, d_2 > 0$.

(a) $d_1 \neq d_2$.

In this case there exist precisely four fixed points $(\pm 1, 0)$, $(0, \pm 1)$ and, respectively, four ergodic Dirac measures at these points.

(b) $d_1 = d_2$.

In this case all the points of S^1 are fixed under $\overline{\beta}$ and the ergodic measures are the Dirac measures at the points of S^1.

II. $d_1 > 0, d_2 < 0$.

(a) $|d_1| \neq |d_2|$.

In this case we have two fixed points $(\pm 1, 0)$ and the pair of periodic points $(0, \pm 1)$ of period 2. Thus there exist three ergodic measures: the Dirac measures μ_1 at $(1,0)$, μ_2 at $(-1,0)$ and the measure μ_3 on $\{(0,1),(0,-1)\}$ given by $\mu_3(0,1) = \mu_3(0,-1) = \frac{1}{2}$.

(b) $|d_1| = |d_2|$.

In this situation the points $(\pm 1, 0)$ are fixed and all the other points are periodic of period 2. Thus we have two Dirac measures μ_1 *at*$(1,0)$, μ_{-1} at $(-1,0)$ and the measures μ_{ξ_1} $-1 < \xi_1 < 1$ concentrated on the sets $\{(\xi_1, \xi_2), (\xi_1, -\xi_2)\}$ and such that

$$\mu_{\xi_1}(\xi_1, \xi_2) = \mu_{\xi_1}(\xi_1, -\xi_2) = \frac{1}{2}.$$

III. $d_1 < 0, d_2 > 0$.

This case is analogous to II.

IV. $d_1 < 0, d_2 < 0$.

(a) $d_1 \neq d_2$.

Here we have two pairs of periodic points of period 2 and respectively two ergodic measures

$$\mu_1(\pm 1, 0) = \frac{1}{2}, \ \mu_2(0, \pm 1) = \frac{1}{2}.$$

(b) $d_1 = d_2$.

All the points of the circle S^1 are periodic of period 2 and the corresponding ergodic measures are given by

$$\mu(\xi_1, \xi_2) = \mu(-\xi_1, \xi_2) = \frac{1}{2}.$$

44.26 Now let $\mathscr{b}: H^s(\mathbb{T}^2) \to H^{s-r}(\mathbb{T}^2)$ be a G-PDO of the form

$$\mathscr{b} = a_0(x, D) + a_1(x, D)U \tag{44.42}$$

where $a_i(x, D)$, $i = 0, 1$, are the PDOs of order r and U is defined by

$$(Uf)(x) = f\left(\alpha^{-1}(x)\right) \tag{44.43}$$

with α^{-1} described in 44.24.

The s-symbol of the operator $\mathscr{b}$ is

$$\sigma_s(\mathscr{b}) = a_0(x, \xi) + a_1(x, \xi) \cdot \left\| \partial\alpha^{-1}(x, \xi) \right\|^{-s} \cdot \left| \alpha^{-1\prime}(x) \right|^{-\frac{1}{2}} V \tag{44.44}$$

where V is defined by β^{-1} (44.40) and (40.6).

44.27 To apply Theorem 17.3 for calculating the invertibility conditions of $\sigma_s(\mathscr{b})$ we have to evaluate the meaning of $\left\| \partial\alpha^{-1} \right\|$ and $\left| \alpha^{-1\prime} \right|$ on the supports of ergodic measures with respect to β^{-1}.

Since we have already described all such measures in 44.25 we can easily make the corresponding calculations in all the situations I-IV.

In the cases I-IV (a) we have

$$\left\| \partial\alpha^{-1}\left(t_i, \tau_j; (\pm 1, 0)\right) \right\| = |d_1| = \left| \alpha_1'(t_i) \right|^{-1}$$

$$\left\| \partial\alpha^{-1}\left(t_i, \tau_j; (0, \pm 1)\right) \right\| = |d_2| = \left| \alpha_2'(t_j) \right|^{-1}$$

and in the cases (b) we have

$$\left\|\partial\alpha^{-1}\left(t_i,\tau_j;\xi\right)\right\| = |d_1| = |d_2|\,.$$

The corresponding values of $\left|\alpha^{-1\prime}\right|$ are equal to $\left|\alpha_1'(t_i)\cdot\alpha_2'(\tau_j)\right|$.

Now we have everything we need to write out the Fredholmity conditions for $\mathscr{b}$ (bearing in mind Theorem 17.3 and the foregoing reasoning).

Theorem 44.28 *Let $\mathscr{b}$ be the operator given by (44.42) acting from $H^s\left(\mathbb{T}^2\right)$ into $H^{s-r}\left(\mathbb{T}^2\right)$. This operator is Fredholm iff one of the following two conditions (1) or (2) is satisfied:*

(1) $a_0(x,\xi)\neq 0$ *for every* $(x,\xi)\in S^*\mathbb{T}^2=\mathbb{T}^2\times S^1$ *and for every* $\left(t_i,\tau_j\right)$, $i=\overline{1,m_1}$; $j=\overline{1,m_2}$, *we have*

(a) *if $\xi\in S^1$ is a fixed point of the mapping $\overline{\beta}$ described in 44.25 I-IV then*

$$\left|a_0\left(t_i,\tau_j;\xi\right)\right| > \gamma\left(t_i,\tau_j;\xi\right)^{-s}\cdot\left|\alpha_1'(t_i)\cdot\alpha_2'(\tau_j)\right|^{-\frac{1}{2}}\left|a_1\left(t_i,\tau_j;\xi\right)\right|\,;$$

(b) *if $\xi\in S^1$ is a periodic point for $\overline{\beta}$ (described in 44.25 II-IV) then*

$$\left|a_0\left(t_i,\tau_j\right);\xi\cdot a_0\left(t_i,\tau_j;\overline{\xi}\right)\right|$$

$$>\left[\gamma\left(t_i,\tau_j;\xi\right)\cdot\gamma\left(t_i,\tau_j;\overline{\xi}\right)\right]^{-s}\left|\alpha_1'(t_i)\cdot\alpha_2'(\tau_j)\right|^{-1}\left|a_1\left(t_i,\tau_j;\xi\right)\cdot a_1\left(t_i,\tau_j;\overline{\xi}\right)\right|$$

where $\gamma\left(t_i,\tau_j;\xi\right)=\left\|\partial\alpha^{-1}\left(t_i,\tau_j;\xi\right)\right\|$ is evaluated in 44.27 and $\overline{\xi}=\overline{\beta}(\xi)$.

(2) $a_1(x,\xi)\neq 0$ *for every* (x,ξ) *and all the inequalities mentioned in (1) have the opposite sign.*

45. *G*-PDOs. Examples.

In this section we continue the presentation of various *G*-PDOs, the Fredholmity conditions for which could be written out in explicit form.

45.1 Let $\alpha\colon M\to M$ be a diffeomorphism of an n-dimensional compact smooth manifold M such that the set $\Omega(\alpha)$ of its non-wandering points consists of periodic points (in particular α possesses this property if $|\Omega(\alpha)|<\infty$).

Remark Note that this class of diffeomorphisms contains in particular the Morse-Smale diffeomorphisms which are known to be the simplest class of diffeomorphisms possessing structural stability (see, for example, Nitecki [1], §2.3).

45.2 In this section we present a number of examples of G-PDOs with shifts generated by diffeomorphisms of this type. Since the ergodic measures with respect to α can be concentrated only on the trajectories of periodic points it follows that the support of any ergodic measure with respect to the induced diffeomorphism $\beta: S^*M \to S^*M$ (see (40.3)) is contained in the union of the fibres (spheres S^{n-1}) lying over the trajectory of a certain periodic point.

If a periodic point x_0 is a fixed point (its period is equal to 1) then the action of β on the fibre $S^*M_{x_0} \simeq S^{n-1}$ is induced by the action of $\partial\alpha(x_0)$ via formula (40.4) and all the ergodic measures for β having supports in this fibre can be obtained by means of Theorem 38.9.

If a periodic point x_0 has period $p \neq 1$ then we arrive at the generalization of the problem considered in Section 38. Thus, the problem is posed as follows.

45.3 Let $S_0, S_1, \ldots, S_{p-1}$ be p copies of the sphere S^{n-1}; $L_0, L_1, \ldots, L_{p-1}$ be non-degenerate $n \times n$ matrices; and let $\beta_j: S_j \to S_{j+1}$, $0 \le j \le p-2$, $\beta_{p-1}: S_{p-1} \to S_0$ be the mappings given by the formula

$$\beta_j(\xi) = \frac{L_j \xi}{\|L_j \xi\|}. \tag{45.1}$$

Consider the mapping α of the space

$$Y = \bigcup_{j=0}^{p-1} S_j \tag{45.2}$$

defined by

$$\beta|_{S_j} = \beta_j. \tag{45.3}$$

The problem is to describe the ergodic measures with respect to β on Y.

The reduction of this problem to the one solved in Section 38 will be obtained as a corollary of the general result (Lemma 45.5) concerning ergodic measures having 'periodic' components.

45.4 Let $\beta: Y \to Y$ be a bijection of the set Y and $Y_0 \subset Y$ be a subset invariant with respect to $\gamma = \beta^p$ and such that

(i) $Y = \bigcup_{i=0}^{p-1} \beta^i(Y_0)$

(ii) $Y_0 \cap \beta^i(Y_0) = \varnothing,\ i = \overline{1, p-1}$.

Observe that (ii) is equivalent to the property

$$\beta^i(Y_0) \cap \beta^j(Y_0) = \varnothing$$

when $i \neq j,\ 0 \le i, j \le p-1$.

The set Y_0 is 'periodic' (of period p) with respect to β.

Lemma 45.5 *Let β and β^{-1} be measurable bijections of a measurable space (Y, Ω) and $Y_0 \subset Y$ be a measurable subset with the properties described in 45.4. The ergodic measures with respect to β on (Y, Ω) and the ergodic measures with respect to $\gamma = \beta^p$ on (Y_0, Ω) are related to each other in the following way:*

I. *If μ is an ergodic measure with respect to β on (Y, Ω) then the measure ν given by*

$$\nu(\omega) = p\mu(X_0 \cap \omega) \tag{45.4}$$

is ergodic with respect to γ.

II. *If ν is an ergodic measure with respect to γ on (Y_0, Ω) then the measure*

$$\mu = \frac{1}{p}\left(\nu + \beta^{-1}(\nu) + \ldots + \beta^{-p+1}(\nu)\right) \tag{45.5}$$

is ergodic with respect to β.

Proof Let us prove II first. Since for every $\omega \in \Omega$ we have

$$\beta^{-j}(\nu)(\omega) = \nu\left(\beta^{-j}(\omega)\right)$$

it follows that

$$\mu\left(\beta^{-1}(\omega)\right)=\frac{1}{p}\left(\beta^{-1}(\nu)+\ldots+\beta^{-p}(\nu)\right)(\omega)$$

$$=\frac{1}{p}\left(\beta^{-1}(\nu)+\ldots+\beta^{-p+1}(\nu)+\nu\right)(\omega)$$

$$=\mu(\omega)$$

which means that μ is invariant with respect to β and is normalized (since ν is).

Now let Y' and Y'' be two measurable sets invariant with respect to β and such that

$$Y=Y'\cup Y'' \text{ and } Y'\cap Y''=\varnothing . \tag{45.6}$$

Let

$$\mu'=\mu|_{Y'},\quad \mu''=\mu|_{Y''},\quad \nu'=\nu|_{Y'},\quad \nu''=\nu|_{Y''} . \tag{45.7}$$

Then

$$\begin{aligned}\mu' &=\frac{1}{p}\left(\nu'+\beta^{-1}(\nu')+\ldots+\beta^{-p+1}(\nu')\right)\\ \mu'' &=\frac{1}{p}\left(\nu''+\beta^{-1}(\nu'')+\ldots+\beta^{-p+1}(\nu'')\right) .\end{aligned} \tag{45.8}$$

The sets Y' and Y'' are both invariant with respect to γ and as ν is ergodic we conclude that either $\nu'=0$ or $\nu''=0$ which implies that either $\mu'=0$ or $\mu''=0$ and thus μ is ergodic and II is proved.

Now let us check I. In view of the properties of Y_0 it follows that $\mu(Y_0)=\frac{1}{p}$ and thus the measure ν defined by (45.4) is normalized and is invariant with respect to γ (since μ is).

Let Y_0' and Y_0'' be two measurable subsets of Y_0 invariant with respect to γ and such that

$$Y_0=Y_0'\cup Y_0'' \text{ and } Y_0'\cap Y_0''=\varnothing .$$

Set

$$Y'=\bigcup_{i=0}^{p-1}\beta^i(Y_0'),\quad Y''=\bigcup_{i=0}^{p-1}\beta^i(Y_0'') .$$

The sets Y' and Y'' are both measurable and invariant with respect to β and in view of the properties of Y_0 satisfy (45.6).

Now using for these sets the notation (45.7) we again obtain (45.8), and as μ is ergodic it follows that either $\mu' = 0$ or $\mu'' = 0$. This in turn, along with (45.8), means that either $\nu' = 0$ or $\nu'' = 0$ which proves the ergodicity of ν and completes the proof as a whole. □

This lemma and Theorem 38.9 allows us to write out the solution of the problem posed in 45.3 which is stated in the next

Theorem 45.6 *Let Y be defined by (45.2) and β by (45.3), (45.1). A measure μ on Y is ergodic with respect to β iff it has the form*

$$\mu = \frac{1}{p}\sum_{j=0}^{p-1}\beta^{-j}(\nu)$$

where ν is a certain ergodic measure on the sphere S_0 with respect to the mapping γ induced by the matrix $L = L_{p-1}\cdot\ldots\cdot L_1 \cdot L_0$.

All such measures ν are described in Theorem 38.9.

Remark 45.7 Theorems 17.3, 38.9, 45.6 along with the description of the symbols of G-PDOs obtained in Section 41 show in particular that if the set $\Omega(\alpha)$ of non-wandering points consists only of periodic points then the Fredholmity conditions for a two-term G-PDO

$$b = a_0 + a_1 T$$

(T is the shift operator generated by α) can be written out in explicit form and the necessary ergodic measures corresponding to periodic points of α can be obtained by means of Theorems 38.9 and 45.6 just as for fixed points.

45.8 Later in this section we present two examples of G-PDOs on the sphere S^2. Since the symbols of these operators are the operators acting in the functional spaces on S^*S^2 let us now say a few words about the structure of S^*S^2 and the action of $\beta = \frac{\partial\alpha(\xi)}{\|\partial\alpha(\xi)\|}$ on S^*S^2 induced by the action of a diffeomorphism α on S^2.

The sphere S^2 does not possess a coordinate system without singularities and the tangent and cotangent bundles of S^2 are not trivial. In order to avoid referring to the local coordinates one can use an embedding into Euclidian space.

Thus, let $M = S^2$ be the unit sphere in $\mathbb{R}^3$. The tangent space TM can be naturally embedded into $\mathbb{R}^6$:

$$TM = \left\{(x,\eta): x \in \mathbb{R}^3, \eta \in \mathbb{R}^3, \|x\| = 1, \langle x,\eta\rangle = 0\right\}$$

and the tangent plane TM_x at a point $x \in M$ has the form

$$TM_x = \{(x,\eta): \langle x,\eta\rangle = 0\} .$$

Every linear function φ on TM_x is defined by

$$\varphi(\eta) = \langle \eta, \xi\rangle$$

(where ξ is a certain vector of TM_x). Thus the cotangent bundle T^*M as a set in $\mathbb{R}^6$ is given by

$$T^*M = \left\{(x,\xi): x \in \mathbb{R}^3, \xi \in \mathbb{R}^3 : \|x\| = 1, \langle x,\xi\rangle = 0\right\} .$$

As manifolds and vector bundles over M, the spaces TM and T^*M are isomorphic and the difference appears under the action of the induced mappings (differentials and codifferentials).

Any diffeomorphism $\alpha: M \to M \left(M = S^2\right)$ can be extended onto a neighbourhood of the sphere via the formula

$$\overline{\alpha}(x) = \|x\| \cdot \alpha\left(\frac{x}{\|x\|}\right) .$$

The extended map acts on a three-dimensional domain and defines the 3×3 Jacobian matrix $J(x) = \dfrac{\mathrm{d}\overline{\alpha}}{\mathrm{d}x}$. With the help of this matrix the differential α' is defined on TX by the formula

$$\alpha'(x,\eta) = (\alpha(x), J(x)\eta) .$$

The codifferential $\partial\alpha$ acts on the cotangent bundle T^*X via the formula

$$\partial\alpha(x,\xi)=\left(\alpha(x),\left[J(x)^{-1}\right]^{t}\xi\right)$$

where ^{t} is the transpose sign.

The bundle S^*M of the normalized cotangent vectors is realized in the form

$$S^*M=\left\{(x,\xi): x\in\mathbb{R}^3, \|x\|=1, \xi\in\mathbb{R}^3, \|\xi\|=1, \langle x,\xi\rangle=0\right\}$$

and is a three-dimensional manifold.

The induced mapping $\beta: S^*M\to S^*M$ is given via the formula

$$\beta(x,\xi)=\left(\alpha(x),\frac{\left[J(x)^{-1}\right]^{t}\xi}{\left\|\left[J(x)^{-1}\right]^{t}\xi\right\|}\right).$$

When introducing the shift operators and the s-symbol we also used the factors $|\alpha'(x)|$ (see (40.2)) and $\|\partial\alpha(x,\xi)\|$ (see (41.14)) which in our case are evaluated as

$$|\alpha'(x)|=|\det J(x)|,\ x\in S^2$$

$$|\partial\alpha(x,\xi)|=\left\|\left[J(x)^{-1}\right]^{t}\xi\right\|,\ (x,\xi)\in S^*S^2.$$

We note also that the factor $|\beta'(x,\xi)|$ (used in (40.6)) is exploited only to make the operator V (defined by (40.6)) into unitary and does not enter directly into the conditions of the invertibility of the s-symbol. Thus there is no need to write out the explicit form of the function $|\beta'(x,\xi)|$.

The realization of S^*S^2 as a submanifold of $\mathbb{R}^6$ determines a natural Riemann metric on S^2 and it is convenient to consider functional spaces with the norms induced by this Riemann metric.

Now let us consider certain examples of G-PDOs on S^2.

Example I.

45.9 Let the mapping $\alpha: S^2\to S^2$ be defined via the formula

$$\alpha(x)=\frac{x+h}{\|x+h\|} \tag{45.9}$$

where $h=(0,0,\tau)$, $0<\tau<1$. The non-wandering points of α are precisely two fixed points - the 'north' pole $p^+=(0,0,1)$ and the 'south' pole $p^-=(0,0,-1)$. Thus the ergodic measures for the induced mapping

$$\beta(\eta)=\frac{\partial\alpha(\eta)}{\|\partial\alpha(\eta)\|},\ \eta\in S^*S^2$$

can be obtained with the help of Theorem 38.9 applied to the spheres $S_\pm=S^*S^2_{p^\pm}$. In the particular situation considered here their description is elementary.

Let us take as the local coordinates in the neighbourhoods of $p^\pm$ the natural coordinates (x_1,x_2) (the projection onto the plane). Then a routine calculation show that

$$\alpha'(p^\pm)=\begin{bmatrix}(1\pm\tau)^{-1} & 0\\ 0 & (1\pm\tau)^{-1}\end{bmatrix},$$

$$\partial\alpha(p^\pm)=\begin{bmatrix}1\pm\tau & 0\\ 0 & 1\pm\tau\end{bmatrix}.$$

Thus

$$\left|\alpha^{-1\prime}(p^\pm)\right|=(1\pm\tau)^2 \tag{45.10}$$

and

$$\left\|\partial\alpha^{-1}(p^\pm,\xi)\right\|=(1\pm\tau)^{-1}. \tag{45.11}$$

Since the matrix $\partial\alpha^{-1}(p^\pm)$ is a multiple of the identity, the mapping β acts in the fibres over $p^\pm$ trivially and all the ergodic measures for β are precisely the Dirac point measures corresponding to the points of the cotangent circle.

45.10 Let us consider the *G*-PDO $\mathcal{b}:H^s(S^2)\to H^{s-r}(S^2)$ of the form

$$\mathcal{b}=a_0(x,D)+a_1(x,D)U \tag{45.12}$$

where $a_i(x,D)$, $i=0,1$, are PDOs of order r and U is the shift operator given by

$$(Uf)(x)=f(\alpha^{-1}(x)) \tag{45.13}$$

where α is defined by (45.9).

The s-symbol of this operator is equal to

$$\sigma_s(\mathcal{b}) = a_0(x,\xi) + a_1(x,\xi)\left|\alpha^{-1'}(x)\right|^{-\frac{1}{2}}\left\|\partial\alpha^{-1}(x,\xi)\right\|^{-s} V \tag{45.14}$$

where V is given by (40.6).

Now the argument of 45.9 (namely the description of the ergodic measures and (45.10) and (45.11)) together with the standard reasoning (the usage of Theorem 17.3) for the derivation of the invertibility conditions for $\sigma_s(\mathcal{b})$ implies the next

Theorem 45.11 *The operator* $\mathcal{b}$ *(45.12) is Fredholm iff one of the following two conditions (1) or (2) holds.*

(1) $a_0(x,\xi) \neq 0,\ (x,\xi) \in S^*S^2$

and

$$\left|a_0\left(p^{\pm},\xi\right)\right| > \left|a_1\left(p^{\pm},\xi\right)\cdot(1\pm\tau)^{s-1}\right|,\ \xi \in S^*S^2_{p^{\pm}}\ ;$$

(2) $\left|a_1(x,\xi)\right| \neq 0,\ (x,\xi) \in S^*S^2$

and

$$\left|a_0\left(p^{\pm},\xi\right)\right| < \left|a_1\left(p^{\pm},\xi\right)\cdot(1\pm\tau)^{s-1}\right|,\ \xi \in S^*S^2_{p^{\pm}}\ .$$

Remark In this example we obtain again the essential dependence of the Fredholmity conditions on the Sobolev space index.

Now let us look at one more example where the description of the necessary ergodic measures needs a combination of the reasoning exploited in the preceding examples.

Example II

45.12 Let $R(h,\lambda)$ be a matrix of the form

$$R(h,\lambda) = \begin{bmatrix} \cos 2\pi h & -\sin 2\pi h & 0 \\ \sin 2\pi h & \cos 2\pi h & 0 \\ 0 & 0 & \lambda \end{bmatrix} \tag{45.15}$$

$\lambda \neq 0,\ 0 \leq h \leq 1.$

This matrix generates the diffeomorphism $\alpha: S^2 \to S^2$ via the formula

$$\alpha(x) = \frac{R(h,\lambda)x}{\|R(h,\lambda)x\|} . \tag{45.16}$$

For different parameters h,λ one can obtain here qualitatively different diffeomorphisms α.

Let

$$R_h \equiv R(h,1) . \tag{45.17}$$

For any continuous function a on S^*S^2 and a point $(x,\xi) \in S^*S^2$ we denote

$$S_{x,\xi}(a) = \exp \int_0^1 \ln\left|a\left(R_h x, R_h \xi\right)\right| \mathrm{d}h . \tag{45.18}$$

The next theorem gives the Fredholmity condition for a two-term G-PDO with the shifts generated by the diffeomorphisms α introduced above (45.16). In its formulation we shall use the notation

$$\prod_N (a_0)(x,\xi) = \prod_{j=0}^{N-1} a\left(\alpha^j(x), \partial\alpha^j(x,\xi)\right),\ (x,\xi) \in S^*S^2 .$$

Theorem 45.13 *Let $\mathfrak{b}: H^s(S^2) \to H^{s-r}(S^2)$ be a two-term G-PDO of the form (45.12), (45.13) where α is given by (45.16). The operator $\mathfrak{b}$ is Fredholm iff one of the following conditions (depending on (h,λ) holds*:

(1) $\lambda = 1,\ h = 0$.

$$a_0(x,\xi) + a_1(x,\xi) \neq 0 \textit{ for every } (x,\xi) \in S^*S^2 .$$

(2) $\lambda = 1,\ h = \dfrac{M}{N}$ *is an irreducible fraction.*

$$\prod_N (a_0)(x,\xi) \neq \prod_N (-a_1)(x,\xi) \textit{ for every } (x,\xi) \in S^*S^2 .$$

(3) $\lambda = 1$, *h is irrational.*

(3a) $a_0(x,\xi) \neq 0$ *for every* $(x,\xi) \in S^*S^2$ *and*

$$S_{x,\xi}(a_0) > S_{x,\xi}(a_1) \textit{ for every } (x,\xi) ,$$

or

(3b) $a_1(x,\xi) \neq 0$ *for every* $(x,\xi) \in S^*S^2$ *and*

$$S_{x,\xi}(a_0) < S_{x,\xi}(a_1) \text{ for every } (x,\xi) .$$

(4) $\lambda > 0,\ \lambda \neq 1,\ h = 0.$

(4a) $a_0(x,\xi) \neq 0$ *for every* $(x,\xi) \in S^*S^2$ *and*

$$\left|a_0\left(p^{\pm},\xi\right)\right| > \left|a_1\left(p^{\pm},\xi\right)\right|\lambda^{s-1},\ \xi \in \left(S^*S^2\right)_{p^{\pm}}$$

($p^{\pm}$ *are the 'poles' of the sphere* (see 45.9)),

$$\left|a_0(x,\xi)\right| > \left|a_1(x,\xi)\right|\lambda^{\frac{1}{2}}$$

where x belongs to the 'equator' of S^2 *('equator' is the set* $\left\{x \in S^2 : x_3 = 0\right\}$*) and* ξ *is any cotangent to the 'equator' vector (that is, orthogonal to any tangent vector which is normal to the 'equator'),*

$$\left|a_0(x,\xi)\right| > \left|a_1(x,\xi)\right|\lambda^{\frac{1}{2}-s}$$

where x belongs to the 'equator' and ξ *is any cotangent conormal to the 'equator' vector (that is, orthogonal to any tangent to the 'equator' vector),*

or

(4b) $a_1(x,\xi) \neq 0$ *for every* $(x,\xi) \in S^*S^2$

and all the inequalities mentioned in (4a) have the opposite sign.

(5) $\lambda > 0,\ \lambda \neq 1,\ h = \dfrac{M}{N}$ *is an irreducible fraction.*

(5a) $a_0(x,\xi) \neq 0$ *for every* $(x,\xi) \in S^*S^2$ *and*

$$\left|\prod_N(a_0)\left(p^{\pm},\xi\right)\right| > \left|\prod_N(a_1)\left(p^{\pm},\xi\right)\right|\lambda^{N(s-1)},\ \xi \in \left(S^*S^2\right)_{p^{\pm}},$$

$$\left|\prod_N(a_0)(x,\xi)\right| > \left|\prod_N(a_1)(x,\xi)\right|\lambda^{\frac{1}{2}}$$

where x belongs to the 'equator' and ξ *is any cotangent to the 'equator' vector,*

$$\left|\prod_N(a_0)(x,\xi)\right| > \left|\prod_N(a_1)(x,\xi)\right|\lambda^{\frac{1}{2}-s}$$

where x belongs to the 'equator' and ξ *is any cotangent conormal to the 'equator' vector;*

or

(5b) $a_1(x,\xi) \neq 0$ *for every* $(x,\xi) \in S^*S^2$

and all the inequalities mentioned in (5a) have the opposite sign.

(6) $\lambda > 0,\ \lambda \neq 1,\ h$ *is irrational.*

(6a) $a_0(x,\xi) \neq 0$ *for every* $(x,\xi) \in S^*S^2$ *and*

$$S_{p^\pm,\xi}(a_0) > S_{p^\pm,\xi}(a_1)\lambda^{s-1},\ \xi \in \left(S^*S^2\right)_{p^\pm},$$

$$S_{x,\xi}(a_0) > S_{x,\xi}(a_1)\lambda^{\frac{1}{2}}$$

where x belongs to the 'equator' and ξ *is any cotangent to the 'equator' vector,*

$$S_{x,\xi}(a_0)S_{x,\xi}(a_1)\lambda^{\frac{1}{2}-s}$$

where x belongs to the 'equator' and ξ *is any cotangent conormal to the 'equator' vector;*

or

(6b) $a_1(x,\xi) \neq 0$ *for every* $(x,\xi) \in S^*S^2$

and all the inequalities mentioned in (6a) have the opposite sign.

(7) $\lambda = -1,\ h = 0.$

$$\prod_2(a_0)(x,\xi) \neq \prod_2(a_1)(x,\xi) \text{ for every } (x,\xi) \in S^*S^2.$$

(8) $\lambda = -1,\ h = \dfrac{M}{N}$ *is an irreducible fraction.*

$$\prod_{N_1}(a_0)(x,\xi) \neq \prod_{N_1}(a_1)(x,\xi) \text{ for every } (x,\xi) \in S^*S^2$$

where

$$N_1 = \begin{cases} N, & \text{if } N \text{ is even,} \\ 2N, & \text{if } N \text{ is odd.} \end{cases}$$

In particular when $h = \frac{1}{2}$ *we have* $\alpha(x) = -x$ *and the condition*

$$a_0(x,\xi)a_0(-x,-\xi) \neq a_1(x,\xi)a_1(-x,-\xi) .$$

(9) $\lambda = -1$, *h is irrational.*

(9a) $a_0(x,\xi) \neq 0$ *for every* $(x,\xi) \in S^*S^2$ *and*

$$S_{x,\xi}\left(\prod_2(a_0)\right) > S_{x,\xi}\left(\prod_2(a_1)\right) \textit{for every } (x,\xi) ;$$

or

(9b) $a_1(x,\xi) \neq 0$ *for every* $(x,\xi) \in S^*S^2$

and the inequality in (9a) has the opposite sign.

(10) $\lambda < 0$, $\lambda \neq -1$, $h = 0$.

(10a) $a_0(x,\xi) \neq 0$ *for every* $(x,\xi) \in S^*S^2$ *and*

$$\left|\prod_2(a_0)(p^+,\xi)\right| > \left|\prod_2(a_1)(p^+,\xi)\right| |\lambda|^{2(s-1)}, \ \xi \in \left(S^*S^2\right)_{p^+} ,$$

$$|a_0(x,\xi)| > |a_1(x,\xi)| \, |\lambda|^{\frac{1}{2}}$$

when x belongs to the 'equator' and ξ is any cotangent to the 'equator' vector,

$$\left|\prod_2(a_0)(x,\xi)\right| > \left|\prod_2(a_1)(x,\xi)\right| |\lambda|^{1-2s}$$

when x belongs to the 'equator' and ξ is any cotangent conormal to the 'equator' vector';

or

(10b) $a_1(x,\xi) \neq 0$ *for every* $(x,\xi) \in S^*S^2$

and all the inequalities in (10a) have the opposite sign.

(11) $\lambda < 0$, $\lambda \neq -1$, $h = \dfrac{M}{N}$ *is an irreducible fraction.*

(11a) $a_0(x,\xi) \neq 0$ *for every* $(x,\xi) \in S^*S^2$ *and*

$$\left|\prod_{N_1}(a_0)(p^+,\xi)\right| > \left|\prod_{N_1}(a_1)(p^+,\xi)\right| |\lambda|^{N_1(s-1)}, \ \xi \in \left(S^*S^2\right)_{p^+},$$

$$\left|\prod_N(a_0)(x,\xi)\right| > \left|\prod_N(a_1)(x,\xi)\right| |\lambda|^{\frac{N_1}{2}}$$

when x belongs to the 'equator' and ξ is any cotangent conormal to the 'equator' vector',

$$\left|\prod_{N_1}(a_0)(x,\xi)\right| > \left|\prod_{N_1}(a_1)(x,\xi)\right| |\lambda|^{N_1\left(\frac{1}{2}-s\right)}$$

when x belongs to the 'equator' and ξ is any cotangent conormal to the 'equator' vector'; here

$$N_1 = \begin{cases} N, & \text{if } N \text{ is even,} \\ 2N, & \text{if } N \text{ is odd;} \end{cases}$$

or

(11b) $a_1(x,\xi) \neq 0$ *for every* $(x,\xi) \in S^*S^2$

and all the inequalities in (11a) have the opposite sign.

(12) $\lambda < 0, \ \lambda \neq -1,$ *h is irrational.*

(12a) $a_0(x,\xi) \neq 0$ *for every* $(x,\xi) \in S^*S^2$ *and*

$$S_{p^+,\xi}\left(\prod_2(a_0)\right) > S_{p^+,\xi}\left(\prod_2(a_1)\right) |\lambda|^{2(s-1)}, \ \xi \in \left(S^*S^2\right)_{p^+},$$

$$S_{x,\xi}(a_0) > S_{x,\xi}(a_1)|\lambda|^{\frac{1}{2}}$$

when x belongs to the 'equator' and ξ is any cotangent to the 'equator' vector,

$$S_{x,\xi}\left(\prod_2(a_0)\right) > S_{x,\xi}\left(\prod_2(a_1)\right) |\lambda|^{1-2s}$$

when x belongs to the 'equator' and ξ is any cotangent conormal to the 'equator' vector;

or

(12b) $a_1(x,\xi) \neq 0$ *for every* $(x,\xi) \in S^*S^2$

and all the inequalities mentioned in (12a) have the opposite sign.

Proof Without going into details we sketch the general scheme of the proof. To describe the ergodic measures with respect to β on S^*S^2 we use the following considerations.

(I) The support of an ergodic measure is concentrated on the part of S^*S^2 based on the set $\Omega(\alpha)$ of non-wandering points of α.

(II) In the situation (1) b is a PDO without a shift.

(III) In cases (2), (7), (8) the diffeomorphism α is periodic and the written-out statement is the explicit record of the condition of Theorem 43.8.

(IV) If $|\lambda| \neq 1$ (in the situations (4), (5), (6), (10), (11), (12)) the set of non-wandering points consists of the poles $p^{\pm}$ and the equator.

The restriction of S^*S^2 to the 'equator' is the torus and the action of β on this torus is diagonal:

(a) on the 'equator' it is the rotation;

(b) on the cotangent circle it is the mapping generated by the Jacobian matrix $\begin{bmatrix} 1 & 0 \\ 0 & \lambda \end{bmatrix}$.

Thus the non-wandering points are the pairs (x,ξ) where ξ is either 'horizontal' (cotangent to the 'equator') or 'vertical' (conormal to the 'equator').

$S^*S^2_{p^{\pm}}$ are circles and the ergodic measures here can be obtained by means of Theorem 38.9 (in fact here it can be used in an elementary way).

(V) If $|\lambda| = 1$ the diffeomorphism α preserves the Riemann metric and thus

$$\left|\alpha^{-1'}(x)\right| = 1 \text{ and } \left\|\partial\alpha^{-1}(x,\xi)\right\| = 1 .$$

(VI) If $|\lambda| \neq 1$ it is enough to calculate the values of these functions on the set of non-wandering points. Direct calculation shows that

$$\left|\alpha^{-1'}\left(p^{\pm}\right)\right| = |\lambda|^2 ,$$

$$\left\|\partial\alpha^{-1}\left(p^{\pm},\xi\right)\right\| = |\lambda|^{-1} ,$$

and on the 'equator'

$$\left|\alpha^{-1'}(x)\right| = \frac{1}{|\lambda|}$$

and

$\left\|\partial\alpha^{-1}(x,\xi)\right\| = 1$ when ξ is a cotangent to the 'equator',

$\left\|\partial\alpha^{-1}(x,\xi)\right\| = |\lambda|$ when ξ is a conormal to the 'equator'.

Thus we obtain the factors

$|\lambda|^{s-1}$ at $p^{\pm}$,

$|\lambda|^{\frac{1}{2}}$ when (x,ξ) is a cotangent to the 'equator',

$|\lambda|^{\frac{1}{2}-s}$ when (x,ξ) is a conormal to the 'equator'.

Bearing the considerations (III)-(VI) in mind we see that the conditions (3)-(6) and (9)-(12) are particular cases of the general invertibility conditions for a two-term operator $\sigma_s(\mathcal{b})$ written out in Theorem 17.3. □

46. *G*-singular integral operators with continuous coefficients

One-dimensional singular integral operators (SIOs) on a curve Γ are in fact a particular case of PDOs on a (one-dimensional) manifold. But usually these operators are considered in a separate way since they possess a number of specific features in comparison with multidimensional PDOs. These specific features are the following.

1. The tangent (and cotangent) sphere is not connected - it consists of two points. In particular the dimension of the cospherical bundle $S^*\Gamma$ coincides with the dimension of the initial manifold - the curve Γ.

2. Any complex bundle over a curve is trivial. The cotangent bundle $T^*\Gamma$ is isomorphic to $\Gamma \times \mathbb{R}$. The isomorphism here is defined 'up to the sign' and the fixation of the orientation and of the unit 'positive' cotangent vector $(x,1)$ determines the isomorphism uniquely.

3. The curve Γ is not an abstract manifold but is given as a subset of the complex plane $\mathbb{C}$. The plane $\mathbb{C}$ induces in a natural way the Riemann metric on Γ and in particular the bundle of the unit cotangent vectors is defined in a canonical way.
   The embedding of Γ into $\mathbb{C}$ also gives us the possibility of establishing the interrelations between the theory of SIOs and the theory of boundary value problems for analytic functions on the plane.

All these features also play an essential role in the investigation of *G*-SIOs.

We now proceed to the precise description of the objects and problems of this section.

Singular integral operators with continuous coefficients. Elementary properties.

46.1 We shall say that Γ is a *closed (Lyapunov) curve* on the complex plane if it is an oriented closed curve without self-crossings and satisfies the Lyapunov condition. The latter means that the angle $\theta_r(t)$ $(t \in \Gamma)$ between the tangent to the curve at the point *t* and the positive real axis satisfies the Hölder condition

$$|\theta_\Gamma(t_1) - \theta_\Gamma(t_2)| \leq C|t_1 - t_2|^\mu \tag{46.1}$$

where $0 < \mu \leq 1$ and *C* is some constant.

In this section we shall consider only the case when a closed curve consists of a finite number of connected components (*simple closed curves*)

$$\Gamma = \{\Gamma^1, \ldots, \Gamma^l\}$$

(Γ^i, $i = \overline{1,l}$, is a connected component of Γ).

In problems related to the investigation of *G*-SIOs the choice of orientation on Γ does not in fact play a fundamental role. The orientation of Γ related to the theory of analytic functions is a rule defined in the following way.

The curve Γ decomposes the plane into two open sets Ω^+ and Ω^- for each of which Γ is the boundary (by Ω^- we denote the set containing ∞ of the extended plane). The orientation of Γ is chosen from the condition that the domain Ω^+ remains on the left while going around the curve.

As in the multidimensional manifold case in *G*-SIO theory the orientation does not in fact play a significant role and the condition on a mapping $\alpha: \Gamma \to \Gamma$ to preserve (change) the orientation does not depend on the choice of orientation. Certain combinatorial situations arise when considering the case of non-simple curves and several mappings α_k but we shall have little need to consider this theme here.

We recall also that if $\Gamma = \{\Gamma^1, \ldots, \Gamma^l\}$ is a closed curve with fixed orientation and φ is a continuous non-vanishing function on Γ then the *index* of φ is defined by the equality

$$\operatorname{ind}[\varphi] = \frac{1}{2\pi}[\arg \varphi(t)]_\Gamma = \frac{1}{2\pi}\sum_{i=1}^{l}[\arg \varphi(t)]_{\Gamma^i} \tag{46.2}$$

where $[\ ]_{\Gamma^i}$ is the increase of the function around a curved circuit traversed according to the orientation.

46.2 Let S be the *Cauchy singular integral* operator (see 2.49) defined in $L^2(\Gamma)$ via the formula

$$(Sf)(t) = \frac{1}{\pi \mathrm{i}} \int_\Gamma \frac{f(\tau)}{\tau - t} \mathrm{d}\tau, \qquad t \in \Gamma$$

where the integral on the right is understood in the principal value sense.

We recall that the *singular integral operator* (SIO) (with continuous coefficients) in $L^2(\Gamma)$ is the operator of the form

$$\mathcal{a} = a_0 + a_1 S + k \tag{46.3}$$

where a_0 and a_1 are the operators of multiplication by continuous functions and k is a compact operator.

Remark Hereinafter in this chapter we shall consider only operators with continuous coefficients. SIOs and *G*-SIOs with discontinuous coefficients will be investigated in the next chapter.

46.3 It is worth mentioning that the classical general Fredholm theory of SIOs with continuous coefficients in $L^2(\Gamma)$ can be extracted in a rather simple way from Theorem 12.8. Indeed, it is known the operator S possesses the following properties:

(i) $S^2 = I$;

(ii) $Sa \sim aS$ for any continuous function a;

(iii) $S^* \sim S$

where in (ii) and (iii) the sign ~ means that the left-hand and the right-hand terms differ by a compact summand;

(iv) $|||a_0 + a_1 S||| \geq |||a_j|||$, $j = \overline{0,1}$,

where $|||\cdot|||$ is the essential norm (see Theorem 2.46).

One can find the proofs of properties (i)-(iii) in for example Gohberg, Krupnik [4], (Theorems 3.2, 4.3, 7.3) and the proof of (iv) is given in Gohberg [1].

Now if we take as the coefficient algebra $\mathcal{A}$ the algebra of operators of the form

$$a + k, \quad k \in K \tag{46.4}$$

(a is the operator of multiplication by a function $a \in C(\Gamma)$ and K is the set of compact operators) then properties (i)-(iii) mean that the operator S generates (the trivial) automorphism of the algebra

$$[\mathcal{A}] = \mathcal{A}/K$$

via the formula

$$\hat{S}([a]) = \left[SaS^*\right] = [SaS]\,. \tag{46.5}$$

Property (iv) means that for the algebra $C^*(\mathcal{A}, S)$ of singular integral operators (46.3) the corresponding factor algebra

$$C^*([\mathcal{A}],[S]) \cong \left[C^*(\mathcal{A},S)\right] = C^*(\mathcal{A},S)/K$$

possesses the property (*) (in view of (i) we consider here $[S]$ as the operator generating the representation of the group $G = \mathbb{Z}_2$).

All these remarks imply (in view of Theorem 12.8) that

$$[C^*(\mathcal{A},S)] \cong [\mathcal{A}] \underset{I}{\times} \mathbb{Z}_2 \tag{46.6}$$

where I denotes the trivial action of $\mathbb{Z}_2$ on $[\mathcal{A}]$.

Now recalling Theorem 14.19 and Corollary 2.77 we have

$$[\mathcal{A}] \cong C(\Gamma) \tag{46.7}$$

which, along with (46.6), establishes the isomorphism

$$[C^*(\mathcal{A}, S)] \cong C(\Gamma) \underset{I}{\times} \mathbb{Z}_2 \tag{46.8}$$

given by

$$[a_0 + a_1 S] \to a_0 \otimes e + a_1 \otimes g_1 \quad (\mathbb{Z}_2 = \{e, g_1\}) . \tag{46.9}$$

By Theorem 15.2 we can also obtain an isomorphism between $C(\Gamma) \underset{I}{\times} \mathbb{Z}_2$ and the algebra of matrix-valued functions of the form

$$\begin{bmatrix} a_0(t) & a_1(t) \\ a_1(t) & a_0(t) \end{bmatrix} . \tag{46.10}$$

This observation implies in particular that an operator $a_0 + a_1 S + k$ is Fredholm iff $a_0^2(t) - a_1(t)^2 \neq 0$ for any $t \in \Gamma$.

By elementary reasoning one can also see that the algebra of matrix-valued functions of the form (46.10) is isomorphic to the algebra of matrix-valued functions of the form

$$\begin{bmatrix} a_0(t) + a_1(t) & 0 \\ 0 & a_0(t) - a_1(t) \end{bmatrix}$$

which means that we have obtained assertions (i) and (ii) of Theorem 2.54.

46.4 Property (i) mentioned in 46.3 means that the operators P and Q given by

$$P = \frac{I+S}{2} \text{ and } Q = \frac{I-S}{2} = I - P \tag{46.11}$$

are projections. These projections define the decomposition of the space $L^2(\Gamma)$ into the direct sum of subspaces $H^+(\Gamma)$ and $H^-(\Gamma)$ where

$$H^{\pm}(\Gamma) = \{u \in L^2(\Gamma) : (I \mp S)u = 0\} . \tag{46.12}$$

A function u belongs to $H^+(\Gamma)$ if and only if it is the boundary value of a certain analytic function in Ω^+ (see 46.1) which can be represented by a Cauchy integral with a density from $L^2(\Gamma)$. A function v belongs to $H^-(\Gamma)$ if and only if it is the boundary value of an analytic function in Ω^- vanishing at infinity and such that it can be represented by a Cauchy integral with a density from $L^2(\Gamma)$.

In view of (46.11) for any $u \in H^\pm(\Gamma)$ we have $Su = \pm u$. The decomposition

$$L^2(\Gamma) = H^+(\Gamma) \oplus H^-(\Gamma)$$

generates a representation of any operator d acting in $L^2(\Gamma)$ in the operator-matrix form

$$d = \begin{bmatrix} d_{11} & d_{12} \\ d_{21} & d_{22} \end{bmatrix}.$$

In a number of situations the operators d_{12} and d_{21} turn out to be compact. If one finds oneself in such a situation then the operator d is Fredholm iff the operators d_{11} and d_{22} are Fredholm and

$$\operatorname{ind} d = \operatorname{ind} d_{11} + \operatorname{ind} d_{22}.$$

In particular the singular integral operator $\mathfrak{a}$ (46.3) in operator-matrix form is written out as

$$\mathfrak{a} = \begin{bmatrix} P(a_0 + a_1) & P\mathfrak{a}Q \\ Q\mathfrak{a}P & Q(a_0 - a_1) \end{bmatrix} \tag{46.13}$$

where the operators $P\mathfrak{a}Q$ and $Q\mathfrak{a}P$ are compact.

46.5 We define a *Toeplitz operator* generated by a function $a \in L^\infty(\Gamma)$ to be an operator acting in $H^+(\Gamma)$ and defined by

$$u \to P(au). \tag{46.14}$$

In an analogous way one can introduce a Toeplitz operator in $H^-(\Gamma)$.

In view of (46.13) we conclude that the singular integral operator a (with continuous coefficients) is Fredholm if and only if the Toeplitz operators $P(a_0+a_1)$ and $Q(a_0-a_1)$ are both Fredholm.

Thus singular integral operator theory and Toeplitz operator theory are closely related and all the results presented below in this and the next sections could be reformulated in terms of Toeplitz operators (see Section 48).

Remark In connection of this observation we recall also Section 30 and 32.21-32.26, 33.19-33.20.

We now consider the

G-singular integral operators

46.6 Let $\alpha:\Gamma\to\Gamma$ be a diffeomorphism of Γ such that $\alpha'(t)\neq 0$ and $\alpha'(t)$ satisfies the Holder condition

$$\sup_{\tau,t\in\Gamma}\frac{|\alpha'(t)-\alpha'(\tau)|}{|t-\tau|^{\mu}}<\infty\,. \tag{46.15}$$

In all the forthcoming considerations all the diffeomorphisms of Γ will be presupposed to satisfy this condition.

Recall that when investigating *G*-PDOs we assumed the diffeomorphisms to be C^{∞}-smooth. This was due to the fact that all the manifolds under consideration were C^{∞}-smooth. Now since we have passed to Lyapunov curves it is reasonable to relax the smoothness properties of the mappings as well. In fact this relaxation of the smoothness conditions does not play any essential role in the theory (see Theorems 2.48 and 46.12).

For a given diffeomorphism α we denote by T_α the operator in $L^2(\Gamma)$ (or $L^2(\Gamma,\mathbb{C}^n)$) defined by the formula

$$(T_\alpha f)(x)=\left|\alpha^{-1\prime}(x)\right|^{\frac{1}{2}}f\left(\alpha^{-1}(x)\right). \tag{46.16}$$

Hereinafter we denote by $\{\alpha_g\}_{g\in G}$ a certain collection of diffeomorphisms of Γ defining the representation of a group *G* (cf. 40.1). If this collection $\{\alpha_g\}_{g\in G}$ is fixed then we denote by T_g the operator T_{α_g} described by (46.16) with $\alpha=\alpha_g$.

46.7 In the space $L^2(\Gamma)$ or $L^2\left(\Gamma,\mathbb{C}^n\right)$ let us consider operators of the form

$$\mathscr{b} = b_0 + b_1 S + k \tag{46.17}$$

where k is a compact operator and

$$b_j = \sum_{g\in F} a_{jg} T_g, \quad j = 0,1,\ |F| < \infty \tag{46.18}$$

where a_g are the operators of multiplication by continuous scalar (or matrix-valued) functions and T_g are defined in 46.6.

The operators (46.17) will be called *G-singular integral operators* (*G*-SIOs) related to the group G and its representation by the diffeomorphisms α_g of Γ.

We denote by $\mathscr{B}$ here the C^**algebra of G-singular integral operators* generated by the operators (46.17).

In fact, *G*-SIO is a particular case of *G*-PDO (just as SIO is a particular case of PDO) but as we have already mentioned at the beginning of this section these operators have certain specific features.

46.8 The bundle $S^*\Gamma$ of the unit vectors of $T^*\Gamma$ is the union of two copies of Γ:

$$\tilde{\Gamma} = S^*\Gamma = \Gamma_+ \cup \Gamma_- = \{(x,\xi): x \in \Gamma, \xi = \pm 1\} \tag{46.19}$$

(here we presume that we have already fixed a certain orientation on Γ, that is we have fixed the orientation of each connected component Γ_i of $\Gamma = \{\Gamma_1,\dots,\Gamma_l\}$).

For any diffeomorphism $\alpha_g:\Gamma\to\Gamma$ the induced diffeomorphism $\beta_g: S^*\Gamma \to S^*\Gamma$ (see (40.3)) acts via the formula

$$\beta_g(x,\xi) = \left(\alpha_g(x), \operatorname{sign}\alpha_g'(x)\xi\right) \tag{46.20}$$

where $\operatorname{sign}\alpha_g'(x) = 1$ if α_g preserves the orientation of the corresponding connected components and $\operatorname{sign}\alpha_g'(x) = -1$ if α_g changes the orientation.

46.9 We define the *symbol of the G-SIO* of the form (46.16) to be the operator $\sigma(\mathscr{b})$ acting in $L^2(\tilde{\Gamma})$ and given by the formula

$$\sigma(\mathscr{b}) = \sum_{g\in F} a_{0g} T_{\beta_g} + \sum_{g\in F} a_{1g} \operatorname{sign}\xi T_{\beta_g}\ . \tag{46.21}$$

We denote by B the C^*-algebra $\left(\subset L\left(L^2(\tilde{\Gamma})\right)\right)$ generated by the operators (46.21).

If the diffeomorphisms α_g, $g \in F$ in (46.21) preserve the orientation of Γ then the symbol $\sigma(\mathcal{b})$ decomposes into the direct sum of two functional operators

$$b^+ = b_0 + b_1$$

and (46.22)

$$b^- = b_0 - b_1$$

(b_j, $j = \overline{0,1}$, are given by (46.18)).

Remark 46.10 To be more precise the symbol $\sigma(\mathcal{b})$ can be decomposed into the direct sum of two operators $(b_0 + b_1)$ and $(b_0 - b_1)$ under weaker conditions than those above on the action of α_g (the so-called weak preservation of the orientation) which is defined as follows.

We say that the group $\{\alpha_g\}_{g\in G}$ of homeomorphisms of Γ *weakly preserves the orientation* of the (non-simple) closed curve Γ if, for any connected component Γ^i of Γ, the mappings α_g which preserve this component $\left(\alpha_g\left(\Gamma^i\right) = \Gamma^i\right)$ preserve its orientation as well.

It is worth mentioning that while the property of $\{\alpha_g\}_{g\in G}$ of preserving the orientation of a (non-simple) curve depends on the choice of the orientation of the connected components Γ_i, the property of weak preservation of the orientation does not depend on the choice of the orientation of connected components.

We see that the construction of the symbol of a G-SIO is simpler than the corresponding construction for a G-PDO (cf. 40.3). For a G-SIO the symbol is composed from the operators involved in the definition of the initial operator $\mathcal{b}$.

The following statement is a particular case of the general theorems on the symbolic calculus of G-PDOs (Theorems 40.7, 40.12).

Theorem 46.11 *Let G be a discrete amenable group acting topologically freely on Γ via mappings $\{\alpha_g\}_{g\in G}$ (or let G be a countable amenable group acting normally (see 40.11)). Then the mapping*

$$\sigma : \mathcal{b} \to \sigma(\mathcal{b})$$

can be extended up to the *-epimorphism

$$\sigma:\mathcal{B}\to B\,,$$

the kernel of which is the ideal K of compact operators in $L^2(\Gamma)$.

In particular the symbol of a G-SIO possesses properties analogous to the properties of the symbol of an SIO (and enumerated in Theorem 40.7).

Since the symbol of a G-SIO is a functional operator on the curve $\tilde{\Gamma}$ and any connected component of $\tilde{\Gamma}$ is isomorphic to the circle all the previously investigated examples where the conditions of invertibility of the functional operators on the circle were obtained (see 18.4) are in fact examples of the evaluation of the invertibility conditions for the symbols of certain G-SIOs.

While calculating the index of a G-SIO we shall also need (besides the invertibility conditions of the symbol) to construct certain topological invariants related to the index formulae. The following Theorems 46.12 and 46.13 present the information which will be used later when describing these invariants. The first theorem is, on the one hand, a particular case of the general theorem on the transformation of an argument for a PDO (see Theorems 2.48, 2.51) but on the other hand contains certain refinements: here we relax the demand on the smoothness of α_g and we write out the transformed operator not only up to a compact summand but up to an *integral operator with a weak singularity.*

Theorem 46.12 *Let Γ be a Lyapunov closed curve (see 46.1), $\alpha:\Gamma\to\Gamma$ be a diffeomorphism of the class C^1 with α' satisfying the Holder condition (46.15) and V_α be the shift operator in $L^2(\Gamma)$ defined by*

$$(V_\alpha f)(t)=f\left(\alpha^{-1}(t)\right). \tag{46.23}$$

Then we have

$$\operatorname{sign}\alpha' V_\alpha^{-1} S V_\alpha - S = +k$$

where k *is an integral operator with a kernel which is continuous off the diagonal and has a weak singularity on the diagonal.*

Proof Direct computation shows that the operator

$$\operatorname{sign}\alpha' V_\alpha^{-1} S V_\alpha - S$$

is an integral operator with kernel

$$K(t,\tau)=\frac{\alpha'(\tau)}{\alpha(\tau)-\alpha(t)}-\frac{1}{\tau-t}$$

$$=\frac{\alpha'(\tau)(\tau-t)-\alpha(\tau)+\alpha(t)}{[\alpha(\tau)-\alpha(t)](\tau-t)}\,. \tag{46.24}$$

The numerator in (46.24) can be rewritten in the form

$$\int_\tau^t [\alpha'(u)-\alpha'(\tau)]\,du\,. \tag{46.25}$$

Since α' satisfies the Holder condition we have

$$|\alpha'(u)-\alpha'(\tau)|\le C|u-\tau|^\mu,\ 0<\mu\le 1, C>0$$

which in turn in view of (46.25) implies the estimate for the numerator in (46.24)

$$\left|\int_\tau^t [\alpha'(u)-\alpha'(\tau)]\,du\right|\le \int_\tau^t |\alpha'(u)-\alpha'(\tau)|\,|du|$$

$$\le \int_0^{|\tau-t|} Cmr^\mu\,dr$$

$$=\frac{Cm}{\mu+1}|\tau-t|^{\mu+1}$$

and it follows that

$$|K(t,\tau)|\le\frac{Cm}{\mu+1}\;\frac{|\tau-t|^{\mu+1}}{|\alpha(\tau)-\alpha(t)|\,|\tau-t|}\,.$$

As

$$\lim_{\tau\to t}\left|\frac{\tau-t}{\alpha(\tau)-\alpha(t)}\right|=\frac{1}{|\alpha'(t)|}$$

and $\alpha'(t)\ne 0$ on Γ we conclude that the ratio

$$\left|\frac{\tau-t}{\alpha(\tau)-\alpha(t)}\right|$$

is bounded from above.

Thus

$$|K(t,\tau)| \le \frac{C_1}{|\tau - t|^{1-\mu}}$$

which means that $K(t,\tau)$ has a weak singularity on the diagonal.

The continuity of $K(t,\tau)$ off the diagonal is obvious. □

Theorem 46.13 *Let $\alpha : \Gamma \to \Gamma$ be a diffeomorphism preserving orientation and satisfying the Holder condition (46.15). If V_α and T_α are the operators in $L^2(\Gamma)$ given by (46.23) and (46.16) and P and Q are the projections (46.11) then the operator*

$$V_\alpha P + Q$$

is invertible and the operator

$$T_\alpha P + Q$$

is Fredholm with zero index.

Proof I. We start by considering the operator

$$\mathfrak{b} = V_\alpha P + Q \,. \tag{46.26}$$

Along with this operator, let us consider also the operator

$$\mathfrak{b}_1 = V_\alpha^{-1} P + Q \,.$$

The latter is the regularizer for $\mathfrak{b}$ since in view of the previous theorem and the definition of P and Q we have

$$\begin{aligned} \mathfrak{b}\mathfrak{b}_1 &= I + k_1 \\ \mathfrak{b}_1\mathfrak{b} &= I + k_2 \end{aligned} \tag{46.27}$$

where k_i, $i = 1,2$, are compact operators.

Let us show that

$$\operatorname{Ker} \mathscr{b} = \{0\} . \tag{46.28}$$

Indeed suppose that $f \neq 0$ and

$$\mathscr{b} f = 0 . \tag{46.29}$$

If we denote

$$f^{+} = Pf, \; f^{-} = Qf \tag{46.30}$$

then (46.29) implies

$$V_{\alpha} f^{+} = -f^{-} \tag{46.31}$$

and since

$$f^{+} = Sf^{+} \text{ and } f^{-} = -Sf^{-} \tag{46.32}$$

it follows that

$$f^{+} = -V_{\alpha}^{-1} f^{-} = V_{\alpha}^{-1} S f^{-} = -V_{\alpha}^{-1} S V_{\alpha} f^{+} . \tag{46.33}$$

Now summing (46.33) and the left-hand equality in (46.32) one obtains

$$2 f^{+} = \left(S - V_{\alpha}^{-1} S V_{\alpha}\right) f^{+} . \tag{46.34}$$

In view of Theorem 46.12 the operator

$$k = S - V_{\alpha}^{-1} S V_{\alpha}$$

is an integral operator, the kernel of which is continuous off the diagonal and has a weak singularity on the diagonal. This implies:

(1) equation (46.34) may have only a finite number of linearly independent solutions;
(2) all the solutions of (46.34) are continuous functions.

Observe that the set of continuous functions belonging to $H^{+}(\Gamma)$ (see 46.4) forms an algebra; similarly for the set of continuous functions belonging to $H^{-}(\Gamma)$ (this follows from the fact that both $H^{+}(\Gamma)$ and $H^{-}(\Gamma)$ are formed by the boundary values of analytic functions and the analyticity and continuity are preserved under multiplication).

Moreover, the operator V_α maps a continuous function into a continuous function and possesses the property

$$V_\alpha(f_1 \cdot f_2) = [V_\alpha f_1] \cdot [V_\alpha f_2] \,. \tag{46.35}$$

In view of these observations we conclude that if a pair f^{+}, f^{-} satisfies (46.31) then

$$V_\alpha \left(f^{+}\right)^{n} = \left(f^{-}\right)^{n}, \quad n = 1,2,\dots \,. \tag{46.36}$$

Note also that if

$$f^{+} \equiv c = \text{const}$$

then by (46.31)

$$f^{-} = -V_\alpha f^{+} = -c$$

and it follows that

$$f = f^{+} + f^{-} \equiv 0 \,.$$

Now if f^{+} is not constant then all its degrees $\left(f^{+}\right)^{n}$, $n = 1,2,\dots,$ are linearly independent and the foregoing consideration means that equation (46.34) has an infinite number of linearly independent solutions which contradicts (1), thus proving (46.28).

By analogous reasoning one can obtain that

$$\text{Ker}\, \mathscr{b}_1 = \{0\}$$

which, along with (46.28), implies

$$\text{Ker}\, \mathscr{b}\mathscr{b}_1 = \{0\} \ \text{ and } \ \text{Ker}\, \mathscr{b}_1\mathscr{b} = \{0\} \,. \tag{46.37}$$

But in view of (46.27) both the operators $\mathscr{b}\mathscr{b}_1$ and $\mathscr{b}_1\mathscr{b}$ are Fredholm and have zero index. Thus by (46.37) we conclude that they are invertible. Now if d_1 is the inverse of $\mathscr{b}\mathscr{b}_1$ and d_2 is the inverse of $\mathscr{b}_1\mathscr{b}$ then we have

$$I = \mathscr{b}\left[\mathscr{b}_1 d_1\right] = \left[d_2 \mathscr{b}_1\right]\mathscr{b}$$

which implies the invertibility of $\mathscr{b}$.

II. Now let us consider the operator

$$T_\alpha P + Q\ .$$

We have

$$\left(T_\alpha P + Q\right) = \left(\left|\alpha^{-1\prime}\right|^{\frac{1}{2}} P + Q\right)\left(V_\alpha P + Q\right) + k \tag{46.38}$$

where k is a compact operator.

Since

$$\left|\alpha^{-1\prime}(t)\right|^{\frac{1}{2}} > 0 \text{ on } \Gamma$$

it follows (recalling Theorem 2.54 (iii)) that

$$\operatorname{ind}\left[\left|\alpha^{-1\prime}\right|^{\frac{1}{2}} P + Q\right] = 0\ . \tag{46.39}$$

Thus, in view of the invertibility of $\left(V_\alpha P + Q\right)$, (46.38) and (46.39) imply

$$\operatorname{ind}\left(T_\alpha P + Q\right) = 0\ .$$

The proof is complete. □

46.14 We shall end this section with the presentation of a number of examples of concrete G-SIOs in the case when a group of diffeomorphisms of the curve Γ is generated by a single diffeomorphism, that is $G = \mathbb{Z}$.

To begin with let us make some remarks about the case when Γ is a simple closed curve.

I. Recall that the two-term *G*-SIOs related to irrational rotations of the circle were investigated in Theorems 10.7 and 10.8 where in fact we carried out a direct proof showing the structure of the symbols of the operators in the situation being considered.

Remark Observe that if Γ is a circle and α is a rotation of the circle then $T_\alpha = V_\alpha$ and the statement of Theorem 46.13 turns out to be trivial, as both the spaces $H^+(\Gamma)$ and $H^-(\Gamma)$ are invariant with respect to T_α.

If Γ is a circle and $\alpha: \Gamma \to \Gamma$ is a diffeomorphism of class C^2 with an irrational rotation number (see Corollary 5.5) then α is conjugate to an irrational rotation (see Theorems 5.7, 5.8) and one can easily obtain the Fredholmity conditions and the index formulae for two-term *G*-SIOs by modifying (with the help of Theorem 46.11) the argument of Theorems 10.7 and 10.8 (replacing the geometric mean with respect to Lebesque measure by the geometric mean with respect to ergodic measure under α (cf. Remark 5.12 (c))).

Remark Observe that even in this (to a certain extent) simplest situation the usage of Theorem 46.11 is essential. This is due to the fact that a homeomorphism h establishing the conjugation between α and irrational rotation (see Theorem 5.8) may not even be absolutely continuous which means that we cannot use it to construct the *spatial conjugation* in $L^2(\Gamma)$ between these operators and those generated by irrational rotation (recall also Theorem 9.5).

Thus Theorem 46.11 enables us to obtain a C^*-algebraic (but not spatial) isomorphism between the necessary algebras which in fact does the job.

II. If the rotation number $\tau(\alpha)$ is rational and has the form $\tau(\alpha) = \frac{M}{N}$ ($\frac{M}{N}$ is an irreducible fraction) then by Theorem 5.6 α has periodic points and all of them are of period N.

Remark (1) Observe that the set of fixed points (that is, the set of points of period 1)

$$\Gamma_1 = \{t \in \Gamma : \alpha(t) = t\} \tag{46.40}$$

can be any closed subset of Γ. Indeed, let $\Omega \subset \Gamma$ be any closed set. Then its complement $\Gamma \setminus \Omega$ is a union of a finite or denumerable number of open arcs $(\widehat{\tau_k, \nu_k})$, $k = 1, 2, \ldots$. Let $\varphi \in C^\infty[0,1]$ be a function such that

$$\begin{aligned}
&\varphi(0) = 0, \quad \varphi^{(k)}(0) = 0, \quad k = 1, 2, \ldots\,, \\
&\varphi(1) = 0, \quad \varphi^{(k)}(1) = 0, \quad k = 1, 2, \ldots\,, \\
&\varphi(x) > 0, \quad 0 < x < 1.
\end{aligned}$$

Let s be the natural parameter on Γ. Then the function

$$\alpha(s) = \begin{cases} s, & s \in \Omega \\ s + e^{-\frac{1}{(\nu_k - \tau_k)}} \varphi\left(\dfrac{s - \tau_k}{\nu_k - \tau_k}\right), & s \in (\widehat{\tau_k, \nu_k}) \end{cases}$$

is a C^∞-diffeomorphism of Γ and the set of its fixed points coincides with Ω.

(2) If $\tau(\alpha) = \dfrac{M}{N}$ and $N > 1$ then the set

$$\Gamma_N = \{t \in \Gamma : \alpha^N(t) = t\} \tag{46.41}$$

of periodic points cannot be an arbitrary closed set. For example, if $|\Gamma_N| < \infty$ then the number of its points is a multiple of N.

We note once more that if Γ is a simple closed curve then (since it is diffeomorphic to the circle S^1) the investigation of G-SIOs in $L^2(\Gamma)$ can be reduced to that for the case when Γ is a circle (by applying Theorem 46.11 and bearing in mind the above diffeomorphism $\Gamma \leftrightarrow S^1$). Thus, for example, the results 18.4 can be used to construct and investigate the corresponding examples of G-SIOs.

46.15 The next result (Theorem 46.16) deals with a two-term G-SIO on a non-simple closed curve Γ. It is worth mentioning at once that the main situation to consider here is the case when Γ is α-irreducible (see 18.6) since an arbitrary case can be reduced to the α-irreducible case by decomposing the curve (and the space $L^2(\Gamma)$) into the union of α-irreducible curves (and the direct sum of invariant subspaces, modulo compact operators).

One more distinctive feature of the α-irreducible situation is that under this condition if α preserves the orientation of Γ and has periodic points then all the periodic points are of the same period.

So bearing these observations in mind we prove

Theorem 46.16 *Let Γ be a (non-simple) Lyapunov closed curve (with a fixed orientation) which is α-irreducible with respect to an orientation-preserving diffeomorphism α having periodic points of period N. Let the set Γ_N of periodic points (46.41) be nowhere dense. Then the G-SIO*

$$\mathcal{b} = (a_{00} + a_{01}T_\alpha) + (a_{10} + a_{11}T_\alpha)S + k$$

(where a_{ij} are the operators of multiplication by continuous functions and k is compact) is Fredholm in $L^2(\Gamma)$ iff for every pair of functions

$$\left(d_0^+, d_1^+\right) \text{ and } \left(d_0^-, d_1^-\right)$$

$\left(d_0^\pm(t) = a_{00}(t) \pm a_{10}(t), d_1^\pm(t) = a_{01}(t) \pm a_{11}(t)\right)$ *one of the following conditions (a) or (b) is satisfied.*

(a) $d_0^\pm(t) \neq 0$ *for every* $t \in \Gamma$ *and*

$$\left|\prod_{k=0}^{N-1} d_0^\pm\left(\alpha^k(t)\right)\right| > \left|\prod_{k=0}^{N-1} d_1^\pm\left(\alpha^k(t)\right)\right|, \quad t \in \Gamma_N;$$

(b) $d_1^\pm(t) \neq 0$ *for every* $t \in \Gamma$ *and*

$$\left|\prod_{k=0}^{N-1} d_0^\pm\left(\alpha^k(t)\right)\right| < \left|\prod_{k=0}^{N-1} d_1^\pm\left(\alpha^k(t)\right)\right|, \quad t \in \Gamma_N.$$

If $\mathcal{b}$ *is Fredholm then*

$$\text{ind}\,\mathcal{b} = -\frac{1}{2\pi}[\arg d(x)]_\Gamma \tag{46.42}$$

where $d(t) = \dfrac{d^+}{d^-}$,

$$d^{\pm} = \begin{cases} d_0^{\pm}, & \textit{in the case } (a), \\ d_1^{\pm}, & \textit{in the case } (b), \end{cases}$$

and the right-hand part of (46.42) is described in (46.2).

Proof By the condition of the theorem the group $\mathbb{Z}$ acts (by diffeomorphisms α^n) topologically freely on Γ. Thus applying Theorem 46.11 we see that the symbol $\sigma(\mathcal{B})$ is the direct sum of two operators (acting in $L^2(S^*\Gamma) = L^2(\Gamma) \oplus L^2(\Gamma)$):

$$\sigma(\mathcal{B}) = \left(d_0^+ + d_1^+ T_\alpha\right) \oplus \left(d_0^- + d_1^-\right) T_\alpha .$$

Thus $\mathcal{B}$ is Fredholm iff both the operators $\left(d_0^+ + d_1^+ T_\alpha\right)$ and $\left(d_0^- + d_1^-\right) T_\alpha$ are invertible.

It is clear that in this situation the set $\Omega(\alpha)$ of non-wandering points (see 5.1) coincides with Γ_N and every ergodic measure is based on the trajectory of a certain point from Γ.

Now Theorem 17.3 implies that conditions (a) and (b) stated in the theorem in fact describe the conditions of invertibility of the symbol $\sigma(\mathcal{B})$ and thus the conditions of the Fredholmity of $\mathcal{B}$.

Thus to finish the proof it remains to check the index formula. This formula could be obtained by the implementation of a linear homotopy (preserving the Fredholmity condition), to an operator with known index (the general scheme of this process is given in Theorem 19.35).

The desired homotopy depends on which case (a) or (b) are realized for the pairs $\left(d_0^+, d_1^+\right)$ and $\left(d_0^-, d_1^-\right)$. Thus we have the four possibilities which we shall denote symbolically by

(a)(a); (a)(b), (b)(a), (b)(b).

To construct the homotopy it is convenient to write the operator $\mathcal{B}$ (modulo compact summand) in the form

$$\mathcal{B} = \left(d_0^+ + d_1^+ T_\alpha\right) P + \left(d_0^- + d_1^- T_\alpha\right) Q \qquad (46.43)$$

and define the homotopy $\mathcal{B}_\tau$, $0 \le \tau \le 1$, $\left(\mathcal{B} = \mathcal{B}_1\right)$ by the

$$\mathcal{b}_\tau = \begin{cases} \left(d_0^+ + \tau d_1^+ T_\alpha\right)P + \left(d_0^- + \tau d_1^- T_\alpha\right)Q, & \text{(a)(a);} \\ \left(d_0^+ + \tau d_1^+ T_\alpha\right)P + \left(\tau d_0^- + d_1^- T_\alpha\right)Q, & \text{(a)(b);} \\ \left(\tau d_0^+ + d_1^+ T_\alpha\right)P + \left(\tau d_0^- + \tau d_1^- T_\alpha\right)Q, & \text{(b)(a);} \\ \left(\tau d_0^+ + d_1^+ T_\alpha\right)P + \left(\tau d_0^- + d_1^- T_\alpha\right)Q, & \text{(b)(b).} \end{cases}$$

In all cases the Fredholmity conditions are preserved and thus

$$\operatorname{ind} \mathcal{b} = \operatorname{ind} \mathcal{b}_1 = \operatorname{ind} \mathcal{b}_0 .$$

The operator $\mathcal{b}_0$ has the form

$$\mathcal{b}_0 = \begin{cases} d_0^+ P + d_0^- Q, & \text{(a)(a);} \\ d_0^+ P + d_1^- T_\alpha Q, & \text{(a)(b);} \\ d_1^+ T_\alpha P + d_0^- Q, & \text{(b)(a);} \\ d_1^+ T_\alpha P + d_1^- T_\alpha Q, & \text{(b)(b).} \end{cases}$$

In the case (a)(a), $\mathcal{b}_0$ is an SIO and its index can be calculated with the help of the Theorem 2.54 since

$$\operatorname{ind} \mathcal{b} = \sum \operatorname{ind} \mathcal{b}|_{L^2(\Gamma^i)}$$

($\Gamma = \{\Gamma^i\}_i$ and Γ^i is a connected component of Γ).

In the case (b)(b) we have

$$\operatorname{ind} \mathcal{b}_0 = \operatorname{ind}\left(d_1^+ P + d_1^- Q\right)T_\alpha = \operatorname{ind}\left(d_1^+ P + d_1^- Q\right)$$

and the index of the right-hand term is calculated by the above argument.

In the case (a)(b) (recalling that by Theorem 46.13

$$\operatorname{ind}\left(P + T_\alpha^{-1} Q\right) = 0 \ \ )$$

we obtain

$$\operatorname{ind} \mathscr{b}_0 = \operatorname{ind}\left(d_0^+ P + d_1^- T_\alpha Q\right)\left(P + T_\alpha^{-1} Q\right)$$

$$= \operatorname{ind}\left(d_0^+ P + d_1^- Q\right)$$

thus again passing to the SIO.

Finally in the case (b)(a) we obtain the SIO by multiplying $\mathscr{b}_0$ by $\left(T_\alpha^{-1} P + Q\right)$.

In all situations the indexes of the SIOs we arrive at are calculated via the formula (46.42). Thus the proof is complete. □

46.17 If we consider a G-SIO in the space $L^2\left(\Gamma, \mathbb{C}^n\right)$ of vector-valued functions then its symbol turns out to a functional operator in the space $L^2\left(S^*\Gamma, \mathbb{C}^n\right)$, and passing to the investigation of the invertibility conditions of such operators we arrive at the problems discussed in Chapter 3. Analogous difficulties arise when investigating the multiterm operators $\sum a_g T_g$ (containing more than two terms). Nevertheless in this situation (since $\dim \Gamma = 1$) we obtain certain simplifications.

Let us present, for example, some applications of the hyperbolicity ideas (Theorems 19.13, 19.21). To do this we confine ourselves to the following example.

In the space $L^2\left(\Gamma, \mathbb{C}^n\right)$ we consider a G-SIO $\mathscr{b}$ of the form

$$\mathscr{b} = b^+ P + b^- Q \tag{46.44}$$

where

$$b^\pm = \sum_{j=0}^{m} a_j^\pm T_\alpha^j \tag{46.45}$$

and where $a_j^\pm$ are continuous matrix valued functions (of dimension n) and T_α is given by (46.16).

Let us apply to the operators $\mathscr{b}$, b^+, b^- the construction from 20.1 and A.26, that is we build the new operators $\tilde{\mathscr{b}}$, $\tilde{b}^+$, $\tilde{b}^-$ via the following rule. These operators act in the space

$$\tilde{H} = \bigoplus_{j=0}^{m-1} L^2\left(\Gamma, \mathbb{C}^n\right) \cong L^2\left(\Gamma, \mathbb{C}^{mn}\right)$$

consisting of elements of the form

$$v = \left(v_0, \ldots, v_{m-1}\right), \quad v_j \in L^2\left(\Gamma, \mathbb{C}^n\right).$$

Let W be the unitary operator in $\tilde{H}$ defined by the formula

$$Wv = \left(T_\alpha^m v_0, \ldots, T_\alpha^m v_{m-1}\right) \tag{46.46}$$

(cf. (20.1)).

The operators $\tilde{b}^{\pm}$ are given by

$$\tilde{b}^{\pm} = d_0^{\pm} + d_1^{\pm} W \tag{46.47}$$

where $d_0^{\pm}, d_1^{\pm}$ are the operators of multiplication by block-matrices of the form

$$d_0^{\pm}(t) = \begin{bmatrix} a_0^{\pm}(t) & a_1^{\pm}(t) & \ldots & a_{m-1}^{\pm}(t) \\ 0 & a_0^{\pm}\left(\alpha^{-1}(t)\right) & \ldots & a_{m-1}^{\pm}\left(\alpha^{-1}(t)\right) \\ \ldots & \ldots & \ldots & \ldots \\ 0 & 0 & \ldots & a_0^{\pm}\left(\alpha^{-m+1}(t)\right) \end{bmatrix}$$

$$d_1^{\pm}(t) = \begin{bmatrix} a_m^{\pm}(t) & 0 & \ldots & 0 \\ a_{m-1}^{\pm}\left(\alpha^{-1}(t)\right) & a_m^{\pm}\left(\alpha^{-1}(t)\right) & \ldots & 0 \\ \ldots & \ldots & \ldots & \ldots \\ a_1^{\pm}\left(\alpha^{-m+1}(t)\right) & a_2^{\pm}\left(\alpha^{-m+1}(t)\right) & \ldots & a_m^{\pm}\left(\alpha^{-m+1}(t)\right) \end{bmatrix}.$$

Set

$$\tilde{\mathcal{b}} = \tilde{b}^{+} P + \tilde{b}^{-} Q \tag{46.48}$$

(where P and Q are the corresponding projections in $L^2\left(\Gamma, \mathbb{C}^{mn}\right)$).

Let ρ^{+} and ρ^{-} be the relations built with respect to $\tilde{b}^{+}$ and $\tilde{b}^{-}$ according to the construction of 19.19.

Theorems 46.11, 19.21, Lemma A.27 and Corollary 20.4 imply the next

Theorem 46.18 *Let $\alpha\colon \Gamma \to \Gamma$ be an orientation-preserving diffeomorphism, the set of non-periodic points of which is dense in Γ. If the operators $\mathcal{b}$ and $\tilde{\mathcal{b}}$ are those defined by (46.44) and (46.48) then the following conditions are equivalent:*

(1) *$\mathcal{b}$ is Fredholm;*

(2) *$\tilde{\mathcal{b}}$ is Fredholm;*

(3) *b^{+} and b^{-} are invertible;*

(4) *$\tilde{b}^{+}$ and $\tilde{b}^{-}$ are invertible;*

(5) *the relations ρ^{+} and ρ^{-} are hyperbolic.*

The fact that any complex vector bundle over a curve is trivial gives us the possibility of rewriting the definition of hyperbolicity in a different (perhaps more visual) way, which is stated in Theorem 46.20.

Before proving this result it is reasonable to make the following useful observation.

Lemma 46.19 *Let A be a C^*-algebra isomorphic to the algebra $C(X, \mathrm{Mat}_n\mathbb{C})$ of continuous matrix-valued functions on a compact space X $(= \mathrm{Prim}\, A)$ and let T be a unitary operator such that $TAT^* = A$. Then*

(i) the automorphism $\hat{T} = T(\,\cdot\,)T^*$ determines a continuous one-dimensional subbundle of $X \times \mathrm{Mat}_n\mathbb{C}$ such that each of its fibres consists of matrices proportional to a unitary matrix.

If the mentioned subbundle is trivial then

(ii) there exists an element $\theta \in A$ such that $\theta(x)$ is unitary for every $x \in X$ and the operator

$$V = \theta T$$

has the property

$$(VaV^*)(x) = a(t(x)),\ a \in A$$

where $t: X \to X$ is the homeomorphism defined by the automorphism (see 12.12).

Proof Identifying A with $C(X, \mathrm{Mat}_n\mathbb{C})$ and bearing in mind the fact that $(T(\,\cdot\,)T^*)(x)$ defines an irreducible representation of the algebra $A|_{t^{-1}(x)}$ we conclude that there exists a unique up to a complex multiplier λ, $|\lambda| = 1$ unitary matrix θ_x such that for every $a \in A$

$$(TaT^*)(x) = \theta_x^* a\left(t^{-1}(x)\right)\theta_x\,. \tag{46.49}$$

Let $L_x: \mathrm{Mat}_n\mathbb{C} \to \mathrm{Mat}_n\mathbb{C}$ be an isometric mapping given by

$$L_x(d) = \theta_x^* d\theta_x\,.$$

One can easily verify that L_x depends on x continuously. Indeed if

$$x_k \to x_0$$

is a certain sequence such that

$$L_{x_k} \to L$$

then (since $a(x)$ depends on x continuously) it follows that

$$La\left(t^{-1}(x_0)\right) = L_{x_0}a\left(t^{-1}(x_0)\right),\ a \in A$$

which implies

$$L = L_{x_0}\ . \tag{46.50}$$

Since the mapping L_x defines in a unique way the one-dimensional subspace $H_x = \{\lambda\theta_x\}_{\lambda\in\mathbb{C}}$ of $\mathrm{Mat}_n\mathbb{C}$, the continuity of L_x implies the continuity of the subbundle $F = \{H_x\}_{x\in X} \subset X \times \mathrm{Mat}_n\mathbb{C}$ and (i) is proved.

To prove (ii) recall that if F is a trivial subbundle then there exists a non vanishing cross-section $\tilde{\theta}_x$ of F. Set

$$\theta(x) = \frac{\tilde{\theta}_x}{\left\|\tilde{\theta}_x\right\|}\ .$$

The unitary matrix-valued function $\theta(x)$, $x \in X$ defines the element $\theta \in A$, and for the operator

$$V = \theta T$$

we have (using (46.49)) that

$$(VaV^*)(x) = (\theta TaT^*\theta^*)(x) = \theta_x(TaT^*)(x)\theta_x^* = a\left(t^{-1}(x)\right).$$

Which completes the proof. □

Remarks (1) If $n = 1$ then (ii) in the statement of the lemma is true for any compact space X.

(2) On the other hand, if X is such that any complex bundle over it is trivial, then (ii) is true for any n.

In connection with the latter remark it is worth mentioning that any complex bundle over Γ is trivial.

Bearing in mind the previous remark, we now proceed to the result describing the desired reformulation of hyperbolicity that will itself be of great use to us later.

Theorem 46.20 *Let* $A \cong C(X, \mathrm{Mat}_n\mathbb{C})$ *where* X *is a compact space such that every subbundle of* $X \times \mathbb{C}^n$ *is trivial. If a unitary operator* T *is such that*

$$(TaT^*)(x) = a\left(t^{-1}(x)\right),\ a \in A \tag{46.51}$$

where $t: X \to X$ *is a homeomorphism having a dense set of non-periodic points, then the operator*

$$b = a_0 + a_1 T,\ a_0, a_1 \in A$$

is invertible iff there exist non-degenerate continuous matrix-valued functions s_1, s_2 *(defining the corresponding elements in* A*) such that*

$$s_1^{-1} b s_2 = b_1 = d_0 + d_1 T \tag{46.52}$$

where

$$d_0 = \begin{bmatrix} I_l & 0 \\ 0 & d_{22}^0 \end{bmatrix},\ d_1 = \begin{bmatrix} d_{11}^1 & 0 \\ 0 & I_{n-l} \end{bmatrix} \tag{46.53}$$

and

$$r\left(d_{22}^0 T^{-1}\right) < 1,\ r\left(d_{11}^1 T\right) < 1 \tag{46.54}$$

(the latter operators and spectral radii are considered in the invariant subspaces defined by the projections

$$\begin{bmatrix} 0 & 0 \\ 0 & I_{n-l} \end{bmatrix} \text{ and } \begin{bmatrix} I_l & 0 \\ 0 & 0 \end{bmatrix}).$$

Proof Clearly if the operator T satisfies (46.51) then the projections

$$\overline{P} = \begin{bmatrix} I_l & 0 \\ 0 & 0 \end{bmatrix} \text{ and } I - \overline{P} = \begin{bmatrix} 0 & 0 \\ 0 & I_{n-l} \end{bmatrix}$$

define invariant subspaces with respect to T.

Suppose first that (46.52)-(46.54) are satisfied. Then these projections define invariant subspaces with respect to b_1, that is

$$b_1 = \beta_1 \oplus \beta_2$$

where

$$\beta_1 = I_l + d_{11}^1 T, \quad \beta_2 = d_{22}^0 + I_{n-l} T$$

and both β_1 and β_2 are invertible (by (46.54)).

Thus b_1 is invertible, which in turn implies the invertibility of $b = s_1 b^1 s_2^{-1}$.

Now let b be invertible. In this case (by Theorem 19.21) the corresponding relation ρ is hyperbolic. Thus there exists a projection $p \in A$ defining the subspaces H^s and H^u (see 19.20), that is

$$\text{Im } P(x) = H_x^s, \ \text{Ker } P(x) = H_x^u .$$

The condition (46.51) means that the operator θ_x in 19.5 (here $X \cong \text{Prim } A \cong \hat{A}$) can be taken to be in the form

$$\theta_x : H_x \to H_{t(x)} \left(H_x = \mathbb{C}^n \right)$$

$$\theta_x(h) = h .$$

The subspaces

$$H^s = \bigoplus_{x \in X} H_x^s, \ H^u = \bigoplus_{x \in X} H_x^u$$

define the subbundles $\overline{H}^s$ and $\overline{H}^u$ of $X \times \mathbb{C}^n$ such that

$$\overline{H}^s \oplus \overline{H}^u = X \times \mathbb{C}^n$$

(these subbundles are continuous since they are defined by P).

By the assumption of the theorem both $\overline{H}^s$ and $\overline{H}^u$ are trivial and thus there exist bases in H_x continuously depending on x such that the vectors

$$h_1(x),\dots,h_l(x)$$

form a basis in H_x^s, and the vectors

$$h_{l+1}(x),\dots,h_n(x)$$

form a basis in H_x^u.

Let the matrix-valued functions $s_1(x)$ and $s_2(x)$ be formed by the vector-columns

$$s_1(x)=\left(a_0(x)h_1(x),\dots,a_0(x)h_l(x),a_1(x)h_{l+1}\left(t^{-1}(x),\dots,a_1(x)h_n\left(t^{-1}(x)\right)\right)\right),$$

$$s_2(x)=\left(h_1(x),\dots,h_l(x),h_{l+1}(x),\dots,h_n(x)\right).$$

Now using the definition of the hyperbolic relation 19.20 and looking through the proof of Theorem 19.21 (namely observing that the operator

$$\tilde a_0+\tilde a_1\theta:\begin{matrix}H^s\\ \oplus\\ H^u\end{matrix}\to\begin{matrix}E^s\\ \oplus\\ E^u\end{matrix}$$

decomposes into the direct sum of operators

$$\tilde a_{01}+\tilde a_{11}\theta_1:H^s\to E^s$$

$$\tilde a_{02}+\tilde a_{12}\theta_2:H^u\to E^u$$

where a_{01} and $a_{12}\theta_2$ are invertible and the spectral radii of β_s and β_u are both less than 1) we conclude that (46.52)-(46.54) are true. □

47. The index of a G-SIO

In this section we shall obtain the index formulae for G-SIOs with different types of diffeomorphisms generating transformations of an argument. We confine ourselves to consideration of the situation when the diffeomorphisms preserve the orientation, though the ideas and methods presented here can equally well be applied to the index calculation of G-SIOs with shifts not necessarily preserving the orientation.

We start with the case of the

Topologically freely acting group $\{\alpha^n\}_{n\in\mathbf{Z}}$

47.1 That is, we consider a G-SIO of the form

$$\mathscr{b} = \sum_{k=0}^{m} \mathscr{a}_k T^k \tag{47.1}$$

where $T = T_\alpha$ is defined by (46.16) and α is an orientation-preserving diffeomorphism of Γ having a dense set of non-periodic points.

Theorem A.36 applied to the operator (47.1) gives us

Theorem 47.2 *Let $\mathscr{b}$ be the G-SIO in $L^2(\Gamma)$ described in 47.1. Let $\tilde{\mathscr{b}}$ be the G-SIO in $L^2(\Gamma,\mathbb{C}^m) = [L^2(\Gamma)]^m$ of the form*

$$\tilde{\mathscr{b}} = \tilde{\mathscr{a}}_0 + \tilde{\mathscr{a}}_1 W \tag{47.2}$$

where the operators $\tilde{\mathscr{a}}_0, \tilde{\mathscr{a}}_1$, W are given by the operator matrices

$$\tilde{\mathscr{a}}_0 = \begin{bmatrix} \mathscr{a}_0 & \mathscr{a}_1 & \dots & \mathscr{a}_{m-1} \\ 0 & T\mathscr{a}_0 T^{-1} & \dots & T\mathscr{a}_{m-2}T^{-1} \\ \dots & \dots & \dots & \dots \\ 0 & 0 & \dots & T^{m-1}\mathscr{a}_0 T^{-m+1} \end{bmatrix} \tag{47.3}$$

$$\tilde{\mathscr{a}}_1 = \begin{bmatrix} \mathscr{a}_m & 0 & \dots & 0 \\ T\mathscr{a}_{m-1}T^{-1} & T\mathscr{a}_m T^{-1} & \dots & 0 \\ \dots & \dots & \dots & \dots \\ T^{m-1}\mathscr{a}_1 T^{-m+1} & T^{m-1}\mathscr{a}_2 T^{-m+1} & \dots & T^{m-1}\mathscr{a}_m T^{-m+1} \end{bmatrix} \tag{47.4}$$

$$W = \begin{bmatrix} T^m & 0 & \dots & 0 \\ 0 & T^m & \dots & 0 \\ \dots & \dots & \dots & \dots \\ 0 & 0 & \dots & T^m \end{bmatrix}. \tag{47.5}$$

The operator $\mathscr{b}$ is Fredholm iff $\tilde{\mathscr{b}}$ is Fredholm and

$$\operatorname{ind} \mathscr{b} = \frac{1}{m} \operatorname{ind} \tilde{\mathscr{b}}\ . \tag{47.6}$$

47.3 In view of this theorem the problem of index calculation for the G-SIO $\mathcal{B}$ is reduced to the problem of index calculation of a two-term G-SIO $\tilde{\mathcal{B}}$ acting in the space of vector functions.

Recall that a G-SIO of the form (47.1) could be represented in the form

$$\mathcal{B} = b^+ P + b^- Q \tag{47.7}$$

with $b^{\pm}$ described by (46.45).

The latter expression in the case of the orientation-preserving mapping α leads to

$$b^+ P + b^- Q \sim \left(b^+ P + Q\right)\left(P + b^- Q\right) \tag{47.8}$$

(here and henceforth in this section the sign ~ means that the left-hand and the right-hand terms differ by a compact summand).

The explicit form of the symbol $\sigma(\mathcal{B})$ (see (46.21)) implies that if $\mathcal{B}$ is Fredholm then both $b^+ P + Q$ and $P + b^- Q$ are Fredholm and in this situation in view of (47.8) it follows that

$$\operatorname{ind}\left(b^+ P + b^- Q\right) = \operatorname{ind}\left(b^+ P + Q\right) + \operatorname{ind}\left(P + b^- Q\right). \tag{47.9}$$

Thus it is enough to calculate the indices of the operators in the right-hand part of the equation.

We shall consider now the operator

$$b^+ P + Q \tag{47.10}$$

(the second operator could be coped with in an analogous way).

Bearing in mind Theorem 47.2 and the fact that α acts topologically freely along with α^m and that W is of the same sort at T, we shall instead of the operator (47.10) investigate the operator (in $L^2\left(\Gamma, \mathbb{C}^m\right)$) of the form

$$\bar{\mathcal{B}} = \left(a_0 + a_1 T\right) P + Q \tag{47.11}$$

where a_0 and a_1 are continuous matrix-valued functions of dimension m.

If $\bar{\mathcal{B}}$ is Fredholm then the operator

$$\approx{b} = a_0 + a_1 T \tag{47.12}$$

is invertible and by Theorem 46.20 there exist continuous non-degenerate matrix-valued functions $s_1(x)$ and $s_2(x)$ such that

$$\approx{b} = s_1(d_0 + d_1 T)s_2 \tag{47.13}$$

where d_0 and d_1 are block-diagonal matrices of the form

$$d_0 = \begin{bmatrix} I_l & 0 \\ 0 & d_{22}^0 \end{bmatrix}, \quad d_1 = \begin{bmatrix} d_{11}^1 & 0 \\ 0 & I_{m-l} \end{bmatrix} \tag{47.14}$$

and the matrix-valued functions d_{22}^0 and d_{11}^1 are such that

$$r\left(d_{22}^0 T^{-1}\right) < 1 \quad \text{and} \quad r\left(d_{11}^1 T\right) < 1 . \tag{47.15}$$

The next theorem shows that the index of $\bar{\mathcal{b}}$ can be calculated in terms of these matrix-valued functions s_1 and s_2.

Theorem 47.4 *The index of the operator $\bar{\mathcal{b}}$ (47.11) is calculated via formula*

$$\operatorname{ind} \bar{\mathcal{b}} = -\frac{1}{2\pi}\left[\arg \det s_1(x)s_2(x)\right]_\Gamma \tag{47.16}$$

where s_1 and s_2 are defined by (47.13)-(47.15) and $\frac{1}{2\pi}[\arg \varphi(t)]_\Gamma$ is defined by (46.2).

Proof Clearly (47.11) and (47.13) imply

$$\bar{\mathcal{b}} = s_1(d_0 + d_1 T)s_2 P + Q \sim (s_1 P + Q)\left[(d_0 + d_1 T)P + Q\right](s_2 P + Q) . \tag{47.17}$$

Since s_1 and s_2 are non-degenerate it follows that both the operators

$$s_1 P + Q \quad \text{and} \quad s_2 P + Q$$

are Fredholm and moreover

$$\operatorname{ind}(s_1P+Q)+\operatorname{ind}(s_2P+Q)=\operatorname{ind}(s_1P+Q)(s_2P+Q)$$

$$=\operatorname{ind}(s_1P+Q)(s_2P+Q)=\operatorname{ind}(s_1s_2P+Q)$$

$$=-\frac{1}{2\pi}\left[\arg\det(s_1\cdot s_2)\right]_\Gamma \tag{47.18}$$

where the last equality follows from 2.56.

Now (47.18) and (47.17) mean that

$$\operatorname{ind}\bar{\mathfrak{L}}=\operatorname{ind}\left[(d_0+d_1T)P+Q\right]-\frac{1}{2\pi}\left[\arg\det(s_1\cdot s_2)\right]_\Gamma \tag{47.19}$$

and to complete the proof it is enough to verify the equality

$$\operatorname{ind}\left[(d_0+d_1T)P+Q\right]=0\ . \tag{47.20}$$

In view of (47.14) and (47.15) the family of operators

$$\left(\begin{bmatrix} I_l & 0 \\ 0 & \tau d_{22}^0 \end{bmatrix}+\begin{bmatrix} \tau d_{11}^1 & 0 \\ 0 & I_{m-l} \end{bmatrix}T\right)P+Q, \qquad 0\le\tau\le1\ ,$$

defines a homotopy in the class of Fredholm operators between $\left[(d_0+d_1T)P+Q\right]$ $(\tau=1)$ and the operator

$$\left(\begin{bmatrix} I_l & 0 \\ 0 & 0 \end{bmatrix}+\begin{bmatrix} 0 & 0 \\ 0 & I_{m-l} \end{bmatrix}T\right)P+Q\ .$$

This operator acts on vector-valued functions coordinatewise, that is it can be decomposed into the direct sum of m operators each acting in the space $L^2(\Gamma)$ of scalar functions.

This operator acts on the first l functions as $P+Q$, that is trivially. It acts on the last $m-l$ functions as $TP+Q$ which is the operator having zero index (by Theorem 46.13). Thus the index of the operator being considered is zero and the proof is complete. □

47.5 The theorem just proved means that the problem of the index calculation of G-SIOs in the case when the diffeomorphism α generating the transformation of an argument on Γ acts topologically freely is principally solved.

Observe at the same time that the matrices s_1 and s_2 (used in (47.16)) in general cannot be written out in explicit form.

It is worth mentioning that if the mapping α has isolated periodic points and the coefficients a_0 and a_1 are constant in certain neighbourhoods of these points then the matrices $s_1(x)$ and $s_2(x)$ could be written out explicitly (in terms of linear algebra). But any Fredholm G-SIO with such an α can be approximated by Fredholm G-SIOs with constant coefficients in certain neighbourhoods of the periodic points and having the same index. Thus in the case when α has isolated periodic points all the foregoing argument has a constructive character.

On the other hand, if α has no periodic points then there is no hope of obtaining an explicit form for s_1 and s_2 in the general situation.

A certain clarification of the theorem can be obtained by consideration of the particular case when the matrices a_0 and a_1 have the block-diagonal form

$$a_0 = \begin{bmatrix} a_{11}^0 & 0 \\ 0 & a_{22}^0 \end{bmatrix}, \quad a_1 = \begin{bmatrix} a_{11}^1 & 0 \\ 0 & a_{22}^1 \end{bmatrix}$$

where a_{11}^0 and a_{22}^1 are non-degenerate matrices and

$$r\left(\left[a_{11}^0\right]^{-1} a_{11}^1 T\right) < 1, \quad r\left(\left[a_{22}^1\right]^{-1} a_{22}^0 T^{-1}\right) < 1 .$$

In this situation one can take

$$s_1 = \begin{bmatrix} a_{11}^0 & 0 \\ 0 & a_{22}^1 \end{bmatrix}, \quad s_2 = I$$

and we have

$$\operatorname{ind} \bar{\mathcal{L}} = -\frac{1}{2\pi}\left[\arg \det s_1\right]_\Gamma .$$

So we see that on a certain part of the space the rotation of the argument is due to a_0 (i.e. a_{11}^0) and on the other part it is due to a_1 (i.e. a_{22}^1).

Periodic shifts

47.6 Now we proceed to another 'extremal' case - the case when α is periodic; namely, we have in mind the following situation. The curve Γ is assumed to be α-irreducible with respect to the orientation-preserving mapping α satisfying the condition $\alpha^N = I$ for some natural N (and $\alpha^n \neq I$ for $n = 1, 2, \ldots, N-1$).

As we have already mentioned, in this situation all the points of Γ have one and the same period N with respect to α. We have also observed that α-irreducibility of Γ is not a burden in general since one can always decompose Γ into α-irreducible parts.

Clearly it is reasonable here to consider the group $\{\alpha^n\}$ as the representation of $\mathbb{Z}_N$, and obviously the condition of topologically free action of $\mathbb{Z}_N$ by α^n, $n \in \mathbb{Z}_N$, is satisfied.

Bearing in mind the argument of 47.3 and that the condition $\alpha^N = I$ is equivalent to $T^N = I$ we shall confine ourselves here to the consideration of operators of the form

$$\mathcal{L} = \left(\sum_0^{N-1} a_k T^k \right) P + Q \,. \tag{47.21}$$

Passing from this operator to the operator $\tilde{\mathcal{L}}$ in $L^2(\Gamma, \mathbb{C}^N)$ (see b and $\tilde{b}$ in A.3) we find that the latter has the form

$$\tilde{\mathcal{L}} = bP + Q \tag{47.22}$$

where b is the operator of multiplication by a matrix-valued function and P and Q are the projections in $L^2(\Gamma, \mathbb{C}^N)$ generated by the corresponding projections in $L^2(\Gamma)$,

$$b(x) = \begin{bmatrix} a_0(x) & a_1 & \ldots & a_N \\ a_{N-1}(\alpha^{-1}(x)) & a_0(\alpha^{-1}(x)) & \ldots & a_1(\alpha^{-1}(x)) \\ \ldots & \ldots & \ldots & \ldots \\ a_{N-1}(\alpha^{-N+1}(x)) & \ldots & \ldots & a_0(\alpha^{-N+1}(x)) \end{bmatrix}, \tag{47.23}$$

which means that $\tilde{\mathcal{L}}$ is Fredholm iff $b(x)$ is non-degenerate for every $x \in \Gamma$.

The operator $\tilde{\mathcal{L}}$ is an SIO (without shifts) and by 2.56 we have

$$\operatorname{ind} \tilde{\mathcal{L}} = -\frac{1}{2\pi} [\arg \det b(x)]_\Gamma \,,$$

so to obtain the index of $\mathfrak{b}$ it is enough to find its relation to ind $\tilde{\mathfrak{b}}$ which is described in the next

Theorem 47.7 *(cf. Remark 43.14) The operator* $\mathfrak{b}$ *(47.21) is Fredholm iff* $\tilde{\mathfrak{b}}$ *(47.22) is Fredholm,*

$$\operatorname{ind} \mathfrak{b} = \frac{1}{N} \operatorname{ind} \tilde{\mathfrak{b}} = -\frac{1}{2\pi N} [\arg \det b(x)]_\Gamma .$$

Proof The operator $\mathfrak{b}$ is Fredholm iff its symbol $\sigma(\mathfrak{b})$ (46.21) is invertible, which is equivalent to the invertibility of the matrix-valued function $b(x)$ (47.23) and thus to the Fredholmity of $\tilde{\mathfrak{b}}$.

Thus it remains to check the index formula. As was observed in Theorem A.17, for the operator $\tilde{\mathfrak{b}}$, the decomposition $\tilde{\mathfrak{b}} = \bigoplus_{i=0}^{N-1} \mathfrak{b}_i$ and formula $\operatorname{ind} \tilde{\mathfrak{b}} = \sum_{i=0}^{N-1} \operatorname{ind} \mathfrak{b}_i$ are valid. So to complete the proof it is enough to verify the equality

$$\operatorname{ind} \mathfrak{b}_0 = \operatorname{ind} \mathfrak{b}_k, \quad k = \overline{1, N-1} \tag{47.24}$$

where $\mathfrak{b}_k$, $k = \overline{0, N-1}$, are the operators mentioned in Theorem A.17 which, in the situation considered here, arise from the operator (47.21) and thus have the form

$$\mathfrak{b}_k = b_k P + Q, \quad i = \overline{0, N-1}$$

with

$$b_k = \sum_{n=0}^{N-1} a_n \omega^{kn} T^n, \quad \omega = e^{i\frac{2\pi}{N}}$$

(recall also Example 43.11).

To prove (47.24) it is enough to establish the conjugacy (modulo compact operators) between $\mathfrak{b}_0$ and $\mathfrak{b}_k$, so let us do this.

As a technical instrument we need a continuous non-vanishing function φ on Γ with the property

$$\varphi\big(\alpha^n(x)\big) = \varphi(x)\omega^{-kn} . \tag{47.25}$$

Let us construct such a function.

Consider any connected component Γ^l of Γ. Take any $x_0 \in \Gamma^l$ and let p be the minimal natural exponent such that $\alpha^p(x_0) \in \Gamma^l$. If $\alpha^p(x_0) = x_0$ (which means that $p = N$) then set $\varphi \equiv 1$ on Γ^l and extend it onto Γ by equation (47.25).

If $p < N$ then let x_0' be the first point on Γ^l after x_0 (with respect to the orientation on Γ^l) belonging to the orbit of x_0 under the mapping α^p, that is $x_0' = \alpha^{pj}(x_0)$ for some $0 < j < \dfrac{N}{p}$.

On the arc $[\widehat{x_0, x_0'}]$ we take any φ satisfying the conditions

$$\varphi(x_0) = 1, \quad \varphi(x_0') = \omega^{-kpj}$$

and φ is continuous and non-zero on $[\widehat{x_0, x_0'}]$.

The desired function φ on the whole of Γ is obtained from the part of it already defined, by extension via equation (47.25).

Now if we consider the operator of multiplication by the function constructed above then we have

$$\begin{aligned}
\varphi^{-1}\mathcal{b}_0\varphi &= \varphi^{-1}\left[\left(\sum a_n T^n\right)P + Q\right]\varphi \\
&\sim \sum a_n \varphi^{-1} \cdot \varphi\left(\alpha^{-n}(\cdot)\right)T^n P + Q \\
&= \left(\sum a_n \omega^{kn} T^n\right)P + Q \\
&= \mathcal{b}_k ,
\end{aligned}$$

which proves (47.24) and thus completes the proof of the theorem. □

Now we consider the most complicated situation - the case when

The diffeomorphism α is not periodic and does not act topologically freely

47.8 Recalling that we are still confining ourselves to the case when α is orientation preserving and Γ is α-irreducible, we distinguish this situation by the following property:

The set Γ_N of all the periodic points (of period N) for α has non-empty interior $\overset{\circ}{\Gamma}_N$ and $\Gamma_N \neq \Gamma$.

Let $\Gamma_\infty = \Gamma \setminus \overset{\circ}{\Gamma}_N$.

The complexity of this situation is related to the fact that the invertibility conditions of the symbol operator are qualitatively different on Γ_∞ and $\overset{\circ}{\Gamma}_N$ and the index calculation requires more ingenious reasoning.

As in the foregoing cases it is enough to investigate the operator of the form

$$\mathcal{b} = b^+ P + Q \tag{47.26}$$

where

$$b^+ = \sum_0^{m-1} a_k T^k . \tag{47.27}$$

In this expression (adding if necessary a number of zero summands) one can always take

$$m = pN,\ p \in \mathbb{N}\ , \tag{47.28}$$

and henceforth up to 47.11 we shall presume this condition to be satisfied.

47.9 The coefficients a_k in (47.27) are uniquely defined on Γ_∞ and one can define uniquely on Γ_N only the sums

$$v_l(x) = \sum_k a_{kN+l}(x),\ l = 0,\dots,N-1\ . \tag{47.29}$$

To begin with we shall (taking advantage of an ambiguity of the coefficients a_k on $\overset{\circ}{\Gamma}_N$) choose the coefficients on $\overset{\circ}{\Gamma}_N$ in a special way (which will be used later when constructing the operator $\tilde{\mathcal{b}}$ (see Theorem 47.10)).

The set $\overset{\circ}{\Gamma}_N$ is open and thus consists of a finite or denumerable number of open arcs, any two of which have no common boundary points.

For every $l = 0,\dots,N-1$ consider the sums

$$\begin{cases} a_l(x) + a_{N+l}(x) + \dots + a_{(p-1)N+l}(x) = v_{l,0}(x) \\ a_l(x) + \varepsilon a_{N+l}(x) + \dots + \varepsilon^{p-1} a_{(p-1)N+l}(x) = v_{l,1}(x) \\ \dots\dots\dots\dots\dots\dots\dots\dots\dots\dots\dots\dots \\ a_l(x) + \varepsilon^{p-1} a_{N+l}(x) + \dots + \varepsilon^{(p-1)^2} a_{(p-1)N+l}(x) = v_{l,p-1}(x) \end{cases} \tag{47.30}$$

where $\varepsilon = e^{i\frac{2\pi}{p}}$.

We shall consider equations (47.30) as a system of equations with respect to $a_l, a_{N+l}, \ldots, a_{(p-1)N+l}$. Since the determinant of this system is non-zero (it is the Vandermonde determinant) to define the functions $a_{kN+l}(x)$ uniquely it is enough to set the functions $v_{l,k}(x)$.

In view of (47.29) the functions $v_{l,0}\ (= v_l)$ are defined uniquely by b^+. By the uniqueness of a_k in (47.27) on Γ_∞ we conclude that the functions $v_{l,k}$, $k = 1, \ldots, p-1$, are defined uniquely on Γ_∞ (and in particular on $\partial \overset{\circ}{\Gamma}_N$) and could be extended in an arbitrary way on Γ_N up to functions continuous of Γ. We shall define this extension in a special way (the sense of which will be clarified later in the proof of Theorem 47.10).

Let (x_0', x_0'') be a certain arc forming the set $\overset{\circ}{\Gamma}_N$. Let

$$\psi_0 : [x_0', x_0''] \to [-1,1]$$

be any bijective continuous mapping.

The arcs $[\alpha^j(x_0'), \alpha^j(x_0'')] = [x_j', x_j'']$, $j = 1, \ldots, N-1$, are the images of $[x_0', x_0'']$ under the iterations of α. Let us set the mappings

$$\psi_j : [\alpha^j(x_0'), \alpha^j(x_0'')] \to [-1,1]$$

via formula

$$\psi_j(x) = \psi_0\left(\alpha^{-j}(x)\right).$$

For every $j = 1, \ldots, N-1$ we set the function $v_{l,k,j}$ on the interval [-2,2] via the expression

$$v_{l,k,j}(s) = \begin{cases} v_l\left(\psi_j^{-1}(s)\right), & s \in [-1,1] \\ \sum\limits_{\nu} a_{\nu N+l}\left(x_j'\right) e^{i\left(\frac{2\pi}{N}k\nu + l\right)(-s-1)}, & s \in [-2,-1] \\ \sum\limits_{\nu} a_{\nu N+l}\left(x_j''\right) e^{i\left(\frac{2\pi}{N}k\nu + l\right)(s-1)}, & s \in [1,2] \end{cases}$$

where v_l are given by (47.29).

Now let $\tau \in [0,1]$. On each arc $\left[x_j', x_j''\right]$ we define a family of functions depending on the parameter τ by

$$w_{l,k}(x,\tau) = v_{l,k,j}\left[(2-\tau)\psi_j(x)\right], \quad x \in \left[x_j', x_j''\right] \tag{47.31}$$

and set $v_{l,k}(x)$ (on the same arc) by

$$v_{l,k}(x) = w_{l,k}(x,1)\ . \tag{47.32}$$

These functions are continuous on $\left[x_j', x_j''\right]$ and at the boundary points take values satisfying the system (47.30). Moreover these functions define the continuous extensions of the above-mentioned functions $v_{l,k}(x)$ (47.30) from Γ_∞ onto the whole of Γ (it is obvious in the case when $\overset{\circ}{\Gamma}_N$ consists of a finite number of arcs and a bit less obvious but still quite easy to verify in the case when $\overset{\circ}{\Gamma}_N$ consists of a denumerable number of arcs). The desired construction of the functions $v_{l,k}(x)$ is completed.

Now we can carry out the reduction of the operator $\mathcal{b}$ (47.26) to a two-term operator. The corresponding statement is formulated in an analogous way to that presented in Theorem 47.2 but the proof is more complicated and in fact combines the ideas of the proofs of Theorems 47.2 and 47.7.

Theorem 47.10 *Let α be the diffeomorphism described in 47.8 and $\mathcal{b}$ be the operator (47.26)-(47.28) with a_k reset by (47.30) and (47.32). Let $\tilde{\mathcal{b}}$ be the operator given by (47.2)-(47.5). Then the operator $\mathcal{b}$ is Fredholm iff $\tilde{\mathcal{b}}$ is Fredholm and*

$$\operatorname{ind} \mathcal{b} = \frac{1}{m} \operatorname{ind} \tilde{\mathcal{b}}\ .$$

Proof Let us consider in the space

$$H^m = \left[L^2(\Gamma)\right]^m \cong L^2\left(\Gamma, \mathbb{C}^m\right)$$

the subspaces

$$F_r = \left\{\overline{u} \in H^m : \overline{u} = \left\{u_0, \ldots, u_{N-1}, \delta^r u_0, \ldots, \delta^r u_{N-1}, \delta^{2r} u_0, \ldots,\right.\right.$$

$$\left.\left.\delta^{2r} u_{N-1}, \ldots, \delta^{(p-1)r} u_0, \ldots, \delta^{(p-1)r} u_{N-1}\right\}, \ r = \overline{0, p-1}\right. \tag{47.33}$$

where $\delta = e^{i\frac{2\pi}{p}}$. Each of the spaces F_r is isomorphic to H^N and the isomorphism is given by

$$Q_r : \overline{u} \to (u_0, \ldots, u_{N-1}) . \tag{47.34}$$

These subspaces are invariant with respect to the operator $\tilde{\mathfrak{b}}$ given by (47.2)-(47.5) (where $\mathfrak{a}_0 = a_0 P + Q$, $\mathfrak{a}_k = a_k P$, $k = \overline{1, m-1}$ and $\mathfrak{a}_m = 0$). We denote by $\tilde{\mathfrak{b}}_r$, $r = 0, \ldots, p-1$, the restriction of $\tilde{\mathfrak{b}}$ onto F_r.

After establishing the isomorphism with H^N (47.34) the operators $\tilde{\mathfrak{b}}_r$ are realized as the operators B_r of the form

$$B_r = \left(\sum_{j=0}^{p-1} \delta^{rj} A_j \overline{T}^j \right) P + Q + k_r \tag{47.35}$$

where k_r are compact operators and

$$A_j = \begin{bmatrix} a_{jN}(x) & a_{jN+1}(x) & \cdots & a_{jN+N-1}(x) \\ a_{jN-1}\left(\alpha^{-1}(x)\right) & a_{jN}\left(\alpha^{-1}(x)\right) & \cdots & a_{jN+N-2}\left(\alpha^{-1}(x)\right) \\ \cdots & \cdots & \cdots & \cdots \\ a_{jN-N+1}\left(\alpha^{-N+1}(x)\right) & a_{jN-N+2}\left(\alpha^{-N+1}(x)\right) & \cdots & a_{jN}\left(\alpha^{-N+1}(x)\right) \end{bmatrix},$$

$$\overline{T} = \begin{bmatrix} T^N & 0 & \cdots & 0 \\ 0 & T^N & & \vdots \\ \vdots & & \ddots & \vdots \\ 0 & \cdots & \cdots & T^N \end{bmatrix}.$$

Let us show that the operators B_r are Fredholm iff $\mathfrak{b}$ is Fredholm.

Let us look closer at the symbols of these operators.

The symbol $\sigma(B_r)$ is the operator in the space $\left[L^2(\Gamma)\right]^N \oplus \left[L^2(\Gamma)\right]^N$ of the form

$$B_r^+ \oplus I = \sum_{j=0}^{p-1} \delta^{rj} A_j \overline{T}^j \oplus I . \tag{47.36}$$

The symbol $\sigma(\mathfrak{b})$ is the operator in $L^2(\Gamma) \oplus L^2(\Gamma)$ of the form

$$b^{+} \oplus I = \sum_{0}^{m-1} a_k T^k \oplus I . \tag{47.37}$$

So to prove the above statement it is enough to verify that

B_r^{+} is invertible in $\left[L^2(\Gamma)\right]^N$

iff

b^{+} is invertible in $L^2(\Gamma)$.

Observe that

$$\left[L^2(\Gamma)\right]^N = \left[L^2(\Gamma_\infty)\right]^N \oplus \left[L^2\left(\overset{\circ}{\Gamma}_N\right)\right]^N$$

where the subspaces on the right are both invariant with respect to B_r^{+}.

We have also that

$$L^2(\Gamma) = L^2(\Gamma_\infty) \oplus L^2\left(\overset{\circ}{\Gamma}_N\right)$$

and both the subspaces on the right are invariant with respect to b^{+}.

Let B_r^{∞}, B_r^{N}, b^{∞}, b^{N} be the restrictions of B_r^{+} and b^{+} onto $\left[L^2(\Gamma_\infty)\right]^N$, $\left[L^2\left(\overset{\circ}{\Gamma}_N\right)\right]^N$, $L^2(\Gamma_\infty)$ and $L^2\left(\overset{\circ}{\Gamma}_N\right)$, respectively. By the reasoning of the proof of Lemma A.27 one could show that B_r^{∞} is invertible iff b^{∞} is invertible, so it suffices to consider the operators B_r^{N} and b^{N}.

Since T^N acts on $L^2\left(\overset{\circ}{\Gamma}_N\right) = L^2\left(\overline{\overset{\circ}{\Gamma}_N}\right)$ trivially it follows that B_r^{N} is the operator of multiplication by the matrix-valued function (on $x \in \overline{\overset{\circ}{\Gamma}_N}$) of the form

$$B_r^N(x) = \begin{bmatrix} v_{r,0}(x) & v_{r,1}(x) & \dots & v_{r,N-1}(x) \\ v_{r,N-1}\left(\alpha^{-1}(x)\right) & v_{r,0}\left(\alpha^{-1}(x)\right) & \dots & v_{r,N-2}\left(\alpha^{-1}(x)\right) \\ \dots & \dots & \dots & \dots \\ v_{r,1}\left(\alpha^{-N+1}(x)\right) & v_{r,2}\left(\alpha^{-N+1}(x)\right) & \dots & v_{r,0}\left(\alpha^{-N+1}(x)\right) \end{bmatrix}$$

where $v_{r,l}(x)$ are described in 47.9.

The matrix $B_0^N(x)$ is the matrix constructed via formula (47.23) for the operator b^N. Thus this matrix is invertible iff b^N is invertible.

Let us show that under the choice of $v_{r,l}$, $r=1,\dots,p-1$; $l=0,\dots,N-1$, described in 47.9 the invertibility of b^+ implies the invertibility of $B_r^N(x)$.

Since the operator B_0^∞ has the form

$$B_0^\infty = \sum_j A_j \bar{T}^j$$

it is invertible (in view of Proposition 13.1) iff the operator

$$\sum_j A_j z^j \bar{T}^j \tag{47.38}$$

is invertible for every z, $|z|=1$.

The necessary condition for the invertibility of the operator (47.38) is the condition

$$\det B(x,z) \neq 0 \text{ for every } x \in \partial\overset{\circ}{\Gamma} \text{ and } |z|=1$$

where

$$B(x,z) = \sum_j A_j(x) z^j \ . \tag{47.39}$$

But the structure of the functions $v_{r,l}(x)$ and the explicit form of the matrix $B_r^N(x)$ show that for every $x \in \overline{\overset{\circ}{\Gamma}}_N$ the matrix $B_r^N(x)$ coincides with a certain matrix $B(\tilde{x},z)$ for some $\tilde{x} \in \partial\overset{\circ}{\Gamma}$ or with a certain matrix $B_0^N\left(\overset{\approx}{x}\right)$ for some $\overset{\approx}{x} \in \overset{\circ}{\Gamma}_N$, which means that the invertibility of b^+ implies the invertibility of the matrices $B_r^N(x)$ and consequently of the operators B_r^N.

Thus we have proved that B_r are Fredholm iff $\mathscr{b}$ is Fredholm.

By the construction of B_r it is clear that

$$\operatorname{ind} \tilde{\mathscr{b}} = \sum_{r=0}^{p-1} \operatorname{ind} \tilde{\mathscr{b}}_r = \sum_{r=0}^{p-1} \operatorname{ind} B_r \ .$$

Let us show now that all the operators B_r^+ (47.36) are homotopic to B_0^+ which will imply the equality

$$\operatorname{ind} \tilde{\mathcal{B}} = p \operatorname{ind} B_0 \,. \tag{47.40}$$

The homotopy between B_0^∞ and B_r^∞ is established by the family of invertible operators

$$\sum_j \delta^{rj\tau} A_j \overline{T}^j, \; 0 \le \tau \le 1 \tag{47.41}$$

(recall (47.38)).

We shall define the homotopy of the matrix $B_r^N(x)$ to $B_0^N(x)$, $x \in \overline{\overset{\circ}{\Gamma}}_N$, in the class of non-degenerate matrices of the same form, by the family of matrices

$$\begin{bmatrix} w_{r,0}(x,\tau) & w_{r,1}(x,\tau) & \dots & w_{r,N-1}(x,\tau) \\ w_{r,N-1}\left(\alpha^{-1}(x),\tau\right) & w_{r,0}\left(\alpha^{-1}(x),\tau\right) & \dots & w_{r,N-2}\left(\alpha^{-1}(x),\tau\right) \\ \dots & \dots & \vdots & \dots \\ w_{r,1}\left(\alpha^{-N+1}(x),\tau\right) & w_{r,2}\left(\alpha^{-N+1}(x),\tau\right) & \dots & w_{r,0}\left(\alpha^{-N+1}(x),\tau\right) \end{bmatrix} \tag{47.42}$$

where the functions $w_{l,k}$ are given by (47.31).

For every $x \in \overline{\overset{\circ}{\Gamma}}_N$ and $0 \le \tau \le 1$ this matrix coincides with a certain matrix $B(\tilde{x},z)$ (47.39) for some $\tilde{x} \in \partial \overset{\circ}{\Gamma}_N$ and $|z| = 1$ or with $B_0^N\left(\tilde{\tilde{x}}\right)$ for some $\tilde{\tilde{x}} \in \overline{\overset{\circ}{\Gamma}}_N$ and thus is invertible.

It has to be emphasized that at the points of $\partial \overset{\circ}{\Gamma}_N$ the homotopics (47.41) and (47.42) for the corresponding coefficients (of B_r^∞ and B_r^N) coincide and consequently we have obtained the homotopy of B_r^+ and B_0^+ in the class of operators we are considering, which as we have already mentioned proves (47.40).

Observe now that the operator B_0 does not depend on the existing arbitrariness of the coefficients a_k since it contains only the uniquely defined functions $v_l(x)$ (47.29).

Now let us show that

$$\operatorname{ind} \mathcal{B} = \frac{1}{N} \operatorname{ind} B_0 \tag{47.43}$$

which in view of (47.40) will finish the proof.

Consider the decomposition of the space

$$H^N = \left[L^2(\Gamma)\right]^N$$

into the direct sum of the subspaces H_ν (cf. (A.51)):

$$H_\nu = \left\{\bar{u} \in H^N : \bar{u} = \left(u, \omega^\nu Tu, \omega^{2\nu} T^2 u, \ldots, \omega^{(N-1)\nu} T^{N-1} u\right)\right\}$$

where $\omega = \mathrm{e}^{\mathrm{i}\frac{2\pi}{N}}$, $\nu = 0, \ldots, N-1$.

Under this decomposition, just as in the proof of Theorem 47.7, the operator B_0 decomposes into the direct sum of operators of the form

$$B_{0,j} = \left(\sum_{k=0}^{m-1} a_k \omega^{jk} T^k\right) P + Q + k_j, \; j = 0, \ldots, N-1$$

where k_j are compact operators.

Here $B_{0,0} = \mathscr{b} + k_0$ and all the operators $B_{0,j}$ are Fredholm when $\mathscr{b}$ is Fredholm. Thus

$$\text{ind } B_0 = \sum_{j=0}^{N} \text{ind } B_{0,j} \, . \tag{47.44}$$

Let us verify now that

$$\text{ind } B_{0,j} = \text{ind } B_{0,0} = \text{ind } \mathscr{b}$$

which in view of (47.44) will prove (47.43) and the theorem as a whole.

We shall use the idea already exploited in the proof of Theorem 47.7, namely we shall find a continuous non-vanishing function φ satisfying the condition

$$\varphi\left(\alpha^n(x)\right) = \varphi(x) \cdot \omega^{-jn}, \; x \in \Gamma \tag{47.45}$$

which will imply

$$\varphi^{-1} B_{0,0} \varphi = B_{0,j} + k \tag{47.46}$$

and in turn

$$\operatorname{ind} B_{0,0} = \operatorname{ind} B_{0,j}\,. \tag{47.47}$$

The desired function is constructed as follows.

Fix any connected component Γ^l of Γ and let s be the minimal natural exponent such that $\alpha^s(\Gamma^l) = \Gamma^l$.

If $s = N$ then set $\varphi \equiv 1$ on Γ^l and extend it onto Γ by the equation (47.45).

If $s < N$ then consider any arc $(\widehat{x', x''}) \subset \Gamma^l$ forming the set $\mathring{\Gamma}_N$. Let x_0 be the first point on Γ^l after x'' (with respect to the orientation on Γ^l) belonging to the orbit of the point x' under the mapping α^s, that is $x_0 = \alpha^{si}(x')$ for some $0 < i < \frac{N}{s}$.

On the arc $[\widehat{x', x_0}]$ we take φ satisfying the conditions

$$\varphi(x') = 1, \quad \varphi(x'') = \omega^{-jsi},$$

φ is continuous and non-zero on $[\widehat{x', x''}]$ and $\varphi \equiv \omega^{-jsi}$ on the arc $[\widehat{x'', x_0}]$.

The necessary function on the whole of Γ is obtained from the part of it just defined, by extension via equation (47.45).

Thus in view of property (47.46) the proof is complete. □

47.11 The theorem just proved reduces the index calculation of a G-SIO to that of a two-term operator associated with an (orientation-preserving) mapping having fixed points. So we shall consider now operators of this sort.

Let

$$\mathcal{b} = (a_0 + a_1 T)P + Q \tag{47.48}$$

where a_0, a_1 are continuous m-dimensional matrix-valued functions and T is a unitary operator of the form (46.16) generated by a diffeomorphism $\alpha: \Gamma \to \Gamma$, the fixed-point set Γ_1 of which has a non-empty interior $\mathring{\Gamma}_1 \neq \Gamma$. Thus $\Gamma_\infty = \Gamma \setminus \mathring{\Gamma}_1 \neq \varnothing$.

The operator $\mathcal{b}$ is Fredholm iff the operator

$$b^+ = a_0 + a_1 T \tag{47.49}$$

is invertible. The latter decomposes into the direct sum of operators b^∞ and b^1 (the restrictions onto $L^2(\Gamma_\infty, \mathbb{C}^m)$ and $L^2(\mathring{\Gamma}_1, \mathbb{C}^m)$). So let us write out the invertibility conditions for b^∞ and b^1.

I. By Theorem 46.20 there exist on Γ_∞ non-degenerate continuous matrix-valued functions $c_1(x)$ and $c_2(x)$ such that

$$b^\infty = c_1(x)\left[d_0 + d_1 T\right]c_2(x) \tag{47.50}$$

where

$$d_0 = \begin{bmatrix} I_{p(x)} & 0 \\ 0 & d_{22}^0 \end{bmatrix} \text{ and } d_1 = \begin{bmatrix} d_{11}^1 & 0 \\ 0 & I_{m-p(x)} \end{bmatrix} \tag{47.51}$$

are matrices satisfying the conditions

$$r\left(d_{22}^0 T^{-1}\right) < 1 \text{ and } r\left(d_{11}^1 T\right) < 1 . \tag{47.52}$$

Clearly the dimension $p(x)$ is constant on every connected component of Γ_∞.

Condition (47.51) in the situation we are considering could be rewritten in a more explicit form; namely, in view of the formula (23.23) we have

$$r\left(d_{22}^0 T^{-1}\right) = \max_{\Gamma_1 \cap \Gamma_\infty} r\left(d_{22}^0(x)\right)$$

$$r\left(d_{11}^1 T^{-1}\right) = \max_{\Gamma_1 \cap \Gamma_\infty} r\left(d_{11}^1(x)\right)$$

which means that (47.52) is equivalent to the condition

$$\text{II.}\quad \begin{aligned} &\max_{\Gamma_1 \cap \Gamma_\infty} r\left(d_{22}^0(x)\right) < 1 \\ &\max_{\Gamma_1 \cap \Gamma_\infty} r\left(d_{11}^1(x)\right) < 1 . \end{aligned} \tag{47.53}$$

The invertibility condition for b^1 is

$$\text{III.}\quad \det\left[a_0(x) + a_1(x)\right] \neq 0 \text{ for every } x \in \overline{\mathring{\Gamma}_1} . \tag{47.54}$$

Thus to obtain the index formula we have to first find the topological (homotopic) invariant on the set of pairs of matrix-valued functions (a_0, a_1) satisfying the conditions I-III in terms of which the index will be expressed.

47.12 To start with let us extend the matrix-valued functions $c_1(x)$ and $c_2(x)$ up to continuous matrix-valued functions on the whole of Γ. Observe that due to the connectedness of the set of non-degenerate matrices this continuation does exist on every interval (x', x'') belonging to $\mathring{\Gamma}_1$. Thus if the number of connected components of $\mathring{\Gamma}_1$ is finite then the existence of the desired continuation is obvious.

If the number of connected components of $\mathring{\Gamma}_1$ is infinite then the continuation of $c_1(x)$ and $c_2(x)$ onto every connected component (x', x'') should be constructed in a special way (to obtain the continuity on the whole of Γ). Here one can apply the next statement:

If A and B are non-degenerate matrices then there exists a path (in the set of non-degenerate matrices) connecting A and B and lying in the ball of radius $\|A-B\|$ with the centre at A.

Indeed, consider the family of matrices

$$(1-z)A+zB, \quad |z| \le 1 .$$

This family belongs to the ball mentioned above. Since

$$f(z)=\det[(1-z)A+zB]$$

is a non-zero polynomial it has a finite set of zeros. Thus the points 0 and 1 could be connected by a path $z(t)$, $0 \le t \le 1$ contained in the ball $|z| \le 1$ and not containing the zeros of the polynomial $f(z)$. The desired path connecting A and B can be defined now by

$$(1-z(t))A+z(t)B, \quad 0 \le t \le 1 .$$

If the continuation of the matrix-valued functions $c_1(x)$ and $c_2(x)$ on the arcs $(x', x'') \subset \mathring{\Gamma}_1$ are constructed using this procedure then one can easily verify that the matrix-valued functions $c_1(x)$ and $c_2(x)$ so obtained will be continuous on the whole of Γ.

47.13 Using these functions $c_1(x)$ and $c_2(x)$ let us consider the operator

$$\mathcal{A} = c_1^{-1}[a_0 + a_1 T]c_2^{-1}P + Q .$$

Since

$$d \sim \left(c_1^{-1}P+Q\right)\left[(a_0+a_1T)P+Q\right]\left(c_2^{-1}P+Q\right)$$

it follows that

$$\operatorname{ind} d = \operatorname{ind}\left(c_1^{-1}P+Q\right)+\operatorname{ind}\left[(a_0+a_1T)P+Q\right]+\operatorname{ind}\left(C_2^{-1}P+Q\right)$$

$$= \operatorname{ind}\left[(a_0+a_1T)P+Q\right]+\frac{1}{2\pi}\left[\arg\det(c_1\cdot c_2)\right]_\Gamma \tag{47.55}$$

and the index of $\left[(a_0+a_1T)P+Q\right]$ will be calculated if we calculate the index of d.

Now we shall introduce the topological invariant in terms of which the index of d can be written out.

The coefficients d_0 and d_1 of the operator

$$d^+ = c_1^{-1}(a_0+a_1T)c_2^{-1} = d_0+d_1T \tag{47.56}$$

possess on Γ_∞ the properties described in I and II in 47.11 and are continuous on Γ.

Let

$$D=\{z\in\mathbb{C}:|z|\le 1\}\ .$$

Consider the set $M\subset\Gamma\times D$ given by

$$M=\{(x,z):\quad z=0,\quad \text{if } x\in\mathring{\Gamma}_\infty;$$

$$z=1,\quad \text{if } x\in\mathring{\Gamma}_1;$$

$$z\in D,\ \text{ if } x\in\partial\left(\mathring{\Gamma}_\infty\right)=\partial\left(\mathring{\Gamma}_1\right)\}\ .$$

On the set M we define the matrix-valued function

$$\Phi(x,z)=\begin{cases} I_m, & x\in\mathring{\Gamma}_\infty \\ d_0(x)+d_1(x), & x\in\mathring{\Gamma}_1 \\ \begin{bmatrix} I_{p(x)}+zd_{11}^1 & 0 \\ 0 & I_{m-p(x)}+zd_{22}^0(x)\end{bmatrix}, & x\in\partial\left(\mathring{\Gamma}_\infty\right)\end{cases} \tag{47.57}$$

where d_0, d_1 are from (47.56) and d_{11}^1, d_{22}^0 were described in (47.51), (47.53).

The matrix-valued function $\Phi(x,z)$ is continuous on M and non-degenerate (by (47.53), (47.54)), thus the function $\varphi = \det \Phi(x,z)$ defines a continuous mapping $\varphi: M \to \mathbb{C} \setminus \{0\}$.

In general when the set of boundary points $\partial\left(\overset{\circ}{\Gamma}_\infty\right)$ is infinite the space M does not have the homotopic type of the circle and one cannot exploit the classical definition of the degree of the mapping. But the function φ has a special form - it is a polynomial, the power of which on the disc D is not higher than a certain fixed number. Due to this it is possible to define the (finite) degree of the mapping $\varphi: M \to \mathbb{C} \setminus \{0\}$. One could exploit a number of methods for this purpose.

For example, one could consider φ as a multifunction on Γ and in this case it belongs to the class of functions for which the degree of the mapping is already worked out, say by Borisovic, Gel'man, Myškis, Obuhovskii [1] and Hu [1].

The second variant is to cut out first an ε-neighbourhood of the set of the limit points of $\partial\left(\overset{\circ}{\Gamma}_\infty\right)$, then to define on the rest of the Γ the increment of the argument of φ and after this is done pass to the limit when $\varepsilon \to 0$ (this procedure is used in Kravchenko and Litvinchuk [1]).

The third variant (which we shall apply) is the following. We implement a homotopy of the operator (47.56) (preserving the invertibility conditions of the symbol) to the one for which the corresponding function φ does not depend on z and therefore for this function the classical definition of the degree of the mapping (as the increment of the argument around a curved circuit) is valid.

The explicit description of this procedure will be given in the proof of the next

Theorem 47.14 ind $\mathcal{A} = -$ind Φ *where* $\mathcal{A}$ *and* Φ *are described in 47.13 and the meaning of* ind Φ *will be defined in the proof (47.61).*

Proof Let (x', x'') be any of the intervals (connected components) forming the set $\overset{\circ}{\Gamma}_1$. In the one-sided neighbourhood of x' the matrices d_0 and d_1 are written out in block-diagonal form

$$d_j(x) = \begin{bmatrix} d_{11}^j(x) & d_{12}^j(x) \\ d_{21}^j(x) & d_{22}^j(x) \end{bmatrix}$$

where

$$d_{11}^0(x')=I_{p(x)},\ d_{12}^j(x')=d_{21}^j(x')=0,\ d_{22}^1(x')=I_{m-p(x)}\,.$$

The condition (47.53) gives us

$$\det\begin{bmatrix} d_{11}^0(x)+zd_{11}^1 & d_{12}^0(x)+d_{12}^1(x) \\ d_{21}^0(x)+d_{21}^1(x) & zd_{22}^0(x)+d_{22}^1(x) \end{bmatrix}\neq 0 \tag{47.58}$$

for every $x\in\partial\left(\overset{\circ}{\Gamma}_\infty\right)$ and $|z|\leq 1$.

By uniform continuity there exists $\delta>0$ such that (47.58) is fulfilled in the δ-neighbourhood of the set $\partial\left(\overset{\circ}{\Gamma}_\infty\right)$.

Let us take any smooth function $\psi(x)$ satisfying the conditions

$$\begin{cases} \psi(x)=1 & \text{on a } \delta/2-\text{neighbourhood of } \partial\left(\overset{\circ}{\Gamma}_\infty\right), \\ 0\leq\psi(x)\leq 1, & \\ \psi(x)=0 & \text{outside a } \delta-\text{neighbourhood of } \partial\left(\overset{\circ}{\Gamma}_\infty\right) \end{cases}$$

and consider the family of operators $\mathscr{d}^\tau$, $0\leq\tau\leq 1$, of the form

$$\mathscr{d}^\tau=\left(\begin{bmatrix} d_{11}^0(x) & d_{12}^0(x) \\ d_{21}^0(x) & (1-\tau\psi(x))d_{22}^0 \end{bmatrix}+\begin{bmatrix} (1-\tau\psi(x))d_{11}^1(x) & d_{12}^1(x) \\ d_{21}^1(x) & d_{22}^1(x) \end{bmatrix}T\right)P+Q\,. \tag{47.59}$$

For every τ, $0\leq\tau\leq 1$, the operators $\mathscr{d}^\tau$ satisfy the conditions I-III (in 47.11) and consequently they are Fredholm. Thus the given family establishes a homotopy between the operators $\mathscr{d}^0=\mathscr{d}$ and $\mathscr{d}^1$ which implies

$$\operatorname{ind}\mathscr{d}=\operatorname{ind}\mathscr{d}^1\,. \tag{47.60}$$

Observe that the coefficients d_0^1 and d_1^1 (of the operator $\mathscr{d}^1$) have a special form on $\overset{\circ}{\Gamma}_1$; namely, at every point of $\partial\left(\overset{\circ}{\Gamma}_1\right)$ only one out of eight blocks forming the matrices d_0^1 and d_1^1 is non-zero.

Let us write out the matrix-valued function (47.57) for the operator $\mathscr{d}^1$:

$$\Phi^1(x,z)=\begin{cases} I, & x\in\Gamma_\infty \\ \begin{bmatrix} d_{11}^0+(1-\psi(x))d_{11}^1 & d_{12}^0+d_{12}^1 \\ d_{21}^0+d_{21}^1 & (1-\psi(x))d_{22}^0+d_{22}^1 \end{bmatrix}, & x\in\overset{\circ}{\Gamma}_1 \end{cases}.$$

This function does not depend on z and thus could be considered as a matrix-valued function on Γ for which ind Φ^1 (the index of the function on the curve Γ) is well defined.

Since Φ^1 is homotopic to Φ in the class of functions for which the invertibility conditions are fulfilled we set

$$\text{ind}\,\Phi \overset{\text{def}}{=} \text{ind}\,\Phi^1 . \tag{47.61}$$

The operator $\mathcal{d}^1$ modulo a compact summand can be written in the form

$$\mathcal{d}^1=\left(\left|\alpha^{-1'}(x)\right|^{\frac{1}{2}}\Phi^1 P+Q\right)\left[\left(\overset{\approx}{a}_0+\overset{\approx}{a}_1 T\right)P+Q\right] \tag{47.62}$$

where

$$\text{on } \Gamma_\infty \quad \overset{\approx}{a}_0(x)=\begin{bmatrix} I_{p(x)} & 0 \\ 0 & 0\end{bmatrix},\ \overset{\approx}{a}_1(x)=\begin{bmatrix} 0 & 0 \\ 0 & I_{m-p(x)}\end{bmatrix}$$

and

$$\text{on } \overset{\circ}{\Gamma}_1 \quad \overset{\approx}{a}_0(x)+\overset{\approx}{a}_1(x)=I_m .$$

Remark The operator $\mathcal{d}^1$ does not define the matrix-valued functions $\overset{\approx}{a}_0$ and $\overset{\approx}{a}_1$ uniquely on $\overline{\overset{\circ}{\Gamma}}_1$. For our purposes it is enough to fix any suitable extensions of these functions from Γ_∞. For example one particular extension is dictated by the explicit record of the operator $\mathcal{d}^1$ namely

$$\overset{\approx}{a}_0(x)=\left[\Phi^1(x)\right]^{-1}d_0$$

$$\overset{\approx}{a}_1(x)=\left[\Phi^1(x)\right]^{-1}d_1 .$$

Since the function $\left|\alpha^{-1'}(x)\right|^{\frac{1}{2}}$ is homotopic to the identity it follows that

$$\operatorname{ind}\left(\left|\alpha^{-1\prime}(x)\right|^{\frac{1}{2}}\Phi^1 P+Q\right)=-\operatorname{ind}\Phi^1 \tag{47.63}$$

and thus (in view of (47.60)-(47.63)) the proof will be complete if we show that

$$\operatorname{ind}\left[\left(\approx{a}_0+\approx{a}_1 T\right)P+Q\right]=0 . \tag{47.64}$$

The operator $\left[\left(\approx{a}_0+\approx{a}_1 T\right)P+Q\right]$ acts on vector-valued functions (from $L^2\left(\Gamma,\mathbb{C}^m\right)=\left[L^2(\Gamma)\right]^m$) coordinate-wise and the action of the operator $\approx{a}_0+\approx{a}_1 T$ on the scalar functions can be described in the following way.

Let

$$u(x)=\left(u_1(x),\ldots,u_m(x)\right)$$

and denote by Ω_k, $k=1,\ldots,m$, the set

$$\Omega_k=\left\{x\in\Gamma_\infty : p(x)<k\right\} .$$

The set Ω_k is an open-closed set in Γ_∞ as $p(x)$ is continuous on Γ_∞.

Observe also that the function $p(x)$ may have different values at points x' and x'' of Γ_∞ only when these points are separated by an arc, the length of which is not less than δ (where δ is used in the definition of the function $\psi(x)$ (see (47.59)).

On the coordinate $u_k(x)$ of the vector $u(x)$ the operator $\approx{a}_0+\approx{a}_1 T$ acts via the rule

$$\left(\approx{a}_0+\approx{a}_1 T\right)\left(u_k(x)\right)=\begin{cases}\left|\alpha^{-1\prime}(x)\right|^{\frac{1}{2}} u_k\left(\alpha^{-1}(x)\right), & x\in\Omega_k\\ u_k(x), & x\overline{\in}\Gamma\setminus\Omega_k .\end{cases} \tag{47.65}$$

Let us define the mapping $\alpha_k:\Gamma\to\Gamma$ by the rule

$$\alpha_k(x)=\begin{cases}\alpha(x), & x\in\Omega_k\\ x, & x\in\Gamma\setminus\Omega_k .\end{cases} \tag{47.66}$$

α_k is clearly a homeomorphism of Γ having the same smoothness as α.

Now in terms of α_k (47.65) can be rewritten as

$$\left(\overset{\approx}{a}_0+\overset{\approx}{a}_1\,T\right)u_k = T_{\alpha_k}u_k$$

and the action of the operator $\left[\left(\overset{\approx}{a}_0+\overset{\approx}{a}_1\,T\right)P+Q\right]$ of the component u_k of u coincides with the action of the operator

$$T_{\alpha_k}P+Q\,.$$

But by Theorem 46.13 the index of the latter operator is zero which in view of the foregoing argument means that (47.64) is true, thus completing the proof. □

The theorem just proved along with Theorem 47.10 and (47.55) imply

Theorem 47.15 *If an operator*

$$\mathcal{L}=\left(\sum_{0}^{m-1}a_kT^k\right)P+Q$$

is Fredholm then

$$\text{ind}\,\mathcal{L}=\frac{-1}{m}\left(\text{ind}\,c_1+\text{ind}\,c_2+\text{ind}\,\Phi\right)$$

where c_1, c_2 and Φ are the matrices constructed with respect to the operator $\tilde{\mathcal{L}}$ (47.2) and described in 47.12, 47.13.

Now let us consider certain examples demonstrating the applications of these results.

Example 47.16 Let $T=T_\alpha$ be the operator (46.15) associated with an orientation-preserving diffeomorphism $\alpha:\Gamma\to\Gamma$ such that Γ is α-irreducible and $\overset{\circ}{\Gamma}_1\neq\varnothing$ and $\Gamma_1\neq\Gamma$.

Consider the operator (in $L^2(\Gamma)$) of the form

$$\mathcal{L} = (a_0 + a_1 T)P + Q \ . \tag{47.67}$$

In view of the results of Sections 17 and 18 the invertibility conditions for the operator $a_0 + a_1 T$ can be written out easily in the following way.

The set Γ_∞ decomposes into two non-intersecting open-closed (in Γ_∞) sets $\Gamma_\infty = \Gamma_\infty^1 \cup \Gamma_\infty^2$ where this means that these sets are open and closed with respect to the induced topology on Γ_∞ (one of which may be empty) such that

(i) $a_0(x) \neq 0$ on Γ_∞^1 and

$$|a_0(x)| > |a_1(x)| \text{ for every } x \in \Gamma_1 \cap \Gamma_\infty^1 \ ;$$

(ii) $a_1(x) \neq 0$ on Γ_∞^2 and

$$|a_0(x)| < |a_1(x)| \text{ for every } x \in \Gamma_1 \cap \Gamma_\infty^2 \ ;$$

(iii) $a_0(x) + a_1(x) \neq 0$ for every $x \in \overset{\circ}{\Gamma}_1$.

The functions $c_1(x)$ and $c_2(x)$ in this situation are scalar and could be chosen for example in the following way.

$c_1(x)$ on Γ_∞ is equal to

$$c_1(x) = \begin{cases} a_0(x), & x \in \Gamma_\infty^1 \\ a_1(x), & x \in \Gamma_\infty^2 \end{cases}$$

and extended on to the whole of Γ in an arbitrary way (up to a non-vanishing continuous function)

$$c_2(x) \equiv 1 \ .$$

With these c_1, c_2 we can represent $\mathcal{L}$ in the form

$$\mathcal{L} = (c_1 P + Q)\left[(d_0 + d_1 T)P + Q\right] + k$$

where k is compact and

$$d_0(x)=\begin{cases}1, & x\in\Gamma_\infty^1,\\ 0, & x\in\Gamma_\infty^2,\end{cases}$$

$$d_1(x)=\begin{cases}0, & x\in\Gamma_\infty^1,\\ 1, & x\in\Gamma_\infty^2.\end{cases}$$

The function Φ (47.57) for the operator $d^+=d_0+d_1T$ is written as

$$\Phi(x,z)=\begin{cases}1, & x\in\Gamma_\infty,\ z=0,\\ d_0(x)+d_1(x), & x\in\mathring{\Gamma}_1,\ z=1,\\ d_0(x)+zd_1(x), & x\in\partial\left(\Gamma_\infty^1\right),\ |z|\le 1,\\ zd_0(x)+d_1(x), & x\in\partial\left(\Gamma_\infty^2\right),\ |z|\le 1.\end{cases}$$

Theorem 47.15 gives us

$$\operatorname{ind}\ell=-\operatorname{ind}c_1-\operatorname{ind}\Phi\,.$$

For our example (in view of its certain degeneration) one can simplify the construction exploited in the proof of Theorem 47.14.

The operator ℓ satisfying the conditions (i)-(iii) is homotopic to the operator of the same sort with

$$a_1(x)\equiv 0 \text{ on } \partial\left(\Gamma_\infty^1\right)$$

and (47.68)

$$a_2(x)\equiv 0 \text{ on } \partial\left(\Gamma_\infty^2\right).$$

But if the coefficients of an operator satisfy the conditions (47.68) then we can immediately write out the function

$$\Psi=c_1(x)\Phi^1(x)$$

$$\Psi(x)=\begin{cases} a_0(x), & x\in\Gamma_\infty^1, \\ a_1(x), & x\in\Gamma_\infty^2, \\ a_0(x)+a_1(x), & x\in\overline{\mathring{\Gamma}_1}, \end{cases}$$

in terms of which we have

$$\operatorname{ind} \mathcal{L} = -\operatorname{ind}\Psi .$$

Remark While successively constructing the functions $c_1(x)$ and $\Phi(x)$ we (in a certain sense) reveal first (via c_1) the dependence on the behaviour of the coefficients on the set of moving points and afterwards (via Φ) on the set of fixed points.

Example 47.17 Let us consider the operator (47.67) in the case when Γ is α-irreducible and $\mathring{\Gamma}_N \neq \varnothing$ and $\Gamma_N \neq \Gamma$ for some $N>1$. Following Theorem 47.10 let us find the index of this operator.

Observe that in spite of the visual simplicity of the form of this operator the index calculation cannot be trivial since, as the limiting cases, the index formula should contain the index formula for the topologically freely acting diffeomorphism α (Theorem 47.4) and the index formula for the periodic diffeomorphism (Theorem 47.7).

If we consider the operator

$$\mathcal{L} = (a_0 + a_1 T)P + Q$$

as an operator of the form

$$\mathcal{L} = \left(\sum_{k=0}^{N-1} a_k T^k\right)P + Q$$

(see (47.26)-(47.28)) then as $a_2 = \ldots = a_{N-1}$ the functions $v_{l,k} = v_{0,k}$ (see (47.30)) satisfy the requirements imposed in the proof of Theorem 47.10 and therefore the coefficients a_0 and a_1 satisfy the system (47.30) and do not need to be reset.

For the operator

$$b^+ = \sum_{k=0}^{N-1} a_k T^k, \; a_2 = \ldots = a_{N-1} = 0,$$

let us construct the operator

$$\tilde{b}^+ = \tilde{a}_0 + \tilde{a}_1 W$$

(see (47.2)-(47.5)).

We have

$$\tilde{a}_0 = \begin{bmatrix} a_0(x) & a_1(x) & 0 & \ldots & 0 \\ & a_0\left(\alpha^{-1}(x)\right) & a_1\left(\alpha^{-1}(x)\right) & \ldots & 0 \\ \ldots & \ldots & \ldots & \ldots & \ldots \\ 0 & 0 & 0 & \ldots & a_1\left(\alpha^{-N+2}(x)\right) \\ 0 & 0 & 0 & \ldots & a_0\left(\alpha^{-N+1}(x)\right) \end{bmatrix},$$

$$\tilde{a}_1 = \begin{bmatrix} 0 & 0 & \ldots & 0 \\ 0 & 0 & \ldots & 0 \\ \ldots & \ldots & \ldots & \ldots \\ 0 & 0 & \ldots & 0 \\ a_1\left(\alpha^{-N+1}(x)\right) & 0 & \ldots & 0 \end{bmatrix},$$

$$W = \begin{bmatrix} T^N & 0 & \ldots & 0 \\ 0 & T^N & \ldots & 0 \\ \ldots & \ldots & \ldots & \ldots \\ 0 & 0 & \ldots & T^N \end{bmatrix}.$$

Let us recall the invertibility conditions for the symbol of $\mathscr{b}$ which coincide (see the proof of Theorem 47.10) with the invertibility conditions for $\tilde{b}^+$. They can be written out in the following way.

(i) The set Γ_∞ decomposes into two open-closed sets Γ_∞^1 and Γ_∞^2 such that on Γ_∞^1 we have

$$a_0(x) \neq 0$$

and

$$\left|\prod_{k=0}^{N-1} a_0\left(\alpha^k(x)\right)\right| > \left|\prod_{k=0}^{N-1} a_1\left(\alpha^k(x)\right)\right|, \ x \in \Gamma_N \cap \Gamma_\infty^1 ;$$

on Γ_∞^2 we have

$$a_1(x) \neq 0$$

and

$$\left|\prod_{k=0}^{N-1} a_0\left(\alpha^k(x)\right)\right| < \left|\prod_{k=0}^{N-1} a_1\left(\alpha^k(x)\right)\right|, \ x \in \Gamma_N \cap \Gamma_\infty^2 .$$

(ii) On the set $\overline{\overset{\circ}{\Gamma}_N}$ we have

$$\prod_{k=0}^{N-1} a_0\left(\alpha^k(x)\right) - \prod_{k=0}^{N-1} a_1\left(\alpha^k(x)\right) \neq 0 .$$

Since (i) means the invertibility of the restriction of $\tilde{b}^+$ onto Γ_∞ it follows (by Theorem 19.21) that the corresponding relation is hyperbolic.

In the situation we are considering we can write out the stable and unstable subspaces (entering the definition of the hyperbolic relation) explicitly and by means of them write out the matrices $c_1(x)$ and $c_2(x)$.

On Γ_∞^1 the whole of $H\left(\Gamma_\infty^1\right) = \bigoplus_{x \in \Gamma_\infty^1} H_x$, $H_x = \mathbb{C}^N$ is stable. So we set here

$$c_1(x) = d_0(x), \ \ c_2(x) = I .$$

On Γ_∞^2 the stable and unstable subspaces are non-trivial:

$$H^s\left(\Gamma_\infty^2\right) = \bigoplus_{x \in \Gamma_\infty^2} H_x^s$$

where $H_x^s = \left\{\xi = (\xi_0, \dots, \xi_{N-1}) \in \mathbb{C}^N : \xi_0 = 0\right\}$.

Thus the dimension of H_x^s here is $N-1$.

The unstable subspaces H_x^u are all one-dimensional and H_x^u consists of the vectors proportional to the vector

$$\xi_x^0 = \left(1, -\frac{a_0(x)}{a_1(x)}, \frac{a_0(x)a_0\left(\alpha^{-1}(x)\right)}{a_1(x)a_1\left(\alpha^{-1}(x)\right)}, \ldots, (-1)^{N-1}\prod_{k=0}^{N-2}\frac{a_0}{a_1}\left(\alpha^{-k}(x)\right)\right).$$

The matrices $c_1(x)$ and $c_2(x)$ could be defined here via the construction used in the proof of Theorem 46.20.

Let us set the basis in $H_x = \mathbb{C}^N$ by

$$\left.\begin{array}{l} e_1(x) = (0,1,0,\ldots,0) \\ e_2(x) = (0,0,1,\ldots,0) \\ \cdots\cdots\cdots\cdots\cdots\cdots \\ e_{N-1}(x) = (0,0,0,\ldots,1) \end{array}\right\} \text{the basis in } H_x^s$$

$e_0(x) = \xi_x^0$ - the base in H_x^u.

We have

$$\left.\begin{array}{l} \tilde{a}_0(x)e_1(x) = \left(a_1(x), a_0\left(\alpha^{-1}(x)\right), 0, \ldots, 0, 0\right) \\ \tilde{a}_0(x)e_2(x) = \left(0, a_1\left(\alpha^{-1}(x)\right), a_0\left(\alpha^{-2}(x)\right), \ldots, 0, 0\right) \\ \cdots\cdots\cdots\cdots\cdots\cdots\cdots\cdots\cdots\cdots \\ \tilde{a}_0(x)e_{N-1}(x) = \left(0,0,0,\ldots, a_1\left(\alpha^{-N+2}(x)\right), a_0\left(\alpha^{-N+1}(x)\right)\right) \end{array}\right]$$

$$\tilde{a}_1(x)e_0(x) = \left(0,0,\ldots, a_1\left(\alpha^{-N+1}(x)\right)\right).$$

Thus using the construction written out in the proof of Theorem 46.20 we obtain on Γ_∞^2

$$c_1(x) = \begin{bmatrix} a_1(x) & 0 & 0 & \ldots & 0 \\ a_0\left(\alpha^{-1}(x)\right) & a_1\left(\alpha^{-1}(x)\right) & 0 & \ldots & 0 \\ 0 & a_0\left(\alpha^{-2}(x)\right) & a_1\left(\alpha^{-2}(x)\right) & \ldots & 0 \\ \ldots & \ldots & \ldots & \ldots & \ldots \\ 0 & 0 & 0 & \ldots & a_1\left(\alpha^{-N+1}(x)\right) \end{bmatrix}$$

$$c_2(x)=\begin{bmatrix} 0 & \dots & 0 & 1 \\ 1 & \dots & 0 & -\dfrac{a_0(x)}{a_1(x)} \\ \vdots & & \vdots & \vdots \\ 0 & \cdots & 1 & (-1)^{N-1}\prod\limits_{k=0}^{N-2}\dfrac{a_0}{a_1}\left(\alpha^{-k}(x)\right) \end{bmatrix}$$

and for the operator $\tilde{b}^+$ we obtain a representation in the form

$$\tilde{b}^+ = c_1(x)\left(d_0 + d_1 W\right)c_2^{-1}(x)$$

where on Γ_∞^2

$$d_0 = \begin{bmatrix} 1 & & & & \vdots & 0 \\ & 1 & & & \vdots & 0 \\ & & \ddots & & \vdots & \vdots \\ & & & 1 & \vdots & 0 \\ \dots & \dots & \dots & \dots & \vdots & \dots \\ 0 & 0 & \dots & 0 & \vdots & \prod\limits_{k=0}^{N-1}\dfrac{a_0}{a_1}\left(\alpha^{-k}(x)\right) \end{bmatrix} = \begin{bmatrix} I_{N-1} & \vdots & 0 \\ \dots & \vdots & \dots \\ 0 & \vdots & \tilde{p}_N \end{bmatrix}$$

$$d_1 = \begin{bmatrix} 0 & \vdots & 0 \\ \dots & \vdots & \dots \\ 0 & \vdots & 1 \end{bmatrix}.$$

Let us assume that the matrices $c_1(x)$ and $c_2(x)$ are extended from $\Gamma_\infty = \Gamma_\infty^1 \cup \Gamma_\infty^2$ onto Γ by continuity. After this is done the operator

$$d = c_1^{-1}\tilde{b}^+ c_2$$

has the form $d_0 + d_1 W$ on Γ_∞ and is the operator of multiplication by the matrix-valued function $\left[d_0(x) + d_1(x)\right]$ on $\mathring{\Gamma}_N$.

The function $\Phi(x,z)$ (47.57) now can be written out as

$$\Phi(x,z)=\begin{cases} I_N, & x\in\Gamma_\infty,\ z=0,\\ d_0(x)+d_1(x), & x\in\mathring{\Gamma}_N,\ z=1,\\ \begin{bmatrix} I_{N-1} & \vdots & 0\\ \dots & \vdots & \dots\\ 0 & \vdots & 1+z\tilde{p}_N \end{bmatrix}, & x\in\partial\left(\mathring{\Gamma}_\infty\right),\ |z|\le 1. \end{cases}$$

Thus Theorem 47.15 implies

$$\operatorname{ind} \ell = -\frac{1}{N}\left(\operatorname{ind} c_1 - \operatorname{ind} c_2 + \operatorname{ind}\Phi\right).$$

48. *G*-Toeplitz operators with continuous coefficients

According to the general ideas developed in Chapter 4 (Sections 30, 32, 33) the theory of the Fredholmity of the Toeplitz (and *G*-Toeplitz) operators is in fact a part of the theory of the Fredholmity of the singular integral operators (and *G*-singular integral operators). Nevertheless in the mathematical literature Toeplitz operator theory occupies a special (singled-out) place which is reflected, for example, by the monograph by Böttcher and Sibberman [1]. Apparently this is due, on the one hand, to the accessibility of the deep structure of these objects and, on the other hand, to their various applications (especially in physics).

Thus for these reasons (and demonstrating our respect for Toeplitz operators) we present in this section the results concerning the Fredholm properties of the *G*-Toeplitz operators with continuous coefficients.

In fact (as we have already emphasized) these results are simply the corollaries of the corresponding results derived in Sections 46 and 47.

48.1 Let Γ be a closed Lyapunov curve consisting of a finite number of connected components $\Gamma^1,\dots,\Gamma^l$. Let P and Q be the projections (46.11) defined via the Cauchy singular integral operator and let $H^{\pm}$ be the subspaces of $L^2(\Gamma)$ given by (46.12).

We shall say that $d(a)$ is a *Toeplitz operator in* H^+ associated with a function $a\in C(\Gamma)$ if

$$d(a)=P\circ a \tag{48.1}$$

where on the right we consider a as the operator of multiplication by a given function a.

Remarks (1) We denote here the Toeplitz operator by $d(a)$ and not by T_a (which is the standard notation and was used in 33.19) in order not to confuse this operator with the shift operators T_α (46.16).
(2) If $\Gamma = S^1 = \{t \in \mathbb{C}: |t| = 1\}$ then H^+ is the Hardy space H^2 (see 33.19) and that is why we think of H^+ as an abstract version of the Hardy space (related to $L^2(\Gamma)$).

We denote by $\mathcal{A}$ the C^*-algebra of operators in H^+ generated by the operators (48.1) and compact operators, the elements of $\mathcal{A}$ will be denoted by $\mathfrak{a}$.

Considering, along with the operator $d(a)$, the singular integral operator $d(a)+Q$ and applying Theorem 2.54 we obtain

Theorem 48.2 *The mapping* $\sigma(\,\cdot\,)$ *defined by*

$$\sigma(d(a)) = a$$

can be extended up to the mapping $\sigma: \mathcal{A} \to C(\Gamma)$ *which has the following properties:*

(i) $\sigma(\mathfrak{a}) = 0$ *iff* $\mathfrak{a} \in \mathcal{K}(H^+)$

($\mathcal{K}(H^+)$ is the set of compact operators in H^+);

(ii) $\sigma: \mathcal{A}/\mathcal{K}(H^+) \to C(\Gamma)$

*establishes an algebraic *-isomorphism;*

(iii) *if $\mathfrak{a}$ is Fredholm (that is $\sigma(\mathfrak{a})$ is invertible) then*

$$\operatorname{ind} \mathfrak{a} = -\operatorname{ind} \sigma(\mathfrak{a}) = -\frac{1}{2\pi}[\arg \sigma(\mathfrak{a})(t)]_\Gamma\,. \tag{48.2}$$

48.3 Let $\{\alpha_g\}_{g\in G}$ be a group of diffeomorphisms of the curve Γ preserving the orientation and satisfying the Holder condition 2.50 (henceforth we shall consider here only such diffeomorphisms). We denote by B the algebra of operators in $L^2(\Gamma)$ generated by the operators T_{α_g}, $g \in G$, (defined by (46.16)), and the operators of multiplication by continuous functions. The elements of B will be denoted by b.

48.4 Let $b \in B$ (see 48.3). The operator

$$d(b) \overset{\text{def}}{=} P \circ b \tag{48.3}$$

acting in H^+ will be called the *G-Toeplitz operator with continuous coefficients* (in H^+).

In this situation the operator b will be called the *symbol of the G-Toeplitz operator* $d(b)$.

We denote by $\mathscr{B}$ the C^*-àlgebra generated by the operators $d(b)$ (48.3).

Now the Theorem 46.11 (applied to the operators of the form $d(b)+Q$) gives us the symbolic calculus of G-Toeplitz operators.

Theorem 48.5 *Let G be a discrete amenable group acting normally on Γ via diffeomorphisms $\{\alpha_g\}_{g \in G}$ preserving the orientation. Then the mapping*

$$\sigma : \mathscr{B} \to B$$

defined by

$$\sigma(d(b)) = b$$

*is a *-epimorphism, the kernel of which is the ideal $\mathscr{K}(H^+)$ of compact operators in H^+.*

In particular the G-Toeplitz operator $d(b)$ is Fredholm iff b is invertible.

48.6 In view of this theorem all the examples considered in Sections 46 and 47, where the explicit conditions for the invertibility of the operators $b \in B$ were written out, can serve as the examples where the explicit conditions of G-Toeplitz operators $d(b)$ to be Fredholm could be obtained.

Moreover since the index calculation of a G-SIO was reduced to that of an operator of the form $bP+Q$, the reasoning of 46.4 implies

$$\operatorname{ind}(bP+Q) = \operatorname{ind} d(b) \tag{48.4}$$

(in the notation of 46.4 the operator d_{11} in the decomposition of an arbitrary operator d is exactly the corresponding Toeplitz operator), which means that the results on the index calculation of $bP+Q$ (Section 47) can be reinterpreted as results concerning the index calculation of the G-Toeplitz operator $d(b)$.

Remark It is worth mentioning that in Section 47 we considered the case of the orientation-preserving diffeomorphisms $\{\alpha_g\}_{g\in G}$ not for certain mathematical reasons but simply to shorten the exposition (the methods developed there could be successfully applied to the situation when α_g changes the orientation). On the other hand, if we pass to the Fredholm theory of G-Toeplitz operators then we do not have to consider changing the orientation diffeomorphisms since in this case we have (by Theorem 2.51)

$$d(T_\alpha) = PT_\alpha \in \mathcal{K}(H^+) \, .$$

Let us now write out a number of results for G-Toeplitz operators in explicit form.

Theorem 48.7 *Let* $\Gamma = \{z \in \mathbb{C}: |z| = 1\}$ *and* $\alpha: \Gamma \to \Gamma$ *be a rotation having the form*

$$\alpha(z) = \mathrm{e}^{\mathrm{i}2\pi h} z \, .$$

Then

I. *If h is an irrational number then the G-Toeplitz operator*

$$\mathcal{b} = P(a_0 + a_1 T_\alpha), \quad a_i \in C(\Gamma) \tag{48.5}$$

is Fredholm in $H^+ (= H^2)$ *iff one of the following conditions is satisfied:*

(i) $a_0(z) \neq 0$ *on* Γ *and* $S(a_0) > S(a_1)$; *or*

(ii) $a_1(z) \neq 0$ *on* Γ *and* $S(a_1) > S(a_0)$

where $S(a) = \exp \frac{1}{2\pi} \int_0^{2\pi} \ln\left|a\left(\mathrm{e}^{\mathrm{i}t}\right)\right| \mathrm{d}t$ *is the geometric mean of a over* Γ.

The index of $\mathcal{b}$ *is expressed by*

$$\operatorname{ind} \mathcal{b} = \begin{cases} -\operatorname{ind} a_0 & \textit{in the case (i)} \\ -\operatorname{ind} a_1 & \textit{in the case (ii),} \end{cases}$$

(recall (48.2)).

II. *If* $h=\dfrac{m}{N}$ *is an irreducible fraction then the operator* $\mathcal{T}$ *(48.5) is Fredholm iff*

$$\varphi(z)\equiv\prod_{k=0}^{N-1}a_0\left(ze^{i2\pi kh}\right)+(-1)^{N-1}\prod_{k=0}^{N-1}a_1\left(ze^{i2\pi kh}\right)\neq 0$$

and under this condition

$$\operatorname{ind}\mathcal{T}=-\frac{1}{N}\operatorname{ind}\varphi\,.$$

Proof Follows from Theorems 48.5, 10.8, 47.7, Corollary 15.6 and (48.4). □

Theorem 48.8 *Let* Γ *be a simple closed Lyapunov curve and* α *be an orientation-preserving diffeomorphism which has periodic points of period N and let the set* Γ_N *of periodic points be nowhere dense.*

The G-Toeplitz operator (48.5) is Fredholm in H^+ *iff one of the following conditions (i) or (ii) is satisfied:*

(i) $a_0(x)\neq 0,\ x\in\Gamma$ *and*

$$\left|\prod_{k=0}^{N-1}a_0\left(\alpha^k(x)\right)\right|>\left|\prod_{k=0}^{N-1}a_1\left(\alpha^k(x)\right)\right|,\qquad x\in\Gamma_N\,,$$

(ii) $a_1(x)\neq 0,\ x\in\Gamma$ *and*

$$\left|\prod_{k=0}^{N-1}a_0\left(\alpha^k(x)\right)\right|<\left|\prod_{k=0}^{N-1}a_1\left(\alpha^k(x)\right)\right|,\qquad x\in\Gamma_N\,.$$

Under these conditions we have

$$\operatorname{ind}\mathcal{T}=\begin{cases}-\operatorname{ind}a_0 & \textit{in the case (i)}\\ -\operatorname{ind}a_1 & \textit{in the case (ii).}\end{cases}$$

Proof Follows from Theorem 46.16 and (48.4). □

Theorem 48.9 *Let* Γ *be a simple closed Lyapunov curve and* $\alpha:\Gamma\to\Gamma$ *be an orientation-preserving diffeomorphism such that the set* Γ_1 *of its fixed points has a non-*

empty interior $\overset{\circ}{\Gamma}_1$ *and* $\Gamma_1 \neq \Gamma$. *The G-Toeplitz operator* $\mathscr{b}$ *(48.5) is Fredholm in* H^+ *iff the following conditions are satisfied:*

(i) $a_0(x) + a_1(x) \neq 0, \quad x \in \overset{\circ}{\Gamma}_1,$

(ii) *the set* $\Gamma_\infty = \Gamma \setminus \overset{\circ}{\Gamma}_1$ *decomposes into two non-intersecting open-closed sets (in* Γ_∞*)* $\Gamma_\infty = \Gamma_\infty^1 \cup \Gamma_\infty^2$ *(one of which may be empty) such that*

(a) $a_0(x) \neq 0$ *on* Γ_∞^1 *and*

$$|a_0(x)| > |a_1(x)| \quad \textit{for every } x \in \Gamma_1 \cap \Gamma_\infty^1,$$

(b) $a_1(x) \neq 0$ *on* Γ_∞^2 *and*

$$|a_0(x)| < |a_1(x)| \quad \textit{for every } x \in \Gamma_1 \cap \Gamma_\infty^2.$$

Proof Follows from Example 47.16. □

48.10 Let

$$D = \{z \in \mathbb{C} : |z| \leq 1\}$$

and $M \subset \Gamma \times D$ be the set (introduced in 47.13) of the form

$$M = \{(x,z): \; z = 0, \text{ if } x \in \overset{\circ}{\Gamma}_\infty; \quad z = 1, \text{ if } x \in \overset{\circ}{\Gamma}_1; \quad z \in D, \text{ if } x \in \partial\left(\overset{\circ}{\Gamma}_\infty\right)\}.$$

With every operator $\mathscr{b}$ (48.5) satisfying the conditions of Theorem 48.9 we shall associate the function (non-vanishing on M)

$$\Phi_{\mathcal{L}}(x,z)=\begin{cases} a_0(x)+a_1(x), & x\in\overset{\circ}{\Gamma}_1,\ z=1, \\ a_0(x), & x\in\Gamma_\infty^1,\ z=0, \\ a_1(x), & x\in\Gamma_\infty^2,\ z=0, \\ a_0(x)+za_1(x), & x\in\partial\left(\Gamma_\infty^1\right),\ |z|\le 1, \\ za_0(x)+a_1(x), & x\in\partial\left(\Gamma_\infty^2\right),\ |z|\le 1. \end{cases} \tag{48.6}$$

Following the procedure presented in the proof of Theorem 47.14 we shall define ind $\Phi_{\mathcal{L}}$; namely, we shall implement the homotopy of $\Phi_{\mathcal{L}}$ (preserving the continuity and invertibility conditions) up to a function Φ^1 not depending on z which therefore could be considered as a continuous invertible function on Γ and thus ind Φ^1 is well defined.

Let us describe this homotopy. At the points $x\in\partial\left(\overset{\circ}{\Gamma}{}_\infty^{j}\right)$, $j=1,2$, we have

$$|a_0(x)|\ne|a_1(x)|$$

and consequently there exists $\delta>0$ such that this inequality is true in the δ-neighbourhood of the set $\partial\left(\overset{\circ}{\Gamma}_\infty\right)$. Let ψ_j, $j=1,2$, be continuous functions on Γ satisfying the conditions

$$\begin{cases} \psi_j(x)=1 & \text{on a } \delta/2-\text{neighbourhood of } \partial\left(\overset{\circ}{\Gamma}{}_\infty^{j}\right), \\ 0\le\psi_j(x)\le 1, & \\ \psi_j(x)=0 & \text{outside a } \delta-\text{neighbourhood of } \partial\left(\overset{\circ}{\Gamma}{}_\infty^{j}\right). \end{cases}$$

Now the family of operators b_τ, $0\le\tau\le 1$, of the form

$$b_\tau=\left[1-\tau\psi_2(x)\right]a_0(x)+\left[1-\tau\psi_1(x)\right]a_1(x)T_\alpha$$

consists of invertible operators (since, for every τ, the invertibility conditions for these operators are satisfied) and depends continuously on τ.

We have $b_0=\sigma(\mathcal{L})$ - the symbol of $\mathcal{L}$ - and the function Φ^1 constructed from the operator b_1 has the form

$$\Phi^1(x,z)=\begin{cases}[1-\psi_2(x)]a_0(x)+[1-\psi_1(x)]a_1(x), & x\in\mathring{\Gamma}_1,\\ a_0(x), & x\in\overline{\Gamma^1_\infty},\\ a_1(x), & x\in\overline{\Gamma^2_\infty},\end{cases}$$

and does not depend on z and is continuous on x.

Thus ind Φ^1 is well defined. So we set

$$\operatorname{ind}\Phi_{\mathfrak{b}} \overset{\text{def}}{=} \operatorname{ind}\Phi^1\,. \tag{48.7}$$

Now Theorem 47.14 implies

Theorem 48.11 *Let the operator* $\mathfrak{b}$ *(48.5) satisfy the conditions mentioned in the statement of Theorem 48.9. Then*

$$\operatorname{ind}\mathfrak{b}=-\operatorname{ind}\Phi_{\mathfrak{b}}$$

where ind $\Phi_{\mathfrak{b}}$ *is defined by (48.7).*

48.12 If one deals with the classical Toeplitz operators (in H^2) then in a number of cases it is not only possible to obtain the Fredholmity conditions and index formula but to find explicitly the *defect numbers*, that is dim Ker $\mathfrak{a}$ and dim Ker $\mathfrak{a}^*$. A typical situation is when these numbers have minimal values (allowable by the meaning of ind $\mathfrak{a}$), that is

$$\dim\operatorname{Ker}\mathfrak{a}=0 \text{ when } \operatorname{ind}\mathfrak{a}\le 0$$

and (48.8)

$$\dim\operatorname{Ker}\mathfrak{a}^*=0 \text{ when } \operatorname{ind}\mathfrak{a}\ge 0\,.$$

If we deal with *G*-Toeplitz operators then calculation of the defect numbers turns out to be a more complicated problem. Let us present now certain examples of *G*-Toeplitz operators where it is possible to calculate the defect numbers and property (48.8) holds.

Example 48.13 Consider the operator

$$\mathfrak{b}=P\left(t^{k_0}u(t)+\lambda t^{k_1}u\left(e^{i2\pi h}t\right)\right),\qquad k_0,k_1\in\mathbb{Z},\lambda\in\mathbb{C} \tag{48.9}$$

acting in the Hardy space H^2 (that is $\Gamma = \{t \in \mathbb{C} : |t| = 1\}$).

If h is an irrational number then by Theorem 48.7, I, $\mathfrak{b}$ is Fredholm iff $|\lambda| \neq 1$ (since $S(t^k) = 1$) and under this condition

$$\operatorname{ind} \mathfrak{b} = \begin{cases} -k_0, & \text{if } |\lambda| < 1, \\ -k_1, & \text{if } |\lambda| > 1. \end{cases}$$

If $h = \dfrac{m}{N}$ is an irreducible fraction then by Theorem 48.7, II, $\mathfrak{b}$ is Fredholm iff

$$\prod_{k=0}^{N-1} \mathrm{e}^{\mathrm{i}2\pi k_0 kh} \neq (-1)^N \lambda^N t^{N(k_1 - k_0)} \prod_{k=0}^{N-1} \mathrm{e}^{\mathrm{i}2\pi k_1 kh}, \qquad t \in \Gamma . \tag{48.10}$$

If $k_1 \neq k_0$ then the set of values of the function $t^{N(k_1 - k_0)}$ coincides with the whole of the unit circle and thus if $|\lambda| = 1$ then condition (48.10) is violated at some $t \in \Gamma$. This means that the operator $\mathfrak{b}$ in the case when h is rational and $k_1 \neq k_0$ is Fredholm iff $|\lambda| \neq 1$ and under this condition Theorem 48.7, II, implies

$$\operatorname{ind} \mathfrak{b} = \begin{cases} -k_0, & \text{if } |\lambda| < 1, \\ -k_1, & \text{if } |\lambda| > 1. \end{cases}$$

If $k_1 = k_0$ then condition (48.10) is equal to $(-\lambda)^N \neq 1$ and again by Theorem 48.7, II, we have

$$\operatorname{ind} \mathfrak{b} = -k_0 .$$

Theorem 48.14 *Let the operator $\mathfrak{b}$ (48.9) be Fredholm. Then*

$$\dim \operatorname{Ker} \mathfrak{b} = 0 \text{ *when* } \operatorname{ind} \mathfrak{b} \leq 0$$

and

$$\dim \operatorname{Ker} \mathfrak{b}^* = 0 \text{ *when* } \operatorname{ind} \mathfrak{b} \geq 0 .$$

Proof As is known (see 33.19) the Taylor series expansion

$$u(t) = \sum_{k=0}^{\infty} c_k t^k$$

establishes an isomorphism between H^2 and the space l^2 of unilateral sequences by the rule

$$H^2 \ni u \xrightarrow{\mathcal{J}} \{c_0, c_1, \ldots\} \in l^2 .$$

Under this isomorphism the operator $(T_h u)(t) = u\left(e^{i2\pi h} t\right)$ is transformed into the operator of multiplication by the sequence $\tilde{T}_h = \left\{e^{i2\pi jh}\right\}_{j=\overline{0,\infty}}$ and the operator $P \circ t^k$ is transformed into the shift operator V_k which shifts a sequence by $|k|$ positions to the right if $k > 0$ and by $(-k)$ positions to the left if $k < 0$. Therefore the operator ℓ under the isomorphism $\mathcal{J}$ is transformed into the operator

$$\tilde{\ell} = V_{k_0} + \lambda V_{k_1} \tilde{T}_h . \tag{48.11}$$

The properties of these operators are well studied (see, for example, Karapetyants, Samko [3], §15, 3°) and the statement of the theorem follows from the known results. Let us consider one of the possible cases. Let, say, $|\lambda| > 1$ and $k_0 < 0 < k_1$.

The reasoning of Example 48.13 shows that in this situation we have

$$\operatorname{ind} \ell = -k_1 < 0 . \tag{48.12}$$

Set

$$\nu = k_1 - k_0 .$$

Then the operator $\tilde{\ell}$ (48.11) maps a sequence $\xi = \{c_0, c_1, \ldots\} \in l^2$ into the sequence

$$\tilde{\ell}\xi = \left\{c_{-k_0}, c_{-k_0+1}, \ldots, c_\nu + \lambda c_0, c_{\nu+1} + \lambda e^{i2\pi h} c_1, \ldots, c_{\nu+k} + \lambda e^{i2\pi kh} c_k, \ldots\right\} . \tag{48.13}$$

Consider the equation

$$\tilde{\ell}\xi = 0 .$$

For this equation (48.13) implies

$$c_{-k_0} = c_{-k_0+1} = \ldots = c_{\nu-1} = 0 \tag{48.14}$$

and

$$c_{\nu+k} + \lambda e^{i2\pi kh} c_k = 0, \quad k = 1, 2, \ldots . \tag{48.15}$$

Equations (48.15) in turn imply

$$c_{\nu+k} = -\lambda^{i2\pi kh} c_k$$

which, along with (48.14), means that

$$c_{n\nu+k} = 0, \ k = -k_0, -k_0+1, \ldots, \nu-1$$

and (48.16)

$$c_{n\nu+k} = (-\lambda)^n \prod_{j=0}^{n-1} e^{i2\pi(j\nu+k)h} c_k, \quad k = 0, \ldots, -k_0 - 1 .$$

Since $|\lambda| > 1$ it follows that if $c_k \neq 0$ for some k, $0 \leq k \leq -k_0 - 1$, then for this k the sequence $c_{n\nu+k}$ is unbounded and thus ξ does not belong to l^2. Thus $c_k = 0, k = 0, \ldots, \nu-1$ and in view of (48.16) we conclude that $c_k = 0$ for all k (that is, Ker $\tilde{\mathfrak{b}} = 0$). By (48.12) the theorem is proved for the situation considered.

Other possibilities could be examined in an analogous way. □

49. Boundary value problems with strong non-localness

In this section we consider the Fredholm theory of boundary value problems for elliptic equations in which the boundary condition contains transformation-of-an-argument operators. The typical simplest example of such conditions is the boundary condition of the form

$$\left[a_0(x)u(x) + a_1(x)u(\gamma(x))\right]_{x\in\Gamma} = f_1(x) \tag{49.1}$$

$$\left[\alpha_0(x,D)u(x) + \alpha_1(x,D)u(\gamma(x))\right]_{x\in\Gamma} = f_2(x) \tag{49.2}$$

where γ is a given mapping of the boundary Γ of a domain Ω into Γ or into the interior of Ω, a_0 and a_1 are given functions, and α_0 and α_1 are given pseudodifferential operators.

Remark It is worth mentioning that non-local problems with conditions of the form (49.1) and (49.2) are a natural generalization of the problem, posed by T. Carleman [1], of finding an analytic function satisfying the condition

$$\Phi(z) + G(z)\Phi(\gamma(z)) = \varphi(z)$$

on the boundary of a domain (recall the Introduction).

The properties of non-local problems with conditions of the form (49.1) or (49.2) depend in an essential way on the properties of the mapping γ. If γ acts from the boundary into the domain, then the condition (49.1) connects a point x only with the point $\gamma(x)$. If γ maps the boundary Γ into itself and is a periodic mapping (that is, $\gamma^n(x) \equiv x$ for some $n \in \mathbb{N}$) then the group $\{\gamma^k\}$ is finite and, as will be seen below, the non-local problem can be investigated by means of certain modifications of the usual methods of the theory of elliptic problems, and from this viewpoint the non-localness is weak. Non-localness in a boundary condition is said to be *strong* if this condition connects the point x with an infinite number of other points. For example, the condition (49.1) gives strong non-localness if the group $\{\gamma^k\}$ of the mappings of Γ is infinite (γ is not periodic).

In this section we obtain the Fredholm theory for a certain class of boundary value problems with strong non-localness for elliptic equations. On the basis of this problem we construct an auxiliary *G*-PDO (*G*-Calderon operator), the Fredholmity of which is equivalent to the Fredholmity of the problem being investigated, and since the Fredholm theory for *G*-PDOs has already been built (in sections 40-42) it gives us a tool to complete the job.

This section also contains certain examples of the explicit calculation of the Fredholmity conditions of non-local boundary value problems.

Remark We would like to emphasize at once that the ideas presented in this section constitute in a sense the 'elementary' Fredholm theory of non-local boundary value problems.

The general theory of non-local boundary value problems (involving the Boutet de Monvel operators and boundary conditions with shifts 'concentrated' only on the boundary) is obtained in Chapter 7.

49.1 So let us discuss the precise formulation of the problem. Let Ω be a closed bounded domain in $\mathbb{R}^{n+1}$ with smooth boundary Γ and let $\mathcal{P}$ be an elliptic differential operator of order m in Ω with smooth coefficients. Let

$$\gamma_k : \Gamma \to \Gamma, \quad k = \overline{1,l}\,, \tag{49.3}$$

be diffeomorphisms of the boundary Γ and let U_{γ_k} be the operators generated by them acting via the formula

$$\left[U_{\gamma_k} f\right](x) = f(\gamma_k(x)) . \tag{49.4}$$

Let Y_k, $k = \overline{l+1,\ldots,l_1}$, be given n-dimensional compact smooth submanifolds of the interior $\overset{\circ}{\Omega}$ of Ω and

$$\gamma_k : \Gamma \to Y_k, \quad k = \overline{l+1, l_1} \tag{49.5}$$

be given diffeomorphisms and

$$\left(U_{\gamma_k} f\right)(x) = f(\gamma_k(x)) \tag{49.6}$$

be the operators generated by them (here U_{γ_k} acts from a space of functions on Y_k into a space of functions on Γ).

Let $b_{kj}(x, D)$, $k = 1,\ldots, l$; $j = 1,\ldots, \nu$, be differential operators of order m_j defined on Γ and $b_{kj}(x, D)$, $k = l+1,\ldots, l_1$, be differential operators of order m_j, $j = 1,\ldots, \nu$, defined on Y_k.

We adjoin to the elliptic equation

$$pu = f \tag{49.7}$$

the non-local conditions of the form

$$b_j u \equiv b_j^0 u + b_j^1 u = f_j, \quad j = 1,\ldots,\nu \tag{49.8}$$

where

$$b_j^0 u \equiv \sum_{k=1}^{l} U_{\gamma_k}\left(b_{kj}(x, D) u|_{\Gamma}\right) \tag{49.9}$$

$$b_j^1 u \equiv \sum_{k=l+1}^{l_1} U_{\gamma_k}\left(b_{kj}(x, D) u|_{Y_k}\right) \tag{49.10}$$

and pose the non-local problem:

find a solution of equation (49.7) satisfying the boundary conditions (49.8).

We shall consider this problem in Sobolev spaces: assuming that $f \in H^{s-m}(\Omega)$ and $f_j \in H^{s-m_j-\frac{1}{2}}(\Gamma)$, where $s \geq m, s \geq m_j+1$, we seek a solution in $H^s(\Omega)$ (thus the notation $u|_\Gamma$ (and $u|_{Y_k}$) means the application of the operator of taking the boundary values of a function u (and the values of u over Y_k respectively)).

Let us introduce the spaces

$$\tilde{H}(\Gamma) = \bigoplus_{j=1}^{\nu} H^{s-m_j-\frac{1}{2}}(\Gamma)$$

$$\tilde{H}^s(\Omega) = H^{s-m}(\Omega) \oplus \tilde{H}(\Gamma) .$$

We fix the numbers s and, to the boundary problem (49.7), (49.8), we assign the operator

$$L_s u \equiv \left(\mathcal{p}u, \mathcal{b}_1 u, \ldots, \mathcal{b}_\nu u\right) \tag{49.11}$$

acting from $H^s(\Omega)$ into $\tilde{H}^s(\Omega)$. This operator is bounded.

We call the *boundary value problem* (49.7), (49.8) *Fredholm* if the operator L_s corresponding to it is Fredholm.

G-Calderon operator (of a non-local boundary value problem)

49.2 When investigating the Fredholmity conditions of ordinary boundary value problems for elliptic equations one uses the method of reduction to the boundary. In this method, on the basis of the boundary value problem a pseudodifferential operator is constructed on the boundary; this operator is called the *Calderon operator* which is Fredholm iff the initial problem is Fredholm (see, for example, Treves [1], Chapter III, §3). We shall now carry out the analogous constructions for the non-local problem (49.7), (49.8) using a number of preparatory facts from Treves [1], Chapter III, §2.

Let Z be some open tube neighbourhood of the boundary Γ in $\mathbb{R}^{n+1}$. We denote by t a certain transversal to the Γ coordinate such that Γ is given by the equation $t=0$ and $t>0$ on $Z \cap \mathring{\Omega}$. Let us take a number $\tau_0 > 0$ such that the inequality

$$|t| \leq \tau_0$$

defines a compact tube neighbourhood of Γ contained in Z. We denote this neighbourhood by Ω_0.

We shall presuppose that the operator p and the boundary operators b_{kj}, $k=\overline{1,l}$, satisfy the conditions pointed out in Treves [1], Chapter III, §3, and we assume that in the neighbourhood Ω_0 the operator p has the form

$$p = \partial_t^m + \sum_{r=1}^{m} p_r(x,t,D_x)\partial_t^{m-r}$$

where $p_r(x,t,D_t)$ is a classical pseudodifferential operator on Γ of order r for each t (see Treves [1], Chapter I, §5).

The principal symbol of the operator p in the neighbourhood Ω_0 has the form

$$\sigma(p) = z^m + \sum_{r=1}^{m} p_r(x,t,\xi)z^{m-r}$$

where $p_r(x,t,\xi)$ is the principal symbol of the operator $p_r(x,t,D_x)$.

The ellipticity condition of p can be written out in the following form: there exist integers m^+ and m^- satisfying the equation

$$m^+ + m^- = m, \quad m^- \geq 1,$$

such that $\sigma(p)$, as the polynomial on z, has m^+ roots with positive real part and m^- roots with negative real part for every (x,t) and every cotangent vector $\xi \neq 0$.

Henceforth we assume that $\nu = m^-$ (ν is the number of the conditions (49.8)).

As is shown in Treves [1], Chapter III, §3, the symbol $\sigma(p)$ decomposes into the factors

$$\sigma(p) = \sigma(M^+)\sigma(M^-)$$

where

$$\sigma(M^\pm) = z^{m^\pm} + \sum_{r=1}^{m^\pm} M_r^\pm(x,t,\xi)z^{m^\pm - r}$$

and the coefficients $M_r^\pm(x,t,\xi)$ are infinitely differentiable and all the roots of $\sigma(M^+)$ (as a polynomial in z) have positive real part and all the roots of $\sigma(M^-)$ have negative real part.

The decomposition of $\sigma(p)$ generates the decomposition

$$\mathscr{p} = M^{+}M^{-} + K$$

where $M^{\pm}$ are operators with the principal symbols $\sigma(M^{\pm})$ and K is a smoothing operator (of order $-\infty$).

We presuppose also that the operators $\mathscr{b}_{kj}$ have the similar form

$$\mathscr{b}_{kj} = \sum_{r=0}^{m_j} b_{kjr}(x, D_x)\partial_t^r, \quad k = 1, \ldots, l;\ j = 1, \ldots, \nu,$$

where $b_{kjr}(x, D_x)$ is a classical pseudodifferential operator on Γ of order $m_j - r$ with smooth coefficients. If k is fixed then the operators $\mathscr{b}_{kj}$, $j = \overline{1, \nu}$, define the local boundary value problem for the equation (49.7).

According to the construction written out in Treves [1], Chapter III, §3, on the basis of each of these boundary value problems we build the Calderon operator $\mathscr{a}_k$ on the boundary.

We recall this construction now.

49.3 First we 'divide' $\mathscr{b}_{kj}$ by $\mathscr{p}$, that is we represent $\mathscr{b}_{kj}$ in the form

$$\mathscr{b}_{kj} = \mathscr{q}'_{kj}\mathscr{p} + \mathscr{b}'_{kj}$$

where the degree of $\mathscr{b}'_{kj}$ (as a polynomial on ∂_t) is less than m.

Second we 'divide' the operator $\mathscr{b}'_{kj}$ by M^{-}, that is we represent it in the form

$$\mathscr{b}'_{kj} = \mathscr{q}''_{kj}M^{-} + \mathscr{b}''_{kj}$$

where the degree of $\mathscr{b}''_{kj}$ as a polynomial on ∂_t satisfies the conditions

$$\deg \mathscr{b}''_{kj} \le m^{-} - 1 = \nu - 1\,,$$

$$\deg \mathscr{q}''_{kj} \le m^{+} - 1\,.$$

The operators $\mathscr{b}''_{kj}$ have the form

$$\mathscr{b}''_{kj} = \sum_{r=1}^{\nu} a_{kjr}(x, t, D_x)\partial_t^{r-1}\,,$$

where a_{kjr} are pseudodifferential operators of order $m_j - r + 1$.

Let Λ be pseudodifferential operator of order 1 with principal symbol $\|\xi\|$ which establishes an isomorphism between the spaces $H^s(\Gamma)$ and $H^{s-1}(\Gamma)$ for any s (see Treves [1], Chapter III, §3). In view of Theorem 2.48 the operator $U_{\gamma_k}^{-1}\Lambda^{m_j}U_{\gamma_k}$ is a pseudodifferential operator of order m_j and therefore the operator

$$R_{kj} = U_{\gamma_k}^{-1}\Lambda^{m_j}U_{\gamma_k}\Lambda^{-m_j} \tag{49.12}$$

is a pseudodifferential operator of order 0 with the symbol

$$\sigma\left(R_{k_j}\right) = \|\partial\gamma_k(\eta)\|^{m_j}, \quad \eta \in S^*\Gamma . \tag{49.13}$$

Let us define the pseudodifferential operator of order 0 acting in $\left[H^{s-\frac{1}{2}}(\Gamma)\right]^\nu$ by means of the operator matrix

$$\left[a_k\right]_{jr} = R_{kj}a_{kjr}(x,0,D_x)\Lambda^{r-m_j+1}, \quad j=\overline{1,\nu};\ r=\overline{1,\nu} . \tag{49.14}$$

The operator a_k is the *Calderon operator* (in fact one of several possible ones, since a Calderon operator is defined (Treves [1], Chapter III, §3) up to an elliptic factor) corresponding to the local boundary value problem for equation (49.7) with the boundary operators b_{kj}.

49.4 The *G-Calderon operator* of the non-local boundary problem (49.7), (49.8) is the operator

$$b_s = \sum_{k=1}^{l} U_{\gamma_k}a_k \tag{49.15}$$

acting in the space $\left[H^{s-\frac{1}{2}}(\Gamma)\right]^\nu$.

Remark As will be shown below (in the proof of Theorem 49.5) the non-localness generated by the mappings acting into the interior of the domain turns out to be weak from another point of view: the corresponding terms in the boundary conditions do not affect the Fredholmity of the problem. Therefore, in particular, these terms are not contained in the construction of the G-Calderon operator.

We denote by Q_s the operator from $H^s(\Omega)$ into $H^{s-m}(\Omega) \oplus \left[H^{s-\frac{1}{2}}(\Gamma)\right]^\nu$ corresponding to the usual boundary value problem for equation (49.7) with the boundary conditions

$$\Lambda^{1-j} \partial_t^{j-1}\Big|_\Gamma = f_j, \quad j = \overline{1,\nu}\,. \tag{49.16}$$

The next result is the main technical result of this section. It shows that the investigation of the Fredholmity conditions of the boundary value problems considered here can be reduced to that for the G-PDOs on the boundary (and thus gives us the possibility of applying the theory already developed (see Sections 40-42)).

Theorem 49.5 *Let the above-mentioned assumptions regarding the domain Ω, the mappings γ_k and the operators $\not{p}$ and ℓ_{kj} be satisfied. The operator L_s (49.11) corresponding to the non-local boundary value problem (49.7), (49.8) is Fredholm iff the G-Calderon operator ℓ_s (49.15) is Fredholm. Moreover*

$$\operatorname{ind} L_s = \operatorname{ind} \ell_s + \operatorname{ind} Q_s\,.$$

Proof Let

$$E_2 = \left\{u \in H^s(\Omega) : \not{p} u = 0\right\}$$

and let E_1 be the orthogonal complement to E_2 in $H^s(\Omega)$.

The decompositions

$$H^s(\Omega) = E_1 \oplus E_2$$

$$\tilde{H}^s(\Omega) = H^{s-m} \oplus \tilde{H}^s(\Gamma)$$

(recall 49.2) generate a representation of L_s by means of the triangular operator matrix

$$L_s = \begin{bmatrix} L_{11} & 0 \\ L_{12} & L_{22} \end{bmatrix} \tag{49.17}$$

where L_{11} is the restriction of the operator $\mathcal{p}$ onto E_1 (L_{11} does not depend on the boundary condition), L_{22} is the restriction of the operator L_s onto E_2 (L_{22} depends on $\mathcal{p}$ and the boundary operators $\mathcal{b}_j$).

The representation of $\mathcal{b}_j$ as a sum

$$\mathcal{b}_j = \mathcal{b}_j^0 + \mathcal{b}_j^1$$

(see (49.8)) generates a representation of the operator L_{22} in the form

$$L_{22} = L_{22}^0 + L_{22}^1 \tag{49.18}$$

where L_{22}^0 is the term generated by the operators $\mathcal{b}_j^0$ and L_{22}^1 is that generated by $\mathcal{b}_j^1$.

To continue the proof we shall first prove the next lemma which is useful in its own right.

Lemma 49.6 *Let Y be an arbitrary n-dimensional compact submanifold of Ω (lying in the interior of Ω), let $\gamma: \Gamma \to Y$ be a diffeomorphism and let $\mathcal{b}$ be a pseudodifferential operator of order r on Ω. Let C be the operator from $H^s(\Omega)$ into $H^{s-r-\frac{1}{2}}(\Gamma)$ of the form*

$$Cu = U_\gamma\left(\mathcal{b}u\big|_Y\right)$$

and let C_0 be the restriction of C onto the subspace E_2 introduced in the proof of the theorem). Then the operator C_0 is compact.

Proof Since $\mathcal{p}$ is an elliptic operator with C^∞-smooth coefficients it follows (Treves [1], Chapter I, §2) that it is hypoelliptic. Therefore all the functions in E_2 are infinitely differentiable inside of Ω.

Let Ω_1 be a subdomain of Ω such that

$$Y \subset \Omega_1 \subset \overline{\Omega}_1 \subset \Omega .$$

We consider on E_2 the operator $\mathcal{J}$ of restriction of functions onto Ω_1. Let us fix a number $q > s$ and consider the operator $\mathcal{J}$ as acting from E_2 into $H^q(\Omega_1)$. This operator is defined on the whole of E_2 and is linear. Let us verify that it is closed.

Let $\{u_k\}$ be a sequence in E_2 such that

$$u_k \to u_0 \text{ in } E_2$$

and

$$\mathcal{J}u_k \to v \text{ in } H^q(\Omega_1) .$$

Since $\mathcal{J}$ acts continuously from E_2 into $H^s(\Omega_1)$ it follows that

$$\mathcal{J}u_k \to \mathcal{J}u_0 \text{ in } H^s(\Omega_1) .$$

Since $\mathcal{J}u_k \to v$ in $H^q(\Omega_1)$ it follows that $\mathcal{J}u_k \to v$ in $H^s(\Omega_1)$ and in view of the foregoing observation this means that

$$\mathcal{J}u_0 = v ,$$

and the closedness of $\mathcal{J}$ is proved.

Now by the closed graph theorem we conclude that $\mathcal{J}$ is bounded.

The operator C can be represented as a product by means of the following commutative diagram:

$$\begin{array}{ccccc}
\mathcal{J}E_2 & & \xrightarrow{\qquad\qquad C \qquad\qquad} & & H^{s-r-\frac{1}{2}}(\Gamma) \\
\downarrow & & & & \uparrow W \\
H^q(\Omega_1) & \xrightarrow{b|_Y} & H^{q-r-\frac{1}{2}}(Y) & \xrightarrow{U_\gamma} & H^{q-r-\frac{1}{2}}(\Gamma)
\end{array}$$

where all the operators are bounded and the operator W (the operator of embedding of $H^{q-r-\frac{1}{2}}(\Gamma)$ into $H^{s-r-\frac{1}{2}}(\Gamma)$) is compact (as $q > s$).

Thus C is compact and the lemma is proved. □

Now we continue the proof of the theorem.

Let us show that the terms in the boundary conditions containing the operators b_{kj} generated by the mappings γ_k acting into the interior of the domain ($k = \overline{l+1, l_1}$) do not affect the Fredholmity and the index of the problem.

Under the presupposed assumptions the operator L_{11} arising in the representation of L_s by means of the operator-matrix (49.17) is Fredholm. Indeed, for the operator p there exists a Fredholm boundary value problem (for example the problem with conditions (49.16)). Thus the image of p contains a closed subspace of $H^{s-m}(\Omega)$ having (in

$H^{s-m}(\Omega)$) a finite-dimensional complement. Therefore the image of $\mathcal{P}$ itself is closed and has a finite-dimensional complement. Since, by the definition of E_2 we have Ker $L_{11}=0$ it follows that L_{11} is Fredholm.

Now as the operator-matrix (49.17) is triangular and the operator L_{11} (on the diagonal) is Fredholm we conclude that L_s is Fredholm iff the operator $L_{22}=L_{22}^0+L_{22}^1$ (49.18) is Fredholm. But by Lemma 49.6 the operator L_{22}^1 is compact and therefore L_{22} is Fredholm iff the operator L_{22}^0 is Fredholm.

Remark Note that in general the Fredholmity of a triangular operator-matrix does not imply the Fredholmity of the diagonal elements and thus in the foregoing reasoning the Fredholmity of L_{11} is essential.

Let L_s^0 be the operator from $H^s(\Omega)$ in $\tilde{H}^s(\Omega)$ corresponding to the boundary value problem with boundary operators $\mathcal{b}_j^0$. The above argument shows that L_s is Fredholm iff L_s^0 is Fredholm (though the operator $L_s^1=L_s-L_s^0$ is not compact).

Now let $\overline{\Lambda}$ be the pseudodifferential operator acting from $\left[H^{s-\frac{1}{2}}(\Gamma)\right]^\nu$ into $\tilde{H}^s(\Gamma)$ via the formula

$$\overline{\Lambda}\overline{u}=\left(\Lambda^{m_1}u_1,\dots,\Lambda^{m_\nu}u_\nu\right).$$

The construction of the G-Calderon operator shows that

$$L_s^0=\left(I\oplus\overline{\Lambda}\mathcal{b}_s\right)Q_s+K$$

where K is a smoothing operator.

Thus, L_s is Fredholm iff $\mathcal{b}_s$ is Fredholm and

$$\text{ind } L_s=\text{ind } L_s^0=\text{ind } \mathcal{b}_s+\text{ind } Q_s\,.$$

The theorem is proved. □

This theorem, along with the results of Sections 40-42, makes it quite natural to introduce the

Symbol of a non-local boundary value problem

49.7 Let $\mathcal{b}_s$ (49.15) be the G-Calderon operator of the boundary value problem (49.7), (49.8). Then we shall call the $\left(s-\frac{1}{2}\right)$-symbol of $\mathcal{b}_s$ (see 41.8) the *s-symbol of the non-local boundary value problem* (49.7), (49.8).

Theorem 49.8 *Let the group G generated by diffeomorphisms γ_k, $k=\overline{1,l}$, of the boundary Γ be amenable and act normally on Γ. Then the non-local boundary value problem (49.7), (49.8) is Fredholm iff its s-symbol is invertible as an operator in the space $\left[L^2(S^*\Gamma)\right]^\nu$.*

Proof Just apply Theorems 49.5 and 41.9. □

Remark 49.9 The necessary and sufficient condition for a *local* boundary value problem to be Fredholm is the Lopatinskij condition (see, for example, Agranovich [1], §13) which can be verified locally. This condition is equivalent to the ellipticity of the Calderon operator associated (as the operators a_k) to the local boundary condition and does not depend on s. If the non-localness in the boundary condition (49.8) is strong (the diffeomorphisms γ_k, $k=\overline{1,l}$, generate an infinite group) then (as the Fredholmity conditions of the problem (49.7), (49.8) are linked with the Fredholmity conditions of G-PDOs (G-Calderon operators)) the Fredholmity conditions for the *non-local* boundary value problem depend on s (recall the examples in Sections 44, 45).

Now we proceed to the consideration of the Fredholmity conditions for some concrete examples.

Examples of non-local boundary value problems

49.10 Let

$$\Omega=\left\{x\in\mathbb{R}^3:\|x\|\le 1\right\}$$

be the ball in $\mathbb{R}^3$ and $l_0(x), l_1(x)$ be smooth unit vector fields defined on the boundary $\Gamma=S^2$ of the ball and $a_0(x), a_1(x)$ be given smooth functions and $\gamma:\Gamma\to\Gamma$ be a diffeomorphism.

For the Poisson equation in Ω,

$$\Delta u = f \ , \tag{49.19}$$

let us consider two types of non-local problems with the conditions

$$\left[a_0(x)u(x) + a_1(x)u(\gamma(x))\right]_\Gamma = f_1(x) \tag{49.20}$$

$$\left[a_0(x)\frac{\partial u}{\partial l_0}(x) + a_1(x)\frac{\partial u}{\partial l_1}(\gamma(x))\right]_\Gamma = f_2(x) \tag{49.21}$$

respectively.

For a given $s \geq 2$ we seek a solution $u \in H^s(\Omega)$ under the condition that

$$f \in H^{s-2}(\Omega),\ f_1 \in H^{s-\frac{1}{2}}(\Gamma),\ f_2 \in H^{s-\frac{3}{2}}(\Gamma)\,. \tag{49.22}$$

The manifold $S^*\Gamma$ can be embedded in a natural way into $\mathbb{R}^6$:

$$S^*\Gamma = \left\{(x,\xi) \in \mathbb{R}^3 \times \mathbb{R}^3 : \|x\| = 1, \|\xi\| = 1, \langle x,\xi\rangle = 0\right\}$$

where $\langle x,\xi\rangle = x_1\xi_1 + x_2\xi_2 + x_3\xi_3$.

The G-Calderon operator of the problem (49.19), (49.20) acts in $H^{s-\frac{1}{2}}(\Gamma)$ and has the form

$$\mathscr{b}_s = a_0(x)I + a_1(x)U_\gamma\,. \tag{49.23}$$

The G-Calderon operator of the problem (49.19), (49.21) acts in $H^{s-\frac{3}{2}}(\Gamma)$ and has the form

$$\mathscr{b}_s = a_0(x,D) + U_\gamma a_1(x,D) \tag{49.24}$$

where $a_k(x,D)$ are the pseudodifferential operators of zero order with the symbols

$$a_0(x,\xi) = a_0(x)\langle x + \mathrm{i}\xi, l_0(x)\rangle$$

and (49.25)

$$a_1(x,\xi) = a_1\left(\gamma^{-1}(x)\right)\|\partial\gamma(x)(\xi)\|\langle x + \mathrm{i}\xi, l_1(x)\rangle$$

respectively.

To begin with we consider the situation when γ is a rotation about the x_3 axis by an angle φ_0. Let

$$R_\varphi = \begin{bmatrix} \cos\varphi & \sin\varphi & 0 \\ -\sin\varphi & \cos\varphi & 0 \\ 0 & 0 & 1 \end{bmatrix}, \quad \gamma(x) = R_{\varphi_0}(x) .$$

For a function $a \in C(\Gamma)$ and a point $x \in \Gamma$ we set

$$S_x(a) = \exp\frac{1}{2\pi}\int_0^{2\pi} \ln\left|a\left(R_\varphi(x)\right)\right| d\varphi . \tag{49.26}$$

For a function $a \in C(S^*\Gamma)$ and a point $(x,\xi) \in S^*\Gamma$ we set

$$S_{x,\xi}(a) = \exp\frac{1}{2\pi}\int_0^{2\pi} \ln\left|a\left(R_\varphi x, R_\varphi \xi\right)\right| d\varphi . \tag{49.27}$$

Now for the problem (49.19), (49.20) we have the following

Theorem 49.11 *Suppose that in the condition (49.20) $\gamma(x) = R_{\varphi_0}(x)$, $\varphi_0 \neq 0$. Then for the non-local boundary value problem (49.19), (49.20) the following assertions are equivalent:*

(1) *Problem (49.19), (49.20) is Fredholm for some given s.*

(2) *Problem (49.19), (49.20) is Fredholm for every $s \geq 2$.*

(3) *Problem (49.19), (49.20) is uniquely and unconditionally solvable for any right-hand side for all s.*

(4) *The following conditions (depending on φ_0) are satisfied:*

(a) $\dfrac{\varphi_0}{2\pi} = \dfrac{M}{N}$ *is an irreducible fraction.*

The condition has the form:

$$\prod_{j=0}^{N-1} a_0\left(R_{j\varphi_0}(x)\right) \neq \prod_{j=0}^{N-1} a_1\left(R_{j\varphi_0}(x)\right)(-1)^N$$

for every $x \in \Gamma$.

(b) $\dfrac{\varphi_0}{2\pi}$ *is irrational.*

The condition has the form:

either

$$a_0(x) \neq 0 \text{ and } S_x(a_0) > S_x(a_1) \text{ for every } x \in \Gamma,$$

or

$$a_1(x) \neq 0 \text{ and } S_x(a_0) < S_x(a_1) \text{ for every } x \in \Gamma,$$

where $S_x(a)$ *is defined by (49.26).*

Proof Since the mapping $\gamma = R_\varphi$ preserves the measure and the natural metric on the sphere it follows that in the s-symbol formula for the G-Calderon operator (49.23) the corresponding factors $\|\partial\gamma(\eta)\|$ and $|\gamma'(x)|$ (recall the formula (41.18) and Remarks (2) following it) are equal to 1 and in particular the s-symbol does not depend on s.

The mapping $\beta: S^*\Gamma \to S^*\Gamma$ (generated by γ according to (40.3)) has the form

$$\beta(x,\xi) = \left(R_{\varphi_0}x, R_{\varphi_0}\xi\right).$$

If $\dfrac{\varphi_0}{2\pi} = \dfrac{M}{N}$ then β is periodic (of period N) and defines the topologically free action of a cyclic group $\mathbb{Z}_N$. Thus we can apply Corollaries 15.5, 15.6, and condition (4) (a) is the explicit record of the condition of the invertibility of the s-symbol

$$\det\left[\tilde{b}(x)\right] \neq 0$$

from Corollary 15.6.

If $\dfrac{\varphi_0}{2\pi}$ is irrational then we obtain condition (4) (b) by applying the results of 18.7 (the action $\gamma = R_\varphi$ is included in the action of the group S^1 by rotations R_φ on the ball Ω).

The operator L_s (49.11) corresponding to the problem (49.19), (49.20) can be represented in the form of a product of an operator Q corresponding to the Dirichlet problem and the functional operator

$$b_0 = a_0 + a_1 U_\gamma .$$

If the problem is Fredholm then assertion (4) is satisfied, which in turn implies the invertibility of b_0. But since the operator of the Dirichlet problem for the Poisson equation is invertible the foregoing argument means that if L_s is Fredholm then it is invertible, which means that assertions (1), (2) and (3) are equivalent (the equivalence of (1) and (2) was observed earlier, recall that the s-symbol does not depend on s).

The proof is complete. □

Now let us consider the problem (49.19), (49.21).

Theorem 49.12 *Suppose that in the condition (49.21) $\gamma(x) = R_{\varphi_0}(x)$, $\varphi_0 \neq 0$. Then the non-local problem (49.19), (49.20) is Fredholm for a given s iff the following conditions (depending on φ_0) are satisfied:*

(1) $\dfrac{\varphi_0}{2\pi} = \dfrac{M}{N}$ *is an irreducible fraction.*

The condition has the form

$$\prod_{j=0}^{N-1} a_0\left(R_{j\varphi_0}x, R_{j\varphi_0}\xi\right) \neq \prod_{j=0}^{N-1} a_1\left(R_{j\varphi_0}x, R_{j\varphi_0}\xi\right)(-1)^N .$$

(2) $\dfrac{\varphi_0}{2\pi}$ *is irrational.*

The condition has the form:

either

$$a_0(x,\xi) \neq 0 \text{ and } S_{x,\xi}(a_0) > S_{x,\xi}(a_1) \text{ for every } (x,\xi) \in S^*\Gamma,$$

or

$$a_1(x,\xi) \neq 0 \text{ and } S_{x,\xi}(a_0) < S_{x,\xi}(a_1) \text{ for every } (x,\xi) \in S^*\Gamma$$

where $S_{x,\xi}$ is defined by (49.27) and a_0 and a_1 by (49.25).

Proof Conditions (1) and (2) are the explicit record of the invertibility conditions of the s-symbol of the G-Calderon operator (49.24) (cf. the corresponding argument in the proof of the previous theorem when verifying (4) (a) and (4) (b)). □

49.13 In the examples just considered the Fredholmity conditions of the non-local boundary value problems do not depend on s. Let us present now examples where these conditions do depend on s.

Set

$$R_{\varphi,\lambda} = \begin{bmatrix} \cos\varphi & \sin\varphi & 0 \\ -\sin\varphi & \cos\varphi & 0 \\ 0 & 0 & \lambda \end{bmatrix}, \ \lambda > 0 \tag{49.28}$$

and

$$\gamma(x) = \frac{R_{\varphi_0,\lambda}x}{\left\| R_{\varphi_0,\lambda}x \right\|}. \tag{49.29}$$

Remark If $\lambda = 1$ then $R_{\varphi,\lambda} = R_\varphi$ is the mapping of the preceding example.

We denote by $P^{\pm}$ the 'north' and 'south' poles of the sphere $\Gamma = S^2$, i.e., the points with coordinates $(0,0,\pm 1)$. We denote by Γ_0 the 'equator', that is the circle

$$\Gamma_0 = \{x \in \Gamma : x_3 = 0\}. \tag{49.30}$$

Theorem 49.14 *Let the diffeomorphism γ be given by the formula (49.29) with $\lambda \neq 1$. The non-local boundary value problem (49.19), (49.20) is Fredholm for a given $s \geq 2$ iff the following conditions (depending on φ_0) are satisfied:*

(1) $\frac{\varphi_0}{2\pi} = \frac{M}{N}$ *is an irreducible fraction.*

The condition has the form:

either

(1a) $a_0(x) \neq 0$ *for every* $x \in \Gamma$

and

$$\left|a_0(P^{\pm})\right| > \left|a_1(P^{\pm})\right| \lambda^{s-\frac{3}{2}},$$

$$\prod_{j=0}^{N-1} \left|a_0(R_{j\varphi_0}(y))\right| > \prod_{j=0}^{N-1} \left|a_1(R_{j\varphi_0}(y))\right| \lambda^{\frac{N}{2}},$$

$$\prod_{j=0}^{N-1} \left|a_0(R_j\varphi_0(y))\right| > \prod_{j=0}^{N-1} \left|a_1(R_{j\varphi_0}(y))\right| \lambda^{-Ns}$$

where y belongs to Γ_0;

or

(1b) $a_1(x) \neq 0$ *for every* $x \in \Gamma$

and all the inequalities in (1a) have the opposite sign.

(2) $\frac{\varphi_0}{2\pi}$ *is irrational.*

The condition has the form:

either

(2a) $a_0(x) \neq 0$ *for every* $x \in \Gamma$

and

$$\left|a_0\left(P^{\pm}\right)\right| > \left|a_1\left(P^{\pm}\right)\right|\lambda^{s-\frac{3}{2}},$$

$$S_y(a_0) > S_y(a_1)\lambda^{\frac{1}{2}},$$

$$S_y(a_0) > S_y(a_1)\lambda^{-s}$$

where y belongs to Γ_0;

or

(2b) $a_1(x) \neq 0$ *for every* $x \in \Gamma$

and all the inequalities in (2a) have the opposite sign.

Proof The stated conditions are the explicit record of the invertibility conditions of the s-symbol of the G-Calderon operator (49.23) in the case of a given mapping γ (recall Theorem 45.13). □

An analogous argument applied to the mapping $\beta: S^*\Gamma \to S^*\Gamma$ generated by the diffeomorphism γ (49.29) leads to the following result describing the problem (49.19), (49.21).

Theorem 49.15 *Let the mapping* γ *be given by the formula (49.29) with* $\lambda \neq 1$. *The problem (49.19), (49.21) is Fredholm for a given* $s \geq 2$ *iff the following conditions (depending on* φ_0*) are satisfied:*

(1) $\dfrac{\varphi_0}{2\pi} = \dfrac{M}{N}$ *is an irreducible fraction.*

The condition has the form:

either

(1a) $a_0(x,\xi) \neq 0, \quad (x,\xi) \in S^*\Gamma$

and

$$\prod_{j=0}^{N-1} \left| a_0\left(P^\pm, R_{j\varphi_0}\xi\right) \right| > \prod_{j=0}^{N-1} \left| a_1\left(P^\pm, R_{j\varphi_0}\xi\right) \right| \lambda^{N\left(s-\frac{3}{2}\right)}, \quad \xi \in \left(S^*\Gamma\right)_{P^\pm};$$

$$\prod_{j=0}^{N-1} \left| a_0\left(R_{j\varphi_0}x, R_{j\varphi_0}\xi\right) \right| > \prod_{j=0}^{N-1} \left| a_1\left(R_{j\varphi_0}x, R_{j\varphi_0}\xi\right) \right| \lambda^{\frac{N}{2}}$$

where x belongs to Γ_0 *and* ξ *is any vector cotangent to* Γ_0*;*

$$\prod_{j=0}^{N-1} \left| a_0\left(R_{j\varphi_0}x, R_{j\varphi_0}\xi\right) \right| > \prod_{j=0}^{N-1} \left| a_1\left(R_{j\varphi_0}x, R_{j\varphi_0}\xi\right) \right| \lambda^{-Ns}$$

when x belongs to Γ_0 *and* ξ *is any vector cotangent to* Γ *and conormal to* Γ_0*;*

or

(1b) $a_1(x,\xi) \neq 0, \quad (x,\xi) \in S^*\Gamma$

and all the inequalities in (1a) have the opposite sign.

(2) $\dfrac{\varphi_0}{2\pi}$ *is irrational.*

The condition has the form:

either

(2a) $a_0(x,\xi) \neq 0, \quad (x,\xi) \in S^*\Gamma$

and

$$S_{P^\pm,\xi}(a_0) > S_{P^\pm,\xi}(a_1)\lambda^{s-\frac{3}{2}};$$

$$S_{x,\xi}(a_0) > S_{x,\xi}(a_1)\lambda^{\frac{1}{2}}$$

when x belongs to Γ_0 *and* ξ *is any vector cotangent to* Γ_0*;*

$$S_{x,\xi}(a_0) > S_{x,\xi}(a_1)\lambda^{-s}$$

when x belongs to Γ_0 and ξ is any vector cotangent to Γ and conormal to Γ_0;

or

(2b) $a_1(x,\xi) \neq 0, \quad (x,\xi) \in S^*\Gamma$

and all the inequalities in (2a) have the opposite sign.

The functions $a_k(x,\xi)$ here are defined by (49.25) and $S_{x,\xi}$ by (49.27).

49.16 Now let the mapping $\gamma: \Gamma \to \Gamma$ be given by

$$\gamma(x) = \frac{x+h}{\|x+h\|}, \quad h = (0,0,\tau), 0 < \tau < 1. \tag{49.31}$$

This mapping is of interest in view of the very simple structure of the set of ergodic measures (with respect to it); namely, every such measure is concentrated at one of the points of the form $(P^{\pm},\xi)$ and an argument analogous to that exploited for the proofs of the previous theorems leads to the next result (recall also 45.9 and Theorem 45.11).

Theorem 49.17 *Let the mapping γ be given by the formula (49.31). The non-local boundary value problem (49.19), (49.21) is Fredholm for a given s iff one of the following conditions is satisfied:*

(a) $a_0(x,\xi) \neq 0, \quad (x,\xi) \in S^*\Gamma$

and

$$|a_0(P^{\pm},\xi)| > |a_1(P^{\pm},\xi)|(1 \pm \tau)$$

or

(b) $a_1(x,\xi) \neq 0, \quad (x,\xi) \in S^*\Gamma$

and

$$|a_0(P^{\pm},\xi)| < |a_1(P^{\pm},\xi)|(1 \pm \tau).$$

49.18 Finally let us look at the non-local boundary conditions containing three (more than two) terms:

$$\left[a_0(x)\frac{\partial u}{\partial l_0}(x)+a_1(x)\frac{\partial u}{\partial l_1}(\gamma(x))+a_2(x)\frac{\partial u}{\partial l_2}\left(\gamma^2(x)\right)\right]_\Gamma = \psi(x) \tag{49.32}$$

where l_0, l_1, l_2 are smooth unit vector fields defined on Γ; a_0, a_1, a_2 are given smooth functions, and $\gamma:\Gamma\to\Gamma$ is a diffeomorphism.

Consider the boundary value problem (49.19), (49.32), that is for a given $s\geq 2$ we seek a solution $u\in H^s(\Omega)$ under the condition that

$$f\in H^{s-2}(\Omega),\quad \psi\in H^{s-\frac{3}{2}}(\Gamma)\ .$$

49.19 Let Ω again be the unit ball in $\mathbb{R}^3$, let Γ be its boundary S^2 and let $\gamma=R_\varphi$ be a rotation about the x_3 axis through an angle φ.

If $\frac{\varphi_0}{\pi}$ is rational then the conditions of the Fredholmity of the problem (49.19), (49.32) (the invertibility of the s-symbol of its G-Calderon operator) can be written in the manner presented in Theorem 49.12 (1) (by the usage of Corollary 15.6).

If $\frac{\varphi_0}{\pi}$ is irrational then in the general situation the effective necessary and sufficient conditions for the invertibility of the s-symbol are not known. We single out a special case where such conditions could be obtained due to the additional symmetry.

Before presenting the result we have in mind we shall write out a general C^*-algebraic statement lying at the basis of this result.

Let G be a discrete commutative group with the (compact) dual group $\hat{G}$; let A be a certain C^*-subalgebra of operators acting in a Hilbert space H and containing the unit operator, and let T_g be a unitary representation of G in $L(H)$ such that $T_gAT_g^*=A,\ g\in A$. As usual we denote by $C^*\left(A,T_g\right)$ the C^*-algebra generated by A and $\left\{T_g\right\}_{g\in G}$.

With every operator $b\in C^*\left(A,T_g\right)$ of the form

$$b=\sum_{g\in F}a_gT_g,\quad |F|<\infty \tag{49.33}$$

we associate the A-valued function on $\hat{G}$

$$b(\chi)=\sum a_g\chi(g),\quad \chi\in\hat{G}\ . \tag{49.34}$$

The C^*-algebraic statement mentioned above is the following.

Theorem 49.20 *Let A, $\{T_g\}_{g\in G}$, and $C^*(A,T_g)$ be as described above and let*

(1) *all the operators T_g, $g \in G$, commute with elements of A;*

(2) *$C^*(A,T_g)$ possess the property (*) (see 12.2).*

Then

I. $C^*(A,T_g) \cong A \otimes C^*(G);$ (49.35)

II. *The following statements are equivalent:*

(i) *an operator b of the form (49.33) is invertible;*
(ii) *for every $\chi \in \hat{G}$ the operator $b(\chi)$ (49.34) is invertible;*
(iii) *for every $\pi \in \hat{A}$ ($\hat{A}$ is the spectrum of A) and every $\chi \in \hat{G}$ the operator*

$$\pi[b(\chi)] = \sum_{g\in F} \pi(a_g)\chi(g)$$

is invertible.

Proof Since the operators T_g commute with elements of A all the automorphisms $T_g(\,\cdot\,)T_g^*$ of A are trivial. In view of this and the fact that $C^*(A,T_g)$ possesses the property (*) it follows (by Theorem 12.8) that

$$C^*(A,T_g) \cong A \underset{I}{\times} G \tag{49.36}$$

and by Lemma 11.18 we conclude that (49.35) is true.

Now let us consider II. In view of (49.36) the mapping

$$b = \sum a_g T_g \to \bar{b} = \sum a_g \otimes g \in A \underset{I}{\times} G$$

is a C^*-algebraic isomorphism.

In view of this and Lemma 22.6 (see also 22.7) and bearing in mind that $b(\chi)$ is a continuous operator-function (on χ) on a compact space $\hat{G}$, we conclude that

$$\|b\| = \|\bar{b}\| = \max_{\chi\in\hat{G}} \|b(\chi)\| . \tag{49.37}$$

Let $\tilde{b} \in L\left(L^2(\hat{G},H)\right)$ be the operator defined by the operator-function $b(\chi)$ (49.34). The equality (49.37) and the definition of $b(\chi)$ means that the mapping

$$b \to \tilde{b}$$

establishes a C^*-algebraic isomorphism between $C^*\left(A,T_g\right)$ and the C^*-algebra generated by the operator-functions $b(\chi)$ (or the operators $\tilde{b}$).

But as $b(\chi)$ depends continuously on χ and $\hat{G}$ is compact it follows that $\tilde{b}$ is invertible iff $b(\chi)$ is invertible for every $\chi \in \hat{G}$. This proves the equivalence of (i) and (ii).

Now let us fix some $\chi_0 \in \hat{G}$. The operator $b(\chi_0)$ belongs to A and is invertible iff $\pi\left(b(\chi_0)\right)$ is invertible for every $\pi \in \hat{A}$ (Naimark [1], §25.3).

This proves the equivalence of (ii) and (iii). □

We return to the consideration of boundary value problems.

Theorem 49.21 *Let the vector fields $l_k(x)$, $k=1,2,3$, (mentioned in formula (49.32)) be invariant under rotations about the x_3 axis (i.e. $l_k\left(R_\varphi x\right) = R_\varphi\left(l_k(x)\right)$) and let the coefficients a_k depend only on the coordinate x_3. The non-local boundary value problem (49.19), (49.32), with $\gamma = R_{\varphi_0}$ where $\frac{\varphi_0}{\pi}$ is irrational, is Fredholm iff the condition*

$$a_0(x,\xi) + a_1(x,\xi)z + a_2(x,\xi)z^2 \neq 0 \tag{49.38}$$

for every $(x,\xi) \in S^*\Gamma$ and $z \in \mathbb{C}$, $|z| = 1$.

Proof The G-Calderon operator of the problem (49.19), (49.32) with $\gamma = R_\varphi$ has the form

$$\mathcal{b}_s = a_0(x,D) + U_\gamma a_1(x,D) + U_\gamma^2 a_2(x,D)$$

where $a_k(x,D)$ are pseudodifferential operators of zero order with the symbols

$$a_0(x,\xi) = a_0(x)\langle x + i\xi, l_0(x)\rangle \text{ ,}$$

$$a_1(x,\xi) = a_1\left(\gamma^{-1}(x)\right)\left\|\partial\gamma(x)\xi\right\|\langle x + i\xi, l_1(x)\rangle \text{ ,}$$

$$a_2(x,\xi) = a_2\left(\gamma^{-2}(x)\right)\left\|\partial\gamma^2(x)\xi\right\|\langle x + i\xi, l_2(x)\rangle \text{ .}$$

Recalling that the mapping $\gamma = R_\varphi$ preserves the measure and the natural metric on the sphere we conclude that in the s-symbol of ℓ_s the corresponding factors $\left(\|\partial\gamma(\eta)\|, \|\partial\gamma^2(\eta)\|, |\gamma'(x)|, |\gamma^{2'}(x)|\right)$ are equal to 1 and it has the form (not depending on s)

$$\sigma_s(\ell_s) = a_0(x,\xi) + U_\beta a_1(x,\xi) + U_\beta^2 a_2(x,\xi) \tag{49.39}$$

where

$$a_k(x,\xi) = \sigma(a_k(x,D))$$

and

$$\beta(x,\xi) = \left(R_{\varphi_0}x, R_{\varphi_0}\xi\right). \tag{49.40}$$

The operator (49.39) could be considered to be the element of the C^*-algebra $C^*(A,T)$ (a subalgebra in $L\left(L^2(S^*\Gamma)\right)$ generated by the operators of multiplication by continuous functions $a(x,\xi) \in C(S^*\Gamma)$ and $T = V_\beta$. In view of the explicit form of β (49.40) and the condition that $\frac{\varphi_0}{\pi}$ is irrational it follows that the action of β on $S^*\Gamma$ (= Prim A) is topologically free and consequently (by Theorem 12.14) $C^*(A,T)$ possesses the property (*).

On the other hand, under the conditions of the theorem the coefficients $a_k(x,\xi)$ in (49.39) belong to the subalgebra $\tilde{A} \subset A$ of operators commuting with T.

Thus the operator

$$b = \sigma_s(\ell_s)$$

belongs to the algebra $C^*\left(\tilde{A}, T\right)$ satisfying assumptions (1) and (2) of Theorem 49.20.

Now it is clear that condition (49.38) follows from the assertion II (iii) of Theorem 49.20 (here any π is defined by a point $(x,\xi) \in S^*\Gamma$ and $\{T^n\}_{n\in\mathbb{Z}}$ is a representation of the group $\mathbb{Z}$, the dual of which is $\hat{\mathbb{Z}} = \{z \in \mathbb{C} : |z| = 1\}$ and $z(n) = z^n$, $z \in \hat{\mathbb{Z}}$, $n \in \mathbb{Z}$). □

49.22 To finish the section let us consider the boundary condition (49.32) with γ being defined by (49.31). This mapping has a simpler structure since the trajectory of each point $x \neq P^\pm = (\pm 1, 0, 0)$ is attracted to the 'pole' P^+ when $n \to +\infty$ and to the pole P^- when $n \to -\infty$ ($P^\pm$ are the fixed points). In view of this one can obtain an explicit form of the Fredholmity conditions for the boundary value problem (49.19), (49.32) with γ being given by (49.31).

We introduce the notation

$$d^{\pm}(z,\xi)=a_0\left(P^{\pm},\xi\right)+a_1\left(P^{\pm},\xi\right)(1\pm\tau)^{-s+\frac{3}{2}}z+a_2\left(P^{\pm},\xi\right)(1\pm\tau)^{-2s+3}z^2,\quad z\in\mathbb{C}$$

$$\kappa^{\pm}=\frac{1}{2\pi}\left[\arg d^{\pm}(z,\xi)\right]_{|z|=1}.$$

The s-symbol b of the boundary value problem (49.19), (49.32) has the form

$$b=a_0+a_1\left\|\partial\gamma(x)(\xi)\right\|U_\beta+a_2\left\|\partial\gamma^2(x)(\xi)\right\|U_\beta^2$$
$$=\tilde{a}_0+\tilde{a}_1T+\tilde{a}_2T^2$$

where

$$\tilde{a}_0=a_0(x,\xi)$$
$$\tilde{a}_1=a_1(x,\xi)\left|\gamma'(x)\right|^{-\frac{1}{2}}\left\|\partial\gamma(x)(\xi)\right\|^{-s+\frac{1}{2}}$$
$$\tilde{a}_2=a_2(x,\xi)\left|\gamma^{2'}(x)\right|^{-\frac{1}{2}}\left\|\partial\gamma^2(x)(\xi)\right\|^{-s+\frac{1}{2}}$$

$$T=\left|\beta'(x,\xi)\right|^{\frac{1}{2}}U_\beta,$$

and thus can be considered as an element of the C^*-algebra $C^*(A,T)$ with $A\cong C(S^*\Gamma)$.

We denote by $b_{x,\xi}$, $(x,\xi)\in S^*\Gamma$, the representation of b in $l^2(\mathbb{Z})$ defined by

$$\left(\tilde{a}_k\right)_{x,\xi}\{\eta_z\}=\left\{\tilde{a}_k\left(\beta^z(x,\xi)\right)\eta_z\right\},$$

$$T_{x,\xi}\{\eta_z\}=\{\eta_{z-1}\},\ \{\eta_z\}\in l^2(\mathbb{Z}).$$

Theorem 49.23 *Let $s\geq 2$ be fixed and let γ be given by (49.31).*

The necessary condition for the problem (49.19), (49.32) to be Fredholm is the satisfaction of the following conditions:

(1) $d^{\pm}(z,\xi)\neq 0$, $|z|=1$,

(2) $\kappa^{+}=\kappa^{-}$.

If these conditions are satisfied we have:

(i) *if* $\kappa^+ = \kappa^- = 0$ *then the problem is Fredholm iff* $a_0(x,\xi) \neq 0,\ (x,\xi) \in S^*\Gamma$;

(ii) *if* $\kappa^+ = \kappa^- = 2$ *then the problem is Fredholm iff* $a_2(x,\xi) \neq 0,\ (x,\xi) \in S^*\Gamma$;

(iii) *if* $\kappa^+ = \kappa^- = 1$ *then the problem is Fredholm iff each of the homogeneous equations* $b_{x,\xi}(n) = 0$ *has only the zero solution in* $l^2(\mathbb{Z})$.

Proof Since for the algebra $A = C(S^*\Gamma)$ we have $\hat{A} = S^*\Gamma$ (= Prim A) and the operator T mentioned in 49.22 induces the topologically freely acting automorphism of A such that the points $(p^{\pm},\xi)$ are attracting for each positive (negative) semitrajectory of a point $(x,\xi) \in S^*\Gamma$, it follows that the invertibility conditions of the s-symbol b of the boundary value problem (49.19), (49.32) can be written out by means of Theorem 21.6.

Now conditions (1) and (2) here coincide respectively with conditions (1) and (2) of Theorem 21.6 and direct computation shows that condition (3) of Theorem 21.6 is equivalent respectively to one of the conditions (i), (ii) or (iii). □

Remark 49.24 In all the examples considered in this section the non-local boundary value problems have zero index.

50. The convolution type operators

Under the Fourier transform the shift operator $(T_h f)(x) = f(x-h)$ $\left(f \in L^2(\mathbb{R})\right)$ changes to the operator of multiplication by the function $\mathrm{e}^{ih\xi}$. Therefore the operators of multiplication by functions of the form $\mathrm{e}^{ih\xi}$ possess the basic properties of the shift operators and the equations containing these operators could be dealt with by exploiting the theory built for the investigation of the equations containing the shift operators. Since any periodic or almost-periodic function could be expanded into a series in functions of the form e^{ihx} the methods already developed in this book are applicable to the investigation of a number of equations with periodic or almost-periodic coefficients.

In this section we shall illustrate the potential of the method by the example of the so-called convolution type operators with oscillating coefficients.

Let us start from the description of the objects under consideration.

50.1 The *classical convolution operator* in $L^2(\mathbb{R})$ is an operator Φ of the form

$$(\Phi f)(x) = \int_{-\infty}^{+\infty} f(y)\varphi(x-y)\,\mathrm{d}y = (f * \varphi)(x)$$

where $\varphi \in L^1(\mathbb{R})$.

Since

$$f * \varphi = F^{-1}(\hat{\varphi} \cdot F(f)) \tag{50.1}$$

where F is the Fourier transform and $\hat{\varphi}$ is the image of φ under the Fourier transform it follows that

$$\|\Phi\| = \max_R |\hat{\varphi}(t)| \tag{50.2}$$

(recall that as φ belongs to $L^1(\mathbb{R})$ we have that $\hat{\varphi}$ belongs to the set $C_0(\mathbb{R})$ of continuous functions tending to zero at infinity).

Bearing in mind the equality (50.2) and the fact that the image of $L^1(\mathbb{R})$ by F contains all the rapidly decreasing functions we conclude that the C^*-algebra generated by the classical convolution operators is isomorphic to $C_0(\mathbb{R})$ and the isomorphism is given by (50.1).

In view of this we shall still call the operators defined by (50.1) with $\hat{\varphi} \in C_0(\mathbb{R})$ the *convolution operators.*

Let us add to this C^*-algebra the operators of multiplication by *stabilizing functions,* that is continuous functions having finite limits at $\pm\infty$. The C^*-algebra obtained in this way will be denoted by $\mathcal{A}$.

Clearly this algebra is generated by operators of the form

$$(\mathcal{a}f)(x) = a_0(x)f(x) + \sum_{j=1}^{m} a_j(x) \int_{-\infty}^{+\infty} \varphi_j(x-y)d_j(y)f(y)\,dy \tag{50.3}$$

where $\varphi_j \in L^1(\mathbb{R})$ and a_j, d_j are stabilizing functions.

Since the convolution operator and the operator of multiplication by stabilizing functions commute modulo a compact operator the factor algebra $\mathcal{A}/K$ with respect to the ideal of compact operators is commutative and is isomorphic to the algebra $C(M)$ of continuous functions on its maximal ideal space M.

The structure of M is well known. It has the form

$$M = \mathbb{R}^- \cup \overline{\mathbb{R}} \cup \mathbb{R}^+$$

where $\overline{\mathbb{R}}$ is the real line extended by $\pm\infty$ and $\mathbb{R}^+$ and $\mathbb{R}^-$ are the real lines, and the topology is given by gluing $\pm\infty$ of $\mathbb{R}^+$ to $+\infty$ of $\overline{\mathbb{R}}$ and $\pm\infty$ of $\mathbb{R}^-$ to $-\infty$ of $\overline{\mathbb{R}}$ (Fig. 7).

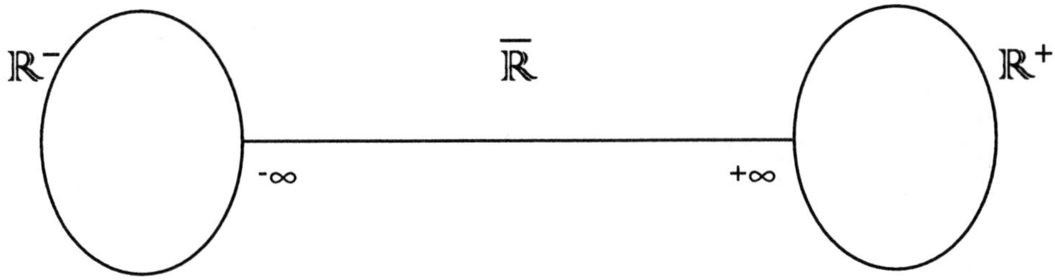

Fig. 7

The isomorphism between $\mathcal{A}/K$ and $C(M)$ is given by the mapping σ of taking the symbol (of an element of $\mathcal{A}$) which is defined by the formulae:

(1) for the operator Φ of convolution by a function φ $\left(\hat{\varphi} \in C_0(\mathbb{R})\right)$

$$\sigma(\Phi)(x) = \begin{cases} 0, & x \in \overline{\mathbb{R}}, \\ \hat{\varphi}(x), & x \in \mathbb{R}^+, \\ \hat{\varphi}(x), & x \in \mathbb{R}^-; \end{cases}$$

(2) for the operator a of multiplication by a stabilizing function a

$$\sigma(a)(x) = \begin{cases} a(x), & x \in \overline{\mathbb{R}}, \\ a(+\infty), & x \in \mathbb{R}^{\prime}, \\ a(-\infty), & x \in \mathbb{R}^-. \end{cases}$$

In particular for the operator $\mathfrak{a}$ of the form (50.3) we have

$$\sigma(\mathfrak{a})(x) = \begin{cases} a_0(x), & x \in \overline{\mathbb{R}}, \\ a_0(+\infty) + \sum_{j=1}^{m} a_j(+\infty) d_j(+\infty) \hat{\varphi}_j(x), & x \in \mathbb{R}^+, \\ a_0(-\infty) + \sum_{j=1}^{m} a_j(-\infty) d_j(-\infty) \hat{\varphi}_j(x), & x \in \mathbb{R}^-. \end{cases}$$

50.2 Let $\mathcal{V}_h$ be the operator in $L^2(\mathbb{R})$ of multiplication by the function e^{ihx}. This operator commutes with the operator of multiplication by stabilizing functions whereas if $h \neq 0$ then it does not commute (even modulo compact operators) with the convolution operators. However $\mathcal{V}_h$ generates an automorphism of the algebra $\mathcal{A}$ which follows from the fact that for the operator Φ of convolution by a function $\varphi \in L^1(\mathbb{R})$ we have

$$\left[\mathcal{V}_h \Phi \mathcal{V}_h^{-1} f\right](x) = \int_{-\infty}^{+\infty} f(y) e^{ixh} \cdot e^{-iyh} \varphi(x-y)\, dy$$

$$= \int_{-\infty}^{+\infty} f(y) \varphi_1(x-y)\, dy \tag{50.4}$$

where $\varphi_1(x) = e^{ixh}\varphi(x)$ belongs to $L^1(\mathbb{R})$ as well.

Bearing this observation in mind we shall consider now the C^*-algebra

$$\mathcal{B} = C^*(\mathcal{A}, \mathcal{V}_h),\ h \in \mathbb{R}$$

generated by the algebra $\mathcal{A}$ and the operators $\mathcal{V}_h$, $h \in \mathbb{R}$. In view of its structure it is reasonable to call the elements of $\mathcal{B}$ *convolution operators with oscillating coefficients.*

The automorphisms $\mathcal{V}_h(\,\cdot\,)\mathcal{V}_h^{-1}$ of $\mathcal{A}$ induce automorphisms of the factor algebra $\mathcal{A}/K \cong C(M)$ described in 50.1 and the corresponding homeomorphisms t_h of the space M (recall 12.12) are given (in view of (50.1) and (50.4)) by

$$t_h(x) = \begin{cases} x, & x \in \overline{\mathbb{R}}, \\ x+h, & x \in \mathbb{R}^+, \\ x+h, & x \in \mathbb{R}^-. \end{cases} \tag{50.5}$$

Remark Clearly the group action here is not topologically free.

50.3 Let us now construct the symbolic calculus of the elements of $\mathcal{B}$ following the general C^*-ideology already developed.

Let

$$L^2(M) = L^2(\overline{\mathbb{R}}) \oplus L^2(\mathbb{R}^+) \oplus L^2(\mathbb{R}^-)$$

where the measure on $\overline{\mathbb{R}}, \mathbb{R}^+, \mathbb{R}^-$ coincides with the standard Lebesque measure.

In the space $L^2(M)$ consider the algebra

$$A \cong C(M)$$

of operators of multiplication by functions from $C(M)$ and the group of unitary operators V_h, $h \in \mathbb{R}$, given by

$$V_h(u_1, u_2, u_3) = \left(e^{ihx} u_1(x), u_2(x-h), u_3(x-h)\right) . \tag{50.6}$$

We denote by $B = C^*(A, V_h)$ the C^*-algebra generated by A and the operators V_h.

The next result shows that B is the algebra of symbols of elements of $\mathcal{B}$.

Theorem 50.4 *Let $\mathcal{B}$ and B be the algebras described in 50.2 and 50.3. Then the mapping*

$$\sigma: \mathcal{B} \to B$$

given by

$$\sigma: \mathfrak{a} \to \sigma(\mathfrak{a}), \qquad \mathfrak{a} \in \mathcal{A}$$

$$\sigma: \mathcal{V}_h \to V_h$$

*(here we denote by $\sigma(\mathfrak{a})$ the operator of multiplication by the function $\sigma(\mathfrak{a}) \in C(M)$ defined in 50.1) is a *-epimorphism, the kernel of which is the ideal K of compact operators in $L^2(\mathbb{R})$.*

In particular an operator $\mathfrak{b} \in \mathcal{B}$ is Fredholm iff its symbol $\sigma(\mathfrak{b}) \in \mathcal{B}$ is invertible.

Proof We have to check that the mapping σ defined by (50.6) establishes the isomorphism

$$\mathcal{B}/K = [\mathcal{B}] \overset{\sigma}{\cong} B . \tag{50.7}$$

The argument in 50.1 shows that

$$\mathcal{A}/K = [\mathcal{A}] \overset{\sigma}{\cong} A = \sigma(\mathcal{A})$$

and the observations in 50.2 along with the definition of V_h (50.6) mean that

$$\sigma\left(\mathcal{V}_h a \mathcal{V}_h^*\right) = V_h \sigma(a) V_h^*, \quad a \in \mathcal{A}. \tag{50.8}$$

Since, as has already been noted, the action of the group of homeomorphisms $\{t_h\}_{h\in\mathbb{R}}$ (see (50.5)) is not topologically free, it is reasonable to apply here Corollary B.30.

For every $h \neq 0$ we have (recall 16.1)

$$X_h = \overline{\mathbb{R}}$$

and consequently

$$I_h \cong C_0(\mathbb{R})$$

is the ideal of the operators of multiplications by functions from $C(M)$ which are zero on $\mathbb{R}^+$ and $\mathbb{R}^-$.

Conditions (i) and (ii) from B.26 in this situation are obviously satisfied (for the algebra $C^*(A, V_h)$).

In view of the definition of the mapping σ on $\mathcal{A}$ (see 50.1) it follows that $\sigma(a) \in I_h$ iff a is (modulo compact operators) an operator of multiplication by a function from $C_0(\mathbb{R})$, which along with the definition of $\mathcal{V}_h$ and V_h means that

$$\sigma([\mathcal{V}_h a]) = V_h \sigma(a), \quad a \in I_h.$$

Thus all the conditions of Corollary B.30 are satisfied and consequently (50.7) is true. □

50.5 The investigation of the invertibility conditions for the symbol $\sigma(\mathcal{b})$ of a convolution operator with oscillating coefficients $\mathcal{b}$ can be carried out by applying the results previously obtained.

In particular let us consider for example the operator $\mathcal{b}$ of the form

$$\mathcal{b} = a_0 + a_1 \mathcal{V}_h \tag{50.9}$$

where

$$(a_k f)(x) = a_{0k}(x) f(x) + \sum_{j=1}^{m} a_{jk}(x) \int_{-\infty}^{+\infty} \varphi_j(x-y) f(y) d_{jk}(y)\, dy + (K_k f)(x) \tag{50.10}$$

and a_{jk}, d_{jk} are stabilizing functions, $\varphi_j \in L^1(\mathbb{R})$ and K_k is a compact operator.

The symbol $\sigma(\mathcal{L})$ is given (in $L^2(M) = L^2(\overline{\mathbb{R}}) \oplus L^2(\mathbb{R}^+) \oplus L^2(\mathbb{R}^-)$) by the formula

$$\begin{aligned}\sigma(\mathcal{L})(u_1, u_2, u_3) = \Big(\big[a_{00}(x) + e^{ihx} a_{10}(x)\big] u_1(x),\\ a_0^+(x) u_2(x) + a_1^+(x) u_2(x-h),\\ a_0^-(x) u_3(x) + a_1^-(x) u_3(x-h)\Big)\\ = \big(\sigma_1(\mathcal{L}) u_1, \sigma_2(\mathcal{L}) u_2, \sigma_3(\mathcal{L}) u_3\big) \qquad (50.11)\end{aligned}$$

where

$$a_k^{\pm}(x) = a_{k0}(\pm\infty) + \sum_{j=1}^{m} a_{kj}(\pm\infty) d_{kj}(\pm\infty) \hat{\varphi}_{kj}(x), \ k = 0,1 .$$

The operators $\sigma_2(\mathcal{L})$ and $\sigma_3(\mathcal{L})$ are two-term functional operators with stabilizing coefficients. Writing out their invertibility conditions (recall 18.2) and bearing in mind that $\sigma_1(\mathcal{L})$ is the operator of multiplication by a function we obtain the next result.

Theorem 50.6 *The operator $\mathcal{L} \in \mathcal{B}$ of the form (50.9) is Fredholm iff*

(i) $a_{00}(x) + e^{ihx} a_{10}(x) \neq 0, \ x \in \mathbb{R}$,

and at each of the points $\pm\infty$ one of the following conditions (ii) or (iii) holds:

(ii) $|a_{00}(\pm\infty)| > |a_{10}(\pm\infty)|$ *and* $a_0^{\pm} \neq 0, \ x \in \mathbb{R}$;

(iii) $|a_{00}(\pm\infty)| < |a_{10}(\pm\infty)|$ *and* $a_1^{\pm}(x) \neq 0, \ x \in \mathbb{R}$.

Remark 50.7 If we consider an operator

$$\mathcal{L} = \sum a_k \mathcal{V}_{h_k}, \qquad a_k \in \mathcal{A}$$

having more than two summands then the invertibility conditions of its symbol $\sigma(\mathcal{L})$ could not be written out in explicit form except in degenerate cases.

One such case is the situation when the coefficients in the symbol are constant in a neighbourhood of ∞ on $\mathbb{R}^+$ and $\mathbb{R}^-$. This condition is realized when the functions $\hat{\varphi}_j$ in (50.10) are finite.

50.8 Along with the operators $\mathcal{V}_h$ there are also some other simple operators generating automorphisms of $\mathcal{A}$; namely, let us consider the shift operators in $L^2(\mathbb{R})$:

$$(\mathcal{W}_\tau f)(x) = f(x-\tau), \quad \tau \in \mathbb{R}. \tag{50.12}$$

A routine calculation shows that $\mathcal{W}_\tau(\,\cdot\,)\mathcal{W}_\tau^{-1}$ is an automorphism of $\mathcal{A}$ which in turn generates an automorphism of $\mathcal{A}/_K$ and the corresponding homeomorphism α_τ of M is given by

$$\alpha_\tau(x) = \begin{cases} x+\tau, & x \in \overline{\mathbb{R}} \\ x, & x \in \mathbb{R}^+ \\ x, & x \in \mathbb{R}^-. \end{cases} \tag{50.13}$$

Remark Here again (recall the remark in 50.2) the group $\{\alpha_\tau\}_{\tau \in \mathbb{R}}$ does not act topologically freely.

Let us consider now the algebra

$$\mathcal{B}_1 = C^*(\mathcal{A}, \mathcal{W}_h), \quad h \in \mathbb{R}$$

generated by the algebra $\mathcal{A}$ and the operators $\mathcal{W}_h$.

50.9 To obtain the symbolic calculus for the elements of $\mathcal{B}'$ we introduce in the space $L^2(M)$ the operators

$$W_\tau(u_1, u_2, u_3) = \left(u_1(x-\tau), \mathrm{e}^{-\mathrm{i}\tau x} u_2(x), \mathrm{e}^{-\mathrm{i}\tau x} u_3(x)\right) \tag{50.14}$$

and consider the algebra $B_1 = C^*(A, W_h)$ where A is that exploited in 50.3.

An argument absolutely analogous to the one used in the proof of Theorem 50.4 shows that B_1 is the algebra of symbols of elements of $\mathcal{B}_1$. Thus we have

Theorem 50.10 *Let $\mathcal{B}_1$ and B_1 be those described in 50.8 and 50.9. Then the mapping*

$$\sigma : \mathcal{B}_1 \to B_1$$

given by

$$\sigma : a \mapsto \sigma(a), \quad a \in \mathcal{A}$$

$$\sigma : \mathcal{W}_h \mapsto W_h$$

*is a *-epimorphism, the kernel of which is the ideal K of compact operators in* $L^2(\mathbb{R})$*, and an operator* $\mathcal{b} \in \mathcal{B}_1$ *is Fredholm iff its symbol* $\sigma(\mathcal{b}) \in B_1$ *is invertible.*

50.11 Let us now extend the algebra $\mathcal{B}$ of convolution operators with oscillating coefficients by adjoining the shift operators $\mathcal{W}_\tau$ (50.12). That is, we consider the C^*-algebra

$$\mathcal{B}_2 = C^*\left(\mathcal{A}, \{\mathcal{V}_h, \mathcal{W}_\tau\}\right), \quad h, \tau \in \mathbb{R}$$

generated by an algebra $\mathcal{A}$ 50.1 and the unitary group $\{\mathcal{V}_h, \mathcal{W}_\tau\}_{\tau,h\in\mathbb{R}}$ generated by the operators $\mathcal{V}_h$ 50.2 and $\mathcal{W}_h$ (50.12).

This situation is a bit more complicated than the previous two cases; namely:

(i) The group $\{\mathcal{V}_h, \mathcal{W}_\tau\}$ is not commutative. However this group is solvable and thus amenable.

(ii) The action of the corresponding (commutative(!)) group $\{t_h, \alpha_\tau\}$ of homeomorphisms of M (see (50.5), (50.13)) is not normal (recall B.25). In fact $G^{\pm\infty}$ $(= G)$ does not act topologically freely. This means that we cannot obtain here the symbolic calculus simply by applying Corollary B.30, but have to exploit a general result such as Theorem 16.13.

50.12 We denote by B_2 the algebra $C^*\left(A, \{V_h, W_\tau\}\right)$ generated by the algebra A (see 50.3) and the unitary group $\{V_h, W_\tau\}$ generated by V_h (50.6) and W_τ (50.14).

This algebra plays the role of symbols for the elements of $\mathcal{B}_2$ which is stated in the next

Theorem 50.13 *Let* $\mathcal{B}_2$ *and* B_2 *be those described in 50.11 and 50.12. Then the mapping*

$$\sigma : \mathcal{B}_2 \to B_2$$

given by

$$\sigma : \mathcal{a} \mapsto \sigma(\mathcal{a}), \quad a \in \mathcal{A}$$
$$\sigma : \mathcal{V}_h \mapsto V_h,$$
$$\sigma : \mathcal{W}_\tau \mapsto W_\tau,$$

*is a *-epimorphism, the kernel of which is the ideal K of compact operators in* $L^2(\mathbb{R})$*; and an operator* $\mathcal{b} \in \mathcal{B}_2$ *is Fredholm iff its symbol* $\sigma(\mathcal{b}) \in B_2$ *is invertible.*

Proof Let $\mathcal{A}_0 \subset \mathcal{A}$ be a subalgebra of elements of $\mathcal{A}$ the symbols of which vanish at $\pm\infty$, and let $A_0 \subset A$ be the corresponding subalgebra of symbols. We denote by $\mathcal{B}_2^0$ and B_2^0 the C^*-algebras

$$\mathcal{B}_2^0 = \underline{C}^*\left(A_0, \{\mathcal{V}_h, \mathcal{W}_\tau\}\right)$$
$$B_2^0 = \underline{C}^*\left(\mathcal{A}_0, \{\mathcal{V}_h, \mathcal{W}_\tau\}\right)$$

where we are using the notation of C.3.

The maximal ideal space M_0 of A_0 is equal to

$$M_0 = \mathbb{R} \cup \mathbb{R} \cup \mathbb{R}$$

and the group action (by homeomorphisms t_h and α_τ) on M_0 is normal.

Now an argument analogous to that used in the proof of Theorem 50.4, along with Corollary C.54, shows that

$$\mathcal{B}_2^0 / K = \left[\mathcal{B}_2^0\right] \overset{\sigma}{\cong} B_2^0$$

which in particular means that the algebras $\left[\mathcal{B}_2\right]$ and B_2 satisfy the condition (16.26) (with σ instead of φ).

So in order to establish the isomorphism

$$\left[\mathcal{B}_2\right] \overset{\sigma}{\cong} B_2$$

it is enough to check (in view of Theorem 16.13) the validity of (16.25), that is

$$P_\infty\left(\left[C^*\left(\mathcal{A}, \{\mathcal{V}_h, \mathcal{W}_\tau\}\right)\right]\right) \cong P_\infty\left(C^*\left(A, \{V_h, W_\tau\}\right)\right). \tag{50.15}$$

Note that the ideal $\mathcal{J}^\infty$ (16.1) for the algebra $[\mathcal{B}_2]$ coincides here with the algebra $[\mathcal{B}_2^0]$ and the corresponding ideal for B_2 (which we shall denote by $\tilde{\mathcal{J}}^\infty$) coincides with B_2^0.

For any element $\alpha \in \mathcal{A}$ we denote by $\bar{\alpha}$ the operator (belonging to $\mathcal{A}$) of multiplication by the function $\sigma(\alpha)|_{\overline{\mathbb{R}}}$.

Let $\bar{\mathcal{A}}$ be the C^*-subalgebra of $\mathcal{A}$ generated by these operators $\bar{\alpha}$. Clearly $\bar{\mathcal{A}} \cong C(\overline{\mathbb{R}})$. We denote by $\bar{\mathcal{B}}_2$ the subalgebra of $\mathcal{B}_2$ generated by $\bar{\mathcal{A}}, \mathcal{V}_h, \mathcal{W}_\tau$, that is

$$\bar{\mathcal{B}}_2 = C^*\left(\bar{\mathcal{A}}, \{\mathcal{V}_h, \mathcal{W}_\tau\}\right).$$

For any finite sum

$$\sum_{h,\tau} \alpha_{h,\tau} \mathcal{V}_h \mathcal{W}_\tau \in \mathcal{B}$$

we have

$$\left[\sum_{h,\tau} \alpha_{h,\tau} \mathcal{V}_h \mathcal{W}_\tau - \sum_{h,\tau} \bar{\alpha}_{h,\tau} \mathcal{V}_h \mathcal{W}_\tau\right] \in \mathcal{J}^\infty, \tag{50.16}$$

which means, in view of Corollary 2.77, that

$$[\mathcal{B}_2] / \mathcal{J}^\infty \cong [\bar{\mathcal{B}}_2] / \mathcal{J}^\infty \cap [\bar{\mathcal{B}}_2]. \tag{50.17}$$

Now for any element $a \in A$ we denote by $\bar{a}$ the operator (belonging to A) of multiplication by the function

$$\bar{a}(x) = \begin{cases} a(x), & x \in \overline{\mathbb{R}} \\ a(+\infty), & x \in \mathbb{R}^+ \\ a(-\infty), & x \in \mathbb{R}^-. \end{cases}$$

Let $\bar{A}$ be the C^*-subalgebra of A generated by the operators $\bar{a}$, and let $\bar{B}_2$ be the C^*-subalgebra of B_2 generated by $\bar{A}, V_h, W_\tau$.

An argument analogous to that used for the derivation of (50.17) shows that

$$B_2 / \tilde{\mathfrak{J}}^\infty \cong \overline{B}_2 / \tilde{\mathfrak{J}}^\infty \cap \overline{B}_2 . \tag{50.18}$$

Now note that

$$\left[\overline{\mathcal{B}}_2\right] \overset{\sigma}{\cong} \overline{B}_2$$

and under this isomorphism the ideal $\mathfrak{J}^\infty \cap \left[\overline{\mathcal{B}}_2\right]$ is mapped into $\tilde{\mathfrak{J}}^\infty \cap \overline{B}_2$, which along with (50.17) and (50.18) means that (50.15) is true and the proof is complete. □

Remark 50.14 It is worth mentioning that for the algebra $\overline{\mathcal{B}}_2$ introduced in the proof we have

$$\left[\overline{\mathcal{B}}_2\right] \cong \overline{\mathcal{B}}_2 .$$

Example 50.15 Let us consider the operator

$$\mathcal{L} = a_0^0 + \sum_{j=1}^{m_0} a_j^0 \Phi_j^0 d_j^0 + \left(a_0^1 + \sum_{j=1}^{m_1} a_j^1 \Phi_j^1 d_j^1 \right) W_\tau$$

$$+ \left(a_0^2 + \sum_{j=1}^{m_2} a_j^2 \Phi_j^2 d_j^2 \right) \mathcal{V}_h + \left(a_0^3 + \sum_{j=1}^{m_3} a_j^3 \Phi_j^3 d_j^3 \right) \mathcal{V}_h W_\tau + k \tag{50.19}$$

where a_j^k, d_j^k are the operators of multiplication by continuous stabilizing functions a_j^k, d_j^k and Φ_j^k are the operators defined by (50.1) (for φ_j^k) and k is a compact operator.

In view of Theorem 50.13 we can write out the symbol $\sigma(\mathcal{L})$ as an operator in $L^2(\overline{\mathbb{R}}) \oplus L^2(\mathbb{R}^+) \oplus L^2(\mathbb{R}^-)$ having the form

$$\sigma(\mathcal{L}) = \sigma_1(\mathcal{L}) \oplus \sigma_+(\mathcal{L}) \oplus \sigma_-(\mathcal{L})$$

$$= \left(a + d W_\tau\right) \oplus \left(a_+ + d_+ \mathcal{W}_h\right) \oplus \left(a_- + d_- \mathcal{W}_h\right) \tag{50.20}$$

where $a, d, a_\pm, d_\pm$ are the operator of multiplication by the functions

$$a(x) = a_0^0(x) + a_0^2 e^{ihx}$$

$$d(x) = a_0^1(x) + a_0^3(x) e^{ihx}$$

$$a_\pm(x) = a_0^0(\pm\infty) + \sum_{j=1}^{m_0} a_j^0(\pm\infty) d_j^0(\pm\infty) \hat{\varphi}_j^0(x) \tag{50.21}$$

$$+ \left[a_0^1(\pm\infty) + \sum_{j=1}^{m_1} a_j^1(\pm\infty) d_j^1(\pm\infty) \hat{\varphi}_j^1(x) \right] e^{-ix\tau}$$

$$d_\pm(x) = a_0^2(\pm\infty) + \sum_{j=1}^{m_2} a_j^2(\pm\infty) d_j^2(\pm\infty) \hat{\varphi}_j^2(x)$$

$$+ \left[a_0^3(\pm\infty) + \sum_{j=1}^{m_3} a_j^3(\pm\infty) d_j^3(\pm\infty) \hat{\varphi}_j^3(x) \right] e^{-ix\tau} .$$

All the operators $\sigma_1(\mathcal{L}), \sigma_+(\mathcal{L}), \sigma_-(\mathcal{L})$ forming the symbol $\sigma(\mathcal{L})$ are two-term functional operators with the standard shift. The coefficients of these operators are not stabilizing but oscillating at infinity.

Let us start, for example, from consideration of the operator $\sigma_1(\mathcal{L})$.

We shall denote by $\mathcal{A}_h$ ($h \neq 0$ is a fixed real number) the C^*-algebra generated by stabilizing functions and the function e^{ihx}. One can easily calculate the maximal ideal space of $\mathcal{A}_h$. Indeed the maximal ideal space of the C^*-algebra of stabilizing functions is $\overline{\mathbb{R}}$ and the maximal ideal space of the C^*-algebra generated by e^{ihx} is the unit circle. Thus the maximal ideal space M_h of $\mathcal{A}_h$ is a subset of $\overline{\mathbb{R}} \times S^1$, that is it could be written out in the form

$$M_h = \Big\{ (x,z) : x \in \overline{\mathbb{R}}, |z| = 1 \text{ and } z = e^{ihx} \text{ when }$$
$$x \in \mathbb{R} \text{ and } z \text{ is arbitrary when } x = \pm\infty \Big\} .$$

That is M_h is the spiral with two attached circles at $\pm\infty$ (Fig. 8).

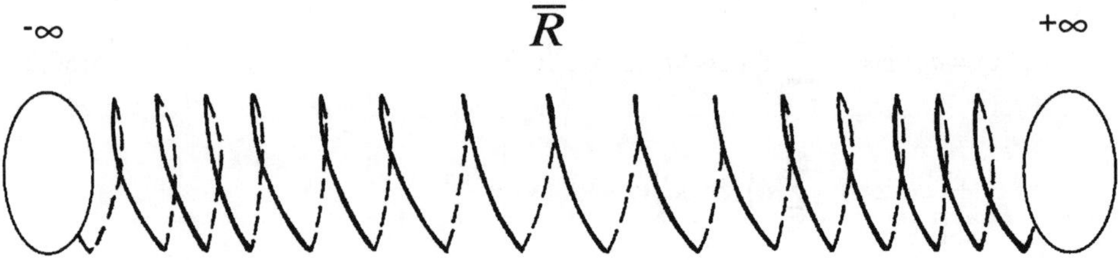

Fig. 8

The operator $\mathcal{W}_\tau$ generates an automorphism of the algebra $\mathcal{A}_h$. The corresponding homeomorphism γ_τ of M_h is given by

$$\gamma_\tau(x,z)=\left(x+\tau,\mathrm{e}^{ih\tau}z\right). \tag{50.22}$$

Thus on the spiral γ_τ acts as a shift and on the infinitely distant circles as a rotation.

Recall that the invertibility conditions of the two-term operators are closely related to the ergodic measures with respect to γ_τ on M_h (Theorem 17.3). In this situation each of these measures is concentrated on one of the infinitely distant circles and has the form described in Example 5.11.

Note also that there is no correlation between the coefficients a and d and $a_\pm$ and $d_\pm$ which means that in general for each of the operators $\sigma_1(\mathcal{b}), \sigma_\pm(\mathcal{b})$ two typical invertibility conditions could be realized ((1) or (2) in Theorem 17.3). Thus we can arrive at $8=2^3$ different invertibility conditions for $\sigma(\mathcal{b})$.

In order not to crowd the exposition we shall write out here only the invertibility conditions for $\sigma_1(\mathcal{b})$ (the invertibility conditions for $\sigma_\pm(\mathcal{b})$ have an analogous form).

Theorem 50.16 *The operator* $\sigma_1(\mathcal{b})=a+d\mathcal{W}_\tau$ *with* a *and* d *defined by (50.21) is invertible if and only iff one of the following conditions (i) or (ii) is satisfied:*

(i) (1) $a(x)=a_0^0(x)+a_0^2(x)\mathrm{e}^{ihx}\neq 0$ *for every* $x\in\mathbb{R}$,

(2) $\left|a_0^0(\pm\infty)\right|\neq\left|a_0^2(\pm\infty)\right|$, *and*
if $\dfrac{h\tau}{2\pi}$ *is irrational then*

(3′) $\max\{|a_0^0(\pm\infty)|,|a_0^2(\pm\infty)|\} > \max\{|a_0^1(\pm\infty)|,|a_0^3(\pm\infty)|\}$,

or

if $\dfrac{h\tau}{2\pi} = \dfrac{m}{N}$ *is an irreducible fraction then*

(3″) $\prod_{k=0}^{N-1} \left|a_0^0(\pm\infty) + a_0^2(\pm\infty)e^{i(\varphi+kh\tau)}\right| > \prod_{k=0}^{N-1} \left|a_0^1(\pm\infty) + a_0^3(\pm\infty)e^{i(\varphi+kh\tau)}\right|,\ 0 \le \varphi < 2\pi;$

(ii) (1) $d(x) = a_0^1(x) + a_0^3(x)e^{ihx} \neq 0$ *for every* $x \in \mathbb{R}$,

(2) $|a_0^1(\pm\infty)| \neq |a_0^3(\pm\infty)|$ *and*

if $\dfrac{h\tau}{2\pi}$ *is irrational then*

(3′) $\max\{|a_0^1(\pm\infty)|,|a_0^3(\pm\infty)|\} > \max\{|a_0^0(\pm\infty)|,|a_0^2(\pm\infty)|\}$,

or

if $\dfrac{h\tau}{2\pi} = \dfrac{m}{N}$ *is an irreducible fraction then*

(3″) $\prod_{k=0}^{N-1} \left|a_0^1(\pm\infty) + a_0^3(\pm\infty)e^{i(\varphi+kh\tau)}\right| > \prod_{k=0}^{N-1} \left|a_0^0(\pm\infty) + a_0^2(\pm\infty)e^{i(\varphi+kh\tau)}\right|,\ 0 \le \varphi < 2\pi.$

Proof Since M_h is connected and the mapping γ_τ acts topologically freely on M_h the invertibility conditions of $a + dW_\tau$ could be obtained by means of Theorem 17.3.

Clearly the conditions (1) and (2) in (i) and (ii) simply express the conditions for invertibility, of the operators a or d, respectively. So we have checked the corresponding inequalities for the geometric means $S(a)$ and $S(d)$ with respect to the ergodic measures.

The support of each ergodic measure lies on an infinitely distant circle and since the action of γ_τ on each of these circles is a rotation (see (50.22)) it follows (recalling Example 5.11) that:

(a) if $\dfrac{h\tau}{2\pi} = \dfrac{m}{N}$ is an irreducible fraction then each such measure is concentrated on N points (Example 5.11 (a)); while

(b) if $\dfrac{h\tau}{2\pi}$ is irrational then the corresponding ergodic measure is the normalized Lebesque measure of the circle (Example 5.11 (b) and Remark 5.12 (ii)).

If we denote by $a_{\pm}(\varphi)$ and $d_{\pm}(\varphi)$, $0 \le \varphi < 2\pi$, the restriction of the elements a and d (as functions of $C(M_h)$) onto the infinitely distant circle at $\pm\infty$ then we have

$$a_{\pm}(\varphi) = a_0^0(\pm\infty) + a_0^2(\pm\infty)\mathrm{e}^{\mathrm{i}\varphi}$$

and

$$d_{\pm}(\varphi) = a_0^1(\pm\infty) + a_0^3(\pm\infty)\mathrm{e}^{\mathrm{i}\varphi}$$

and clearly the inequalities (3″) in (i) and (ii) express precisely the corresponding inequalities for the geometric means mentioned in the statement of Theorem 17.3 (bearing in mind the foregoing observation (a)).

Thus to complete the proof we have to verify that if $\dfrac{h\tau}{2\pi}$ is irrational then the corresponding inequalities for the geometric means are transformed into (3′).

This is stated in the next

Lemma 50.17 *Let*

$$f(\varphi) = d_0 + d_1\mathrm{e}^{\mathrm{i}\varphi}, \quad d_0, d_1 \in \mathbb{C},\ \varphi \in [0, 2\pi]\,.$$

Then the geometric mean is

$$S(f) = \exp\frac{1}{2\pi}\int_0^{2\pi} \ln|f(\varphi)|\,\mathrm{d}\varphi = \max\{|d_0|, |d_1|\}\,.$$

Proof If $d_0 \cdot d_1 = 0$ the statement is obviously true. So let $d_0 \cdot d_1 \ne 0$.

If we rewrite $f(\varphi)$ in the form

$$f(\varphi) = \begin{cases} d_0\left(1 + \dfrac{d_1}{d_0}\mathrm{e}^{\mathrm{i}\varphi}\right), & \text{when } |d_0| \ge |d_1|, \\[2ex] d_1\mathrm{e}^{\mathrm{i}\varphi}\left(1 + \dfrac{d_0}{d_1}\mathrm{e}^{-\mathrm{i}\varphi}\right), & \text{when } |d_0| < |d_1| \end{cases}$$

then we have

$$|f(\varphi)| = \max\{|d_0|, |d_1|\} \cdot \left|1 + r\mathrm{e}^{\mathrm{i}(\psi+\theta)}\right| \tag{50.23}$$

where

$$r=\left|\frac{d_1}{d_0}\right|,\quad \theta=\arg\frac{d_1}{d_0},\quad \psi=\varphi \text{ when } |d_0|\geq|d_1|$$

and

$$r=\left|\frac{d_0}{d_1}\right|,\quad \theta=\arg\frac{d_0}{d_1},\quad \psi=-\varphi \text{ when } |d_0|<|d_1|,$$

and consequently the calculation of the geometric mean of f is reduced to the calculation of the geometric mean of the function $\left|1+re^{i\psi}\right|$.

Since

$$\left|1+re^{i\psi}\right|^2=1+2r\cos\psi+r^2$$

when calculating the geometric mean we deal with the known Poisson integral (see Fihtengolts [1], Volume II, Example 14 in subsection 314)

$$\frac{1}{2\pi}\int_0^{2\pi}\ln\left(1+2r\cos\psi+r^2\right)d\psi=\begin{cases}0, & \text{if } r\leq 1,\\ \ln r, & \text{if } r>1.\end{cases}$$

Since, in the situation considered here, we have $r\leq 1$, this integral is equal to zero, which along with (50.23) implies

$$S(f)=\max\{|d_0|,|d_1|\}$$

thus completing the proof of the lemma and the theorem as a whole. □

Now we consider

Multidimensional convolution operators with oscillating coefficients and shifts

50.18 Let $\mathcal{A}$ be a C^*-algebra of operators in $L^2\left(\mathbb{R}^n\right)$, $n>1$, generated by the operators of multiplication by continuous functions having limit at infinity (that is the functions from $C\left(\overline{\mathbb{R}^n}\right)$ where $\overline{\mathbb{R}^n}$ is the one-point compactification of $\mathbb{R}^n$; as in the previous situation we shall call these functions *stabilizing functions*) and the operators of the form

$$(\Phi f)(x)=\left(F^{-1}\varphi(\xi)Ff\right)(x) \tag{50.24}$$

where $\varphi \in C(\overline{\mathbb{R}}^n)$. The elements of $\mathcal{A}$ will be called *convolution type operators.*

The factor algebra $\mathcal{A}/K$ is commutative and its maximal ideal space M is equal to the pair of extended spaces $\overline{\mathbb{R}}^n$ and $\overline{\mathbb{R}}^n_\infty$ $(\overline{\mathbb{R}}^n_\infty \cong \overline{\mathbb{R}}^n)$ having their infinite points glued together. The symbol $\sigma(a)$ of an element $a \in \mathcal{A}$ is defined by the rule

(1) if a is the operator of multiplication by a function $a \in C(\overline{\mathbb{R}}^n)$ then

$$\sigma(a)(x) = \begin{cases} a(x), & x \in \overline{\mathbb{R}}^n, \\ a(\infty), & x \in \overline{\mathbb{R}}^n_\infty; \end{cases}$$

(2) if Φ is the operator given by (50.24) then

$$\sigma(\Phi)(x) = \begin{cases} \varphi(\infty), & x \in \overline{\mathbb{R}}^n, \\ \varphi(x), & x \in \overline{\mathbb{R}}^n_\infty. \end{cases}$$

50.19 Let us pass at once to the multidimensional analogue of Theorem 50.13.

We adjoin to the algebra $\mathcal{A}$ the operators $\mathcal{V}_h$ and $\mathcal{W}_\tau$, $h, \tau \in \mathbb{R}^n$, defined by the formulae

$$(\mathcal{V}_h f)(x) = e^{i\langle h,x\rangle} f(x)$$

$$(\mathcal{W}_h f)(x) = f(x-\tau)$$

where $\langle h, x\rangle = h_1x_1 + \ldots + h_nx_n$.

Observe that the group $G = \{\mathcal{V}_h, \mathcal{W}_\tau\}_{h,\tau\in\mathbb{R}^n}$ generated by $\mathcal{V}_h$ and $\mathcal{W}_\tau$ is solvable. Indeed G contains a commutative subgroup $G_1 = \{\lambda \mathcal{W}_\tau,\ \tau \in \mathbb{R}^n, |\lambda| = 1\}$ and the factor group G/G_1 is naturally isomorphic to the group $\{\mathcal{V}_h\}_{h\in\mathbb{R}^n}$ which is commutative as well.

The operators $\mathcal{V}_h$ and $\mathcal{W}_\tau$ generate the automorphisms of $\mathcal{A}$.

We now denote by $\mathcal{B} = C^*(\mathcal{A}, \{\mathcal{V}_h, \mathcal{W}_\tau\})$ the C^*-algebra generated by $\mathcal{A}, \mathcal{V}_h, \mathcal{W}_\tau$. As has just been noticed the group $G = \{\mathcal{V}_h, \mathcal{W}_\tau\}$ is amenable.

50.20 In the space

$$L^2(M) = L^2(\overline{\mathbb{R}}^n) \oplus L^2(\overline{\mathbb{R}}^n_\infty) \cong L^2(\mathbb{R}^n) \oplus L^2(\mathbb{R}^n)$$

we introduce the operators

$$V_h(u_1,u_2)=\left(e^{i\langle h,x\rangle}u_1(x),u_2(x-h)\right)$$

$$W_\tau(u_1,u_2)=\left(u_1(x-\tau),e^{i\langle \tau,x\rangle}u_2(x)\right)$$

and consider the algebra

$$B=C^*\left(A,\{V_h,W_\tau\}\right)$$

generated by the algebra $A\cong C(M)$ of operators of multiplication by functions $a\in C(M)$ and the operators $V_h, W_\tau,\ h,\tau\in\mathbb{R}^n$.

An argument analogous to that used in the proof of Theorem 50.13 shows that B is the algebra of symbols of elements of $\mathcal{B}$.

Thus we arrive at

Theorem 50.21 *Let $\mathcal{B}$ and B be as described in 50.18 and 50.19. Then the mapping*

$$\sigma:\mathcal{B}\to B$$

given by

$$\sigma(\mathcal{a}):\mathcal{a}\to\sigma(\mathcal{a}),\quad \mathcal{a}\in\mathcal{A}$$
$$\sigma:\mathcal{V}_h\mapsto V_h,$$
$$\sigma:\mathcal{W}_\tau\mapsto W_\tau,$$

*is a *-epimorphism, the kernel of which is the ideal of compact operators in $L^2\left(\mathbb{R}^n\right)$. An operator $\mathcal{b}\in\mathcal{B}$ is Fredholm iff its symbol $\sigma(\mathcal{b})$ is invertible.*

50.22 The symbol $\sigma(\mathcal{b})$ just described is the direct sum of two operators with standard shifts in $\mathbb{R}^n$ having coefficients which may oscillate at infinity.

Here we can single out a number of situations when the invertibility conditions of $\sigma(\mathcal{b})$ can be written out in explicit form.

For example

(i) the operators forming $\sigma(\mathcal{b})$ are two-term operators;

(ii) one of the two operators forming $\sigma(\mathcal{b})$ is an operator with constant coefficients;

(iii) the coefficients of the operators forming $\sigma(\mathcal{b})$ are stabilizing functions and entering shift operators are generated by shifts forming a rationally independent convex system (recall Section 24).

Every so often we arrive at some interesting maximal ideal spaces. In the situation analogous to Example 50.15, say we have to consider the algebra $\mathcal{A}_h$ generated by continuous stabilizing functions (on $\mathbb{R}^n$) and the function $e^{i\langle h,x\rangle}$ ($h \neq 0$ is a fixed vector). The maximal ideal space M_h of $\mathcal{A}_h$ is the compactification of $\mathbb{R}^n$ by an infinitely distant circle. It could be realized as a subset of $\overline{\mathbb{R}}^n \times S^1$ given by

$$M_h = \Big\{(x,z): x \in \overline{\mathbb{R}}^n, |z| = 1 \text{ and } z = e^{i\langle h,x\rangle} \text{ when } x \in \mathbb{R}^n \text{ and } z \text{ is arbitrary when } x = \infty\Big\}.$$

The corresponding invariant measures and the invertibility conditions of two-term operators here are in perfect analogy to those written out in Theorem 50.16.

50.23 In the algebra $\mathcal{B}_2$ considered in Theorem 50.13 and the algebra $\mathcal{B}$ considered in Theorem 50.20 one could single out a subalgebra containing only the 'shift' operators of the form $(\mathcal{V}_h \mathcal{W}_\tau)^k$ with the fixed $h \neq 0$ and $\tau \neq 0$. This subalgebra, although formed by functional integral operators with oscillating coefficients, is simpler than $\mathcal{B}_2$ ($\mathcal{B}$) from the following point of view.

The operator $\mathcal{T}_{h,\tau} = \mathcal{V}_h \mathcal{W}_\tau$ generates an automorphism of the algebra $\mathcal{A}$ of convolution type operators and the corresponding homeomorphism $\alpha_{h,\tau}$ of the maximal ideal space M of the factor algebra $\mathcal{A}/K$ has the form

$$\alpha_{h,\tau}(x) = (x+\tau, x+h, x+h) \quad (\text{for } \mathcal{B}_2),$$
$$\alpha_{h,\tau}(x) = (x+\tau, x+h) \quad (\text{for } \mathcal{B}).$$

The simplifications arising here are the following:

(i) $\alpha_{h,\tau}$ acts topologically freely, thus we could obtain the proof of the symbolic calculus only by using the isomorphism theorem (Corollary 12.17) not including any checking of extra conditions (arising in Theorem 16.13 and Corollary B.30).

(ii) Although we consider operators with oscillating coefficients the corresponding functional operators forming the symbol appear to be the operators with stabilizing coefficients which again simplifies their investigation.

We come now to the final point of this section.

Discrete convolution operators

50.24 Let us denote by T, the 'left shift' operator in the space $l^2 = l^2(\mathbb{Z})$ of bilateral sequences

$$(Tf)(n) = f(n-1) \ .$$

The operator

$$\sum a_k T^k, \quad a_k \in \mathbb{C}$$

is the discrete convolution operator. These operators generate the C^*-algebra $C^*(\mathbb{C}, T)$ which possesses the property (*) (see 12.2) and is isomorphic to the algebra $C(S^1)$ of continuous functions on the unit circle and the isomorphism is given by the formula

$$\sum a_k T^k \to \sum a_k z^k, \quad |z| = 1 \ .$$

Let $\mathcal{A}$ be some C^*-algebra of bilateral bounded sequences. We consider the elements of this algebra as the operators in l^2 of multiplication by the corresponding sequences. We denote by $\mathcal{B}$ the C^*-algebra generated by $\mathcal{A}$ and T. We call the elements of $\mathcal{B}$ *discrete convolution type operators.*

Let us assume further that $\mathcal{A}$ is shift invariant, that is along with a sequence $\{a(j)\} \in \mathcal{A}$ the sequences $\{a(j-1)\}$ and $\{a(j+1)\}$ belong to $\mathcal{A}$ as well. Under this assumption the operator T generates an automorphism of $\mathcal{A}$ and the investigation of the algebra $\mathcal{B} = C^*(\mathcal{A}, T)$ can be carried out with the help of the theory already developed.

For example, the invertibility conditions of a two-term operator could be obtained by means of the general result (Theorem 17.3) (the action here is always topologically free and the space $M(\hat{T})$ is connected (recall 18.1)).

50.25 More substantive in this situation is the problem of obtaining the corresponding symbolic calculus, that is the investigation of Fredholmity conditions of discrete convolution operators.

Note that the algebra $\mathcal{A}$ may contain compact operators (such operators correspond to sequences tending to zero at $\pm\infty$). The action of the automorphism generated by T on the maximal ideal space M of the factor algebra $A = \mathcal{A}/K$ is not necessarily topologically free (which makes the investigation of the structure of the algebra $C^*(\mathcal{A},T)/K$ non-elementary).

Let $\mathfrak{M}$ be the maximal ideal space of $\mathcal{A}$. The set $\mathbb{Z}$ (the domain of the sequences belonging to $\mathcal{A}$) is naturally everywhere densely embedded in $\mathfrak{M}$ and is an open subset of $\mathfrak{M}$. If $\mathcal{K} = K \cap \mathcal{A}$ is the ideal of compact operators in $\mathcal{A}$ then the foregoing observation shows that

$$k \in \mathcal{K} \Leftrightarrow k(m) = 0, \quad m \in \mathfrak{M} \setminus \mathbb{Z}$$

which implies (recall Theorem 2.83)

$$M = \mathfrak{M} \setminus \mathbb{Z}$$

and

$$A = C(M) .$$

Let us consider now some concrete examples.

Example 50.26 Let $\mathcal{A}$ be the C^*-algebra of stabilizing sequences (that is having limits at $\pm\infty$). In this situation we have

$$M = \{-\infty, +\infty\}$$

and the action of the automorphism of the algebra $A = \mathcal{A}/K$ induced by T is trivial (i.e. not topologically free).

Let

$$b = \sum_{k=-n}^{n} a_k T^k, \quad a_k \in \mathcal{A}$$

be a certain element of $\mathcal{B} = C^*(\mathcal{A},T)$. For any compact operator k there exists a sequence e_{n_j}, $n_j \to \pm\infty$ ($\{e_n\}$ is the canonical orthonormal basis in l^2) such that

$$ke_{n_j} \to 0 .$$

By using this sequence one could easily verify (see, for example, the reasoning in 12.3) that

$$\|\mathcal{b} + k\| \geq \max\{|a_0(+\infty)|, |a_0(-\infty)|\}$$

which means that the algebra

$$[\mathcal{B}] = \mathcal{B}/K = C^*([\mathcal{A}], [T])$$

possesses the property (*) (see 12.2).

This in turn (in view of Corollary 12.9) implies the isomorphism

$$[\mathcal{B}] \to C(S^1) \oplus C(S^1)$$

given by

$$[\mathcal{b}] = \left[\sum a_k T^k\right] \to \left(\sum a_k(+\infty)z^k, \sum a_k(-\infty)z^k\right). \tag{50.25}$$

Remark The property (*) for the functions in the right-hand side of (50.25) is satisfied by the Cauchy inequalities (recall Notes and Remarks to Chapter 2, Section 12).

So the right-hand side of (50.25) is in fact the record of the symbol $\sigma(\mathcal{b})$, $\mathcal{b} \in \mathcal{B}$ and we conclude that

$\mathcal{b}$ is Fredholm iff $\sum a_k(\pm\infty)z^k \neq 0, \quad |z| = 1.$

Example 50.27 Let $\mathcal{A}_1$ be the algebra generated by the algebra $\mathcal{A}$ considered in Example 50.26 and define the sequence

$$v = \{v(n)\} = \{e^{ihn}\}, \quad n \in \mathbb{Z} \tag{50.26}$$

where $\dfrac{h}{2\pi}$ is an irrational number.

The maximal ideal space $\mathfrak{M}_1$ of $\mathcal{A}_1$ is the set $\mathbb{Z}$ compactified by two infinitely distant circles (at $\pm\infty$) (recall the reasoning in Example 50.15) and by the argument of 50.25 we obtain that the maximal ideal space M_1 of the algebra $A_1 = \mathcal{A}_1/K = [\mathcal{A}_1]$ consists precisely of these two infinitely distant circles $S^1_{+\infty} \cup S^1_{-\infty}$ $\left(S^1_{\pm\infty} = \{z \in \mathbb{C} : |z| = 1\}\right)$.

Observe that an isomorphism

$$A_1 \cong C\left(S^1_{+\infty}\right) \oplus C\left(S^1_{-\infty}\right) \tag{50.27}$$

could be established by the formulae

$$\begin{array}{ll} \mathcal{A} \ni a \to \{a(+\infty), a(-\infty)\} \\ v \quad\ \ \to \{z, \qquad z\} \end{array} \tag{50.28}$$

where v is the sequence given by (50.26).

Indeed, clearly A_1 is isomorphic to the direct sum of two algebras $A_1(+\infty) \oplus A_1(-\infty)$ where

$$A_1(+\infty) = A_1(-\infty) = \tilde{A} \subset l^\infty(\mathbb{Z}) \tag{50.29}$$

is the uniform closure of sequences of the form $\sum d_k v^k$, $d_k \in \mathbb{C}$, and the isomorphism is given by

$$\begin{array}{ll} \mathcal{A} \ni a \to \{a(+\infty), a(-\infty)\} \\ v \quad\ \ \to \{v, \qquad v\} \end{array} \tag{50.30}$$

where $a(\pm\infty)$ are constant sequences (with values equal to $a(\pm\infty)$).

Observe now that

$$\left\|\sum d_k v^k\right\|_{l^\infty} \geq |d_0| \,. \tag{50.31}$$

This could be verified directly or one could note that $\left\|\sum d_k v^k\right\|_{l^\infty}$ coincides with the norm of the operator in $l^2(\mathbb{Z})$ of multiplication by this sequence and under the isomorphism $l^2(\mathbb{Z}) \cong L^2\left(S^1\right)$ established by

$$l^2 \ni \{c_n\}_{n\in\mathbb{Z}} \to \sum_{-\infty}^{+\infty} c_n z^n \in L^2\left(S^1\right)$$

the operator of multiplication by $\sum d_k v^k$ is mapped into the operator $\sum d_k U^k_{-h}$ where

$$(U_{-h} f)(z) = f\left(e^{ih} z\right) . \tag{50.32}$$

Since $\frac{h}{2\pi}$ is irrational the automorphism (of $C(S^1)$) generated by U_{-h} acts topologically freely (on S^1) which means by Theorem 12.14 that

$$\left\|\sum d_k U_{-h}^k\right\| \geq |d_0|$$

and consequently (50.31) is true.

Recalling now that by the Cauchy inequalities we have

$$\left\|\sum d_k z^k\right\|_{C(S^1)} \geq |d_0| \tag{50.33}$$

(see Notes and Remarks to Chapter 2, Section 12) we conclude (applying Corollary 12.9) that

$$A_1(+\infty) \oplus A_1(-\infty) \cong C\left(S^1_{+\infty}\right) \oplus C\left(S^1_{-\infty}\right)$$

and the isomorphism is established via the formula

$$\left\{\sum a_k(+\infty)v^k, \sum a_k(-\infty)v^k\right\} \mapsto \left\{\sum a_k(+\infty)z^k, \sum a_k(-\infty)z^k\right\} \tag{50.34}$$

which proves (50.27).

Let T be the 'left shift' operator in $l^2(\mathbb{Z})$ mentioned in 50.24 and let $\mathcal{B}_1 = C^*(\mathcal{A}_1, T)$. By the reasoning of Example 50.15 the homeomorphism of $M_1 = S^1_{+\infty} \cup S^1_{-\infty}$ induced by the automorphism $[T](\,\cdot\,)\left[T^{-1}\right]$ of the algebra A_1 has the form

$$S^1_{\pm\infty} \ni z \mapsto e^{ih} z\,. \tag{50.35}$$

Let $\tilde{B}$ be the subalgebra of operators acting in $L^2\left(S^1\right)$ which is the uniform closure of the set of operators of the form

$$\sum d_k(z) U_h^k$$

where $d_k(\,\cdot\,) \in C\left(S^1\right)$ and U_h is defined by (50.32).

Since the automorphism $U_h(\ \cdot\ )U_h^{-1}$ of $C(S^1)$ defines the homeomorphism (50.35) of S^1 and the latter acts topologically freely it follows (by the isomorphism theorem 12.17) that

$$[\mathcal{B}_1] = C^*([\mathcal{A}_1],[T]) \cong B(+\infty) \oplus B(-\infty) \tag{50.36}$$

where $B(+\infty) = B(-\infty) = \tilde{B}$; and the isomorphism is given by

$$[a] \to (a(+\infty)(z), a(-\infty)(z)), \quad a \in \mathcal{A}_1$$
$$[T] \to (U_h, U_h)$$

where $a(\pm\infty)(z)$ are the operators of multiplication by the functions from $C(S^1_{\pm\infty})$ defined by (50.28).

Observe also that the algebra $\approx{B} \subset L(l^2(\mathbb{Z}))$ which is the uniform closure of the set of operators of the form

$$\sum a_k(n) T^k$$

where a_k belongs to the algebra $\tilde{A}$ (50.29) possesses the property (*) (which could be verified say by the argument of 12.3). Therefore the foregoing reasoning, along with Corollary 12.9 and Theorem 12.14, means that

$$\tilde{B} \cong \approx{B}$$

and consequently (in view of (50.36))

$$[\mathcal{B}_1] \cong \approx{B} \oplus \approx{B}$$

and the isomorphism is given by

$$[a] \to (a(+\infty)(n), a(-\infty)(n)), \quad a \in \mathcal{A}_1$$
$$[T] \to \{T, \qquad T\}$$

where $a(\pm\infty)(n)$ are the operators of multiplication by the sequences defined by (50.30).

Notes and Remarks

Section 36 The facts described in Lemma 36.4 are well known but we do not have at hand a source to quote. Lemma 36.5 is in fact a two space version of the corresponding C^*-result and the idea of the proof is a two space reinterpretation of the corresponding C^*-proof (cf. for example Dixmier [1], 1.8.2). This lemma is borrowed from Belousov, Lebedev [1].

Sections 37, 39 The results presented here have a preparatory character preceding the general construction of the symbol of *G*-PDOs on a manifold. Nevertheless most of the results of these sections are published here for the first time. Among the known results concerning the operators considered here (except those dealing with *G*-PDOs with *constant* coefficients and *standard shifts* in $\mathbb{R}^n$, which can be easily investigated by means of the Fourier transform) we distinguish the work by Rabinovich [1,2] who obtained statements analogous to Theorem 37.4 and Lemma 37.6 for differential difference and pseudodifferential difference equations having a more complicated behaviour of the symbol at infinity in comparison with that considered in the theorem and lemma mentioned above.

The two-term partial differential equations containing a linear transformation of an argument were considered by Antonevich, Lebedev [2] who constructed the symbols for such operators and for particular cases the Fredholmity conditions were calculated.

We would like to recall that the Cauchy problem for partial differential equations in a half-space containing linear transformations of an argument, along with certain boundary value problems for such equations in a strip, have been considered in the work by V.M. Borok and Ya. I. Zhitomirskij and their followers (see for example Borok [1], Borok, Zhitomirskij [1]). The boundary value problems for differential difference equations in bounded domains have been considered by A.L. Skubachevskij, but these problems are of a somewhat different type: here the deviations of an argument in view of the boundness of a domain link the value at a point of the domain with the values only at a *finite* number of other points and locally (at the typical points) the problem is structured (as in the finite group case) as a system of equations of higher order. The complexity of the problems considered by A.L. Skubachevskij is related to the behaviour of the solutions at the points which are the images of the boundary under the shifts contained in the equation. In these points the solution may possess a singularity and the main job here is associated with the investigation of these singularities (see Skubachevskij [1,2], Kamenskij, Skubachevskij [1]).

It is worth mentioning also that the applications of the theory of C^*-algebras to the development of the theory of PDOs (not *G*-PDOs!) have been considered in the work of a number of researchers (see, for example Cordes [1-3], Gramsh [1], Gramsh, Kalb [1], Gramsh, Kaballa [1], Schrohe [1,2]).

Section 38 The problems related to the study of the behaviour of the vectors $L^n x, n \to \infty$, for a given matrix x are quite classical but we do not know the sources containing a total description of this behaviour (like that presented in Theorem 38.6). A result close to the statement of this theorem was obtained, for example, by Godunov, Zabreiko [1] in connection with the subtle analysis of the convergence character of successful approximations. For a particular case of the matrix L the description of the corresponding invariant measures on the sphere was in fact given by Antonevich, Lebedev [2].

The principal results of this section are borrowed from Antonevich, Buraczewski [1].

In connection with the results of Sections 38, 39 we would also like to mention the paper by Antonevich, Serigue Aliu Lo [1] where all the ergodic measures arising in the process of the explicit calculation of the Fredholmity conditions for a two-term *G*-PDO with affine shifts are written out.

Sections 40-42 The construction of the symbol of *G*-PDOs on a manifold in the case of a finite group G was obtained by Antonevich [2,5]. The C^*-approach for this situation was presented in Antonevich [4]. Further, for a number of concrete two-term *G*-PDOs (for $G = \mathbb{Z}$), Antonevich [6,11] has constructed the principal symbol and obtained the Fredholmity conditions by means of explicit bulky computation. Lebedev [4,7] derived the first 'simplest' variant of the isomorphism theorem (Corollary 12.17) (for $G = \mathbb{Z}$ and $A \cong C(X)$), and the construction of the symbol of *G*-PDOs was developed by Antonevich, Lebedev [1] and Lebedev [7] for $G = \mathbb{Z}$ on the basis of the theory of C^*-algebras generated by dynamical systems. Further advances (which have proceeded along with advances in the discovery of results similar to the isomorphism theorem) have been obtained in the work by Antonevich, Brenner [2], Brenner [2,3], Antonevich [13,19,18]. We mention here also the work of Karlovich [4,6] and Batkhin [1-4] concerning *G*-SIOs and multidimensional convolution type operators.

The dependence of the symbol on the exponent of the Sobolev space was first noted by Antonevich, Brenner [2].

The symbolic calculus for *G*-PDOs in the situation when G is not acting topologically freely (but only normally) was described by Antonevich, Lebedev [4] and we would like to emphasize that again it was obtained after the derivation of an isomorphism theorem type result (Corollary B.30) proved by Lebedev [19]. The construction of the 'homogeneous' symbol is also given in Antonevich, Lebedev [4] (the bigraded pseudoalgebra for G-PDOs acting in Sobolev spaces was introduced by Antonevich [20]).

Section 43 This section follows the work by Antonevich [2,5] and its algebraic part *not related to the nature of PDOs* is carried out in Appendix A.

Section 44 The results 44.2-44.6 were obtained by Antonevich [6]. The remaining results are published here for the first time. (With regard to the conditions for the invertibility of the *functional* operators (symbols) considered here, it is reasonable to mention the paper by Lebedev [5].)

Section 45 The spectral properties of the functional operators related to a Morse-Smale diffeomorphism were studied by Lebedev [3], Theorems 45.11 and 45.13 are in essence contained in Antonevich [20].

Lemma 45.5 is probably known but we do not have a source to quote.

Sections 46, 47 Certain information on the history of the research related to G-SIOs was presented in the Introduction to this volume.

The results contained in Section 46 were obtained here as particular cases of the general theorems on G-PDOs. Analogous results have been obtained (by means of the 'SIO nature' methods) in a number of papers by Ju. I. Karlovich and V.G. Kravchenko (see for example Karlovich, Kravchenko [4]). A more detailed review of this direction of the research is given in the monograph by Litvinchuk [2], the survey by Karlovich, Kravchenko, Litvinchuk [1] and the monograph by Kravchenko, Litvinchuk [1]. As we have already mentioned (see Notes and Remarks to Sections 40-42) the C^*-algebraic approach especially oriented to the investigation of G-SIOs has been developed by Karlovich [1,6].

Theorems 46.12, 46.13 are due to Kveselava [1].

Theorem 46.16 was obtained by Kravchenko [1] in the situation when α has a finite set of fixed points.

Theorem 46.20 is borrowed from Antonevich [18] (Theorem 9.4); a close result was proved by Karlovich, Kravchenko [4].

In connection with Theorems 47.2, 47.4 we should again mention the paper by Karlovich, Kravchenko [4] where the index calculation for these operators in the case when α has a finite set of fixed points is written out.

The index of a two-term G-SIO in the space of scalar functions (Example 47.16) was calculated by Karlovich, Kravchenko [1] (see also Kravchenko, Litvinchuk [1]). The corresponding 'vector valued' result (Theorem 47.14) is published here for the first time (here we follow Antonevich [21]). In connection with this result we would like to mention the paper by Antonevich, Gankin [1] where the homotopic classification of Fredholm two-term G-SIOs in the space of vector-valued functions was described and all the homotopical invariants (including the index) were written out.

Section 48 The symbolic calculus for G-Toeplitz operators with continuous coefficients (for the case $G = \mathbb{Z}$) was obtained for the first time by Lebedev [7,9].

Section 49 The name 'non-local boundary conditions' is applied to various classes of boundary conditions as the negation of the localness property. The work by Browder [1], Beals [1], Schehter [1], Grubb [1] and other authors, considered an approach to the investigation of general non-local boundary value problems for elliptic equations with boundary condition operators, the explicit form of which has not been specified. Therefore the results had ineffective character in view of the great generality of the setting of the problems.

In 1969 Bitsadze, Samarskij [1] posed non-local boundary value problems (based on physical applications) where the boundary condition links the value of the solution sought at a point x of the boundary with that at the point $h(x)$, where h is a given mapping of the boundary into the domain. In succeeding years the work of many researchers has been devoted to the study of various aspects of the theory of non-local boundary value problems (recall in this convection the names mentioned in the Introduction to this volume).

In the setting presented in Section 49 the non-local boundary value problem for the case of a periodic mapping of the boundary onto itself for an elliptic equation in a two-dimensional domain was investigated by Kordzadze [1]. Non-local boundary value problems generated by a finite group of the boundary were considered by Antonevich [1,3], and non-local boundary value problems for functional differential equations generated by a finite group of diffeomorphisms of the domain were studied by Antonevich, Dobrushkin [1].

The presentation here follows Antonevich [20].

In connection with the results of this section see also Chapter 7.

Section 50 One-dimensional convolution type operators with oscillating coefficients were considered by Antonevich [8] and by Karapetyants, Samko [3], §21 (where one will also find a description of the Fredholmity conditions for the discrete convolution operators with almost stabilizing coefficients and with coefficients having 'periodic' oscillation). In connection with these results we mention also the work by Karlovich [3,6] which contains a presentation of a systematic investigation of the Fredholm property of certain C^*-algebras of convolution type operators with shifts and coefficients of a much more general sort and in particular a description of the corresponding symbolic calculus in terms of the 'trajectorial ideology' (cf. say Theorem 53.23).

Multidimensional convolution type operators with oscillating coefficients and shifts were studied by Antonevich, Hodzaj [1].

We mention here also the work by Rabinovich [3] where one can find a deep study (exploiting the limit operators method developed by the author) of pseudodifferential operators and multidimensional convolution type operators with coefficients having a complicated behaviour at infinity.

Appendix A
The index calculation in $C^*(A,T_g)$

'...you can't help respecting anybody who can spell TUESDAY, even if he doesn't spell it right...'
'What *is* Learning?
A thing *Rabbit* knows!
You mean this **A** thing is a thing *Rabbit* knew?' A.A. Milne *The House at Pooh Corner.*

In this section we present certain general C^*-algebraic ideas lying at the heart of the index calculation of elements of $C^*(A,T_g)$.

A.1 Let $A \subset L(H)$ be a certain C^*-algebra and $\mathcal{J} \subset L(H)$ be a closed two-sided ideal.
Let

$$G \ni g \to T_g \in L(H)$$

be a unitary representation of a *finite* group G such that

$$T_g A T_g^* = A, \; g \in G, \tag{A.1}$$

and

$$T_g \mathcal{J} T_g^* = \mathcal{J}, \; g \in G, \tag{A.2}$$

(A.1) along with (A.2) implies that the automorphism $\hat{T}_g$ of A, given by

$$\hat{T}_g(a) = T_g a T_g^*, \; a \in A \tag{A.3}$$

generates an automorphism $\left[\hat{T}_g\right]$ of the C^*-algebra

$$[A] = A/\mathcal{J} \, . \tag{A.4}$$

Remark Since (by Corollary 2.77)

$$[A] = A/\mathcal{J} = A+\mathcal{J}/\mathcal{J} = A/A\cap\mathcal{J}$$

we shall henceforth in this section assume that

$$\mathcal{J} \subset A \tag{A.5}$$

(we can always obtain this inclusion passing to the C^*-algebra generated by A and $\mathcal{J}$).

A.2 Let the action of $[\hat{T}_g]$ on Prim$[A]$ be topologically free. Then by Corollary 12.16

$$[C^*(A,T_g)] = C^*([A],[T_g]) \cong [A] \underset{[\hat{T}]}{\times} G \tag{A.6}$$

where by $[\ \cdot\ ]$ we denote the operation of taking the factor-image under the canonical mapping

$$L(H) \to [L(H)] = L(H)/\mathcal{J} \ .$$

Consequently an operator

$$b = \sum_G a_g T_g \in C^*(A,T_g) \tag{A.7}$$

is $\mathcal{J}$-Fredholm (recall 32.9) iff its image

$$[b] = \sum_G [a_g] \otimes g \in [A] \underset{[\hat{T}]}{\times} G \tag{A.8}$$

is invertible.

A.3 Let us introduce now the space

$$\tilde{H} = H^{|G|} = \underset{g \in G}{\oplus} H_g \quad (H_g \approx H) \tag{A.9}$$

($|G|$ is the number of elements in G) and the ideal $\tilde{\mathcal{J}} \subset L(\tilde{H})$ consisting of all the operator matrices $\{d_{h,g}\}_{h,g \in G}$ such that

$$d_{h,g} \in \mathcal{J} \ . \tag{A.10}$$

For any operator b given by (A.7) we denote by $\tilde{b}$ the operator acting in $\tilde{H}$ and defined by the operator matrix $\left\{\tilde{b}_{h,g}\right\}_{h,g\in G}$ where

$$\tilde{b}_{h,g} = \hat{T}_h\left(a_{h^{-1}g}\right). \tag{A.11}$$

Remark In general the operator b (A.7) does not determine uniquely its coefficients, that is the equality

$$\sum_{g\in G} a_g T_g = \sum_{g\in G} a'_g T_g$$

does not imply

$$a_g = a'_g$$

which means that in this case formula (A.11) is not defined correctly. But if the action $\left[\hat{T}_g\right]$ on Prim$[A]$ is topologically free (and this is the main situation we shall deal with in this section) then in view of Theorem 12.14 the coefficients a_g of b are uniquely defined (modulo $\mathcal{J}$) and consequently the operator $\tilde{b}$ is uniquely defined (modulo $\tilde{\mathcal{J}}$).

Bearing this in mind we write out the interrelation between the operators b and $\tilde{b}$ in the following

Lemma A.4 *Let A, $\mathcal{J}$, T_g, $g\in G$ be those mentioned in A.1 and let $\left[\hat{T}_g\right]$ act topologically freely on* Prim$[A]$. *Then the operator b (A.7) is $\mathcal{J}$-Fredholm iff the operator $\tilde{b}$ described in A.3 is $\tilde{\mathcal{J}}$-Fredholm.*

Proof It is enough to prove that the mapping

$$b \to \tilde{b}$$

establishes a C^*-isomorphism between the algebras $C^*(A,T_g)\big/\mathcal{J}$ and $\widetilde{C^*(A,T_g)}\big/\tilde{\mathcal{J}}$ where $\widetilde{C^*(A,T_g)}$ is the C^*-algebra generated by operators of the form $\tilde{b}$. Clearly the mapping

$$a \to \tilde{a},\ a \in A \tag{A.12}$$

establishes an isomorphism between the algebras $[A]$ and $[\tilde{A}]$ where

$$[A] = A/\mathcal{J} \text{ and } [\tilde{A}] = \tilde{A}/\tilde{\mathcal{J}}$$

where $\tilde{A}$ is the C^*-algebra generated by the elements $\tilde{a}$, $a \in A$ (in view of the remark in A.3 and (A.5) it follows that

$$\tilde{\mathcal{J}} \subset \tilde{A}). \tag{A.13}$$

By the definition of $\tilde{T}_g$ and $\tilde{a}$ (see A.3) it follows that

$$\left[\widetilde{T_g a T_g^*}\right] = \left[\tilde{T}_g\right][\tilde{a}](\tilde{T}_g^*)$$

which in view of the isomorphism theorem (Corollary 12.17) completes the proof. □

Remark Looking through the proof one can see that it was enough to assume only that $[T_g]$, $g \in G$, forms a unitary representation of G and $[T_g A T_g^*] = [A]$.

Corollary A.5 *Let* $\mathcal{J}$ *be the ideal of compact operators acting in* H *and let* $[\hat{T}_g]$ *act topologically freely on* Prim$[A]$. *Then the operator* b *(A.7) is Fredholm iff* $\tilde{b}$ *is Fredholm.*

Proof Just note that in this case the ideal $\tilde{\mathcal{J}}$ is the ideal of all the compact operators acting in $\tilde{H}$. Thus, $\mathcal{J}(\tilde{\mathcal{J}})$-Fredholmity coincides with the standard Fredholmity in the space $H(\tilde{H})$. □

Before passing to the index formulae we recall now certain basic facts of the theory of unitary representations of compact and finite groups. The proofs of results A.6-A.10 can be found for example in Dixmier [1], Section 15.

Lemma A.6 *Every irreducible representation of a compact group is finite dimensional.*

A.7 Let v be an irreducible unitary representation of a compact group G. A continuous function

$$g \to \mathrm{Tr}\ \nu(g),\ g \in G$$

is called the *character* (of the representation ν) and is denoted by χ_ν.

If G is compact and commutative then all the irreducible representations are one-dimensional and we obtain the ordinary notion of a character (see 2.24). The character χ_ν depends only on the equivalence class of the representation ν. We denote by $\hat{G}$ the set of equivalence classes of irreducible unitary representations of G (recall again that if G is compact and commutative then $\hat{G}$ coincides with the dual group of G (Theorems 2.92, 2.93)).

The characters can be shown to form an orthonormal system in $L^2(G)$ (with the normalized Haar measure), that is

$$\int_G \chi_{\nu_1} \cdot \overline{\chi}_{\nu_2}\, d\mu = \delta_{\nu_1\nu_2}\ .$$

A.8 Let π be a unitary representation of a compact group G. We denote by P_ν $(\nu \in \hat{G})$ the operator in H_π given by

$$P_\nu = \pi\big((\dim \nu)\cdot \overline{\chi}_\nu\big) \qquad \text{(A.14)}$$

(recall (2.13)).

In particular if G is finite then

$$P_\nu = \frac{\dim \nu}{|G|} \sum_{g\in G} \overline{\chi}_\nu(g)\pi(g)\ . \qquad \text{(A.15)}$$

These operators P_ν play an important role in the theory of the decomposition of unitary representations into direct sums of irreducible representations; namely, we have in mind the following result.

Theorem A.9 *Let π be a unitary representation of a compact group G. Then*

(i) *The operators P_ν $(\nu \in \hat{G})$ defined by (A.14) are mutually orthogonal projections in H_π and*

$$H_\pi = \bigoplus_{\nu\in\hat{G}} P_\nu(H_\pi)\ .$$

(ii) *For every* $\nu \in \hat{G}$ *the subspace* $P_\nu(H_\pi)$ *is either* $\{0\}$ *or is the direct (Hilbert) sum of* n_ν *mutually orthogonal subspaces* $L_{\nu,n}$ *such that* $\dim(L_{\nu,n}) = \dim \nu$ *and on every* $L_{\nu,n}$ π *is equivalent to* ν *(the cardinal number* n_ν *may be finite or infinite). The subspace* $P_\nu(H_\pi)$ *is the minimal closed subspace of* H_π *containing all the invariant subspaces on which* π *is equivalent to* ν.

(iii) *The cardinal numbers* n_ν *are determined by* π *and* ν *uniquely (the number* n_ν *is called the multiplicity of* ν *in* π*).*

Theorem A.10 *If G is a finite group then the set* $\hat{G}$ *is finite and* $|\hat{G}| = p$ *where p is the number of equivalence classes of conjugate elements in G and*

(i) $\sum_{\nu \in \hat{G}} (\dim \nu)^2 = |G|$,

(ii) $\dim \nu$ *is a divisor of* $|G|$ *for every* $\nu \in \hat{G}$.

A.11 Now let A and $\{T_g\}_{g \in G}$, $\tilde{H}$ be those mentioned in A.1 and A.3. Consider the cross-product $A \underset{\hat{T}}{\times} G$. With every element

$$\overline{b} = \sum_G a_g \otimes g \in A \underset{\hat{T}}{\times} G \tag{A.16}$$

we shall link the operator $\tilde{b} \in L(\tilde{H})$ defined by the operator matrix $\{\tilde{b}_{h,g}\}_{h,g \in G}$ given by (A.11). Recall that by Theorem 15.2 the mapping

$$\overline{b} \to \tilde{b}$$

establishes an isomorphism between $A \underset{\hat{T}}{\times} G$ and the C^*-algebra generated by operators of the form $\tilde{b}$.

Note that for any vector

$$\overline{u} = \{u_g\}_{g \in G} \in \tilde{H}$$

and for the element $\overline{b}$ (A.16) we have

$$(\tilde{b}\overline{u})_h = \sum_g \tilde{b}_{h,g} u_g = \sum_g \hat{T}_h\left(a_{h^{-1}g}\right) u_g = \sum_\tau \hat{T}_h(a_\tau) u_{h\tau} . \tag{A.17}$$

A.12 In the space $\tilde{H}$ let us introduce the unitary representation of the group G

$$G \ni g \to \check{T}_g \in L(\tilde{H}) ,$$

where $\check{T}_g$ is the operator defined by the operator matrix $\left\{\left(\check{T}_g\right)_{h,l}\right\}_{h,l \in G}$ given by

$$\left(\check{T}_g\right)_{h,l} = \begin{cases} T_g, & l = g^{-1}h \\ 0, & l \neq g^{-1}h . \end{cases} \tag{A.18}$$

Observe that for every $\overline{u} = \{u_g\}_{g \in G} \in \tilde{H}$

$$\left(\check{T}_g \overline{u}\right)_h = \sum_l \left(\check{T}_g\right)_{h,l} u_l = T_g u_{g^{-1}h} . \tag{A.19}$$

Lemma A.13 *If $\tilde{b}$ and $\check{T}_g$ are as mentioned in A.11, A.12 then*

$$\tilde{b}\check{T}_g = \check{T}_g \tilde{b} .$$

Proof Applying (A.17) and (A.19) we have

$$\left(\tilde{b}\check{T}_g \overline{u}\right)_h = \sum_\tau \hat{T}_h(a_\tau)\left(\check{T}_g \overline{u}\right)_{h\tau} = \sum_\tau T_h a_\tau T_h^{-1} T_g u_{g^{-1}h\tau} . \tag{A.20}$$

On the other hand again applying (A.19) and (A.17) we obtain

$$\left(\check{T}_g \tilde{b}\overline{u}\right)_h = T_g(\tilde{b}\overline{u})_{g^{-1}h} = T_g \sum_\tau T_{g^{-1}h} a_\tau T_{h^{-1}g} u_{g^{-1}h\tau}$$

$$= \sum_\tau T_h a_\tau T_h^{-1} T_g u_{g^{-1}h\tau}$$

which along with (A.20) proves the lemma. □

Remark A.14 The only fact which was exploited in the proof is that the automorphism $\hat{T}_g$ is generated by the operator T_g. So if any other unitary representation

$$G \ni g \to V_g \in L(H)$$

generates the same automorphisms of A (that is $T_g a T_g^* = V_g a V_g^*$, $a \in A$) then for the unitary representation

$$G \ni g \to \check{V}_g \in L(\tilde{H})$$

defined in a way analogous to $\check{T}_g$ we shall also have

$$\tilde{b}\check{V}_g = \check{V}_g \tilde{b} \ .$$

A.15 Let

$$\hat{G} = \{v_i\}_{i=\overline{1,p}}, \qquad d_i = \dim v_i, \qquad \chi_i = \chi_{v_i}$$

(recall A.7 and Theorem A.10). For the representation $\pi : g \mapsto \check{T}_g$ described in A.12 we denote by P_i the projection P_{v_i} given by (A.14) and set

$$F_i = \operatorname{Im} P_i \ .$$

We shall assume here that v_1 is the trivial representation.

Theorem A.9 implies

$$\tilde{H} = \bigoplus_{i=1}^{p} F_i \tag{A.21}$$

where (in view of (A.15))

$$P_i = \frac{d_i}{|G|} \sum_{g \in G} \overline{\chi_i(g)} \check{T}_g \ . \tag{A.22}$$

Observe that the algebra $C^*\left(\Phi, \check{T}_g\right)$ generated by the operators $\check{T}_g$, $g \in G$, and the algebra Φ of operators defined by diagonal matrices possesses the property (*) (which can be easily verified for example by an argument analogous to that presented in the proof of Theorem 15.2) and consequently all the projections P_i, $i = \overline{1, p}$, are non-zero.

For example, the projection P_1 corresponding to the trivial representation v_1 has the form

$$P_1 = \frac{1}{|G|} \sum_{g \in G} \check{T}_g$$

and $F_1 = \operatorname{Im} P_1$ coincides with the subspace of the vectors $\overline{u} = \{u_g\}_{g \in G}$ such that

$$u_g = T_g u_e \ . \tag{A.23}$$

Theorem A.16 *The operator $\tilde{b}$ described in A.11 (corresponding to the operator $\overline{b}$ (A.16)) is the direct sum of the operators $\tilde{b}^i$, $i = \overline{1, p}$, acting in the subspaces F_i $(= \operatorname{Im} P_i)$. Moreover the operator $\tilde{b}^1$ is conjugate to the operator*

$$b = \sum a_g T_g \ , \tag{A.24}$$

namely

$$\tilde{b}^1 = S^{-1} b S \tag{A.25}$$

where S is the operator acting from F_1 into H via the formula

$$S\overline{u} = u_e; \ S^{-1} u = \left(u, T_{g_1} u, T_{g_2} u, \ldots\right)$$

where $\overline{u} = \left\{u_e, u_{g_1}, u_{g_2}, \ldots\right\} \in \tilde{H}$, $u \in H$.

Proof In view of Lemma A.13 and the explicit form of the projections P_i (A.22) all the subspaces F_i are invariant with respect to $\tilde{b}$. Denoting by $\tilde{b}^i$, $i = \overline{1, p}$, the restriction of $\tilde{b}$ onto F_i, we obtain the decomposition

$$\tilde{b} = \bigoplus_{i=1}^{p} \tilde{b}^i \ .$$

Now for any $\overline{u} \in F_1$ we have (recalling (A.23))

$$bS\overline{u} = \sum_g a_g T_g u_e = \sum_g a_g u_g$$

and

$$S\tilde{b}^1\overline{u} = S\tilde{b}\overline{u} = \sum_g a_g u_g$$

which means that

$$bS = S\tilde{b}^1 ,$$

thus completing the proof. □

The argument used in the proof of the theorem can be naturally used to prove the following statement concerning the operators mentioned in A.2-A.3.

Theorem A.17 *Let b be the operator given by (A.7) and* $\tilde{b}$ *be the operator defined in A.3 (via formula (A.11)). Then* $\tilde{b}$ *is the direct sum of the operators* $\tilde{b}^i$, $i = \overline{1,p}$, *acting in the subspaces* F_i $(= \operatorname{Im} P_i$, P_i *are given by (A.22)). Moreover* $\tilde{b}^1$ *is conjugate to the initial operator b and in particular if b (or* $\tilde{b}^1$*) is Fredholm then*

$$\operatorname{ind} b = \operatorname{ind} \tilde{b}^1 . \tag{A.26}$$

Proof Just follow the idea of the proof of Theorem A.16 (since $\tilde{b}$ is given by (A.11) the equality $\tilde{b}\check{T}_g = \check{T}_g\tilde{b}$ for $\check{T}_g$ given by (A.18) is established by the same argument as that used in the proof of Lemma A.13). □

Remarks A.18 (1) We would like to observe that in fact Theorems A.16 and A.17 are *essentially different*; namely, in Theorem A.16 the initial object is the element

$$\overline{b} = \sum a_g \otimes g \in A \underset{\hat{T}}{\times} G$$

and the operators b (A.24) and $\tilde{b}$ (described in A.11) are simply the representations of $\overline{b}$ (in particular here the coefficients a_g, $g \in G$, are uniquely determined).

On the other hand, in Theorem A.17 the initial object is the operator b (A.7) where (recall the remark in A.3) the coefficients a_g are not necessarily uniquely determined (we fix them while defining $\tilde{b}$(!) by (A.11)). Thus in general in this case with the operator $b \in L(H)$ here we link *several* operators $\tilde{b} \in L(\tilde{H})$ (depending on the coefficients a_g chosen to represent b in the form (A.7)). But for any $\tilde{b}$ defined in this way by (A.11) the corresponding operator $\tilde{b}^1$ will be conjugate with b.

(2) It is worth mentioning also that in general the Fredholmity of $b\left(\tilde{b}^1\right)$ does not imply the Fredholmity of $\tilde{b}$ (due to Corollary A.5 this of course could not happen when $\mathcal{J}$ is the ideal of compact operators and $\left[\hat{T}_g\right]$ acts topologically freely on Prim[A]).

(3) Theorem A.17 implies in particular that if $\tilde{b}$ (for any fixed choice of the coefficients a_g in (A.7)) is Fredholm (invertible) then all the corresponding operators $\tilde{b}^i$, $i=\overline{1,p}$, are Fredholm (invertible) and in particular b is Fredholm (invertible).

(4) If $C^*\left(A,T_g\right)$ possesses the property (*) (the coefficients in (A.7) are uniquely determined) then all the operators $\tilde{b}^i$, $i=\overline{1,p}$, are simultaneously Fredholm (invertible) or are simultaneously not Fredholm (not invertible).

A.19 The decomposition

$$\tilde{b} = \bigoplus_{i=1}^{p} \tilde{b}^i$$

established in Theorems A.16 and A.17 makes it possible to obtain the interrelations between the topological invariants of the operators $\tilde{b}$ and $\tilde{b}^i$ and in particular between their indices. Thus (bearing in mind the equality (A.26)) one can find the interrelation between the indices of $\tilde{b}$ and b.

To describe these interrelations let us recall first certain notions and facts.

We denote by $R(G)$ the *group of virtual representations* which is defined in the following way. Let F be the free abelian group generated by all the equivalence classes of finite dimensional representations of G. In F, we single out the subgroup generated by the elements of the form $V \oplus W - V - W$. The factor group of F with respect to this group is denoted by $R(G)$.

The definition of $R(G)$ means that in fact addition here corresponds to the direct addition of representations: $V + W = V \oplus W$.

Since any finite dimensional representation of a compact group G is equivalent to a certain unitary representation (see Pontryagin [1], Section 32) it follows in view of

Theorems A.9 and A.10 that any element of $R(G)$ is a direct sum of a finite number of non-equivalent irreducible representations with integer coefficients (*multiplicities*).

Thus $R(G)$ is in fact the free abelian group generated by the irreducible representations of G.

A.20 Let the operator $\tilde{b}$ (considered in A.3 or A.11) be Fredholm. Then Ker $\tilde{b}$ is a finite dimensional subspace of $\tilde{H}$ invariant in view of Lemma A.13 with respect to the representation $\left\{\check{T}_g\right\}_{g\in G}$. Thus the restriction of $\check{T}_g$ to Ker $\tilde{b}$ is a certain finite dimensional representation of G. Let us denote by [Ker $\tilde{b}$] the equivalence class of this representation in $R(G)$.

The orthogonal complement to the image of $\tilde{b}$ which is denoted by Coker $\tilde{b}$ (and coincides with Ker $\tilde{b}^*$) is also invariant with respect to $\check{T}_g$ and consequently generates the element [Coker $\tilde{b}$] $\in R(G)$.

The *analytical G-index* of an equivariant operator $\tilde{b}$ with respect to $\left\{\check{T}_g\right\}_{g\in G}$ is defined as the element of the virtual representations group $R(G)$:

$$G\text{-}\operatorname{ind}\tilde{b} = [\operatorname{Ker}\tilde{b}] - [\operatorname{Coker}\tilde{b}]\,. \tag{A.27}$$

A.21 Let an element $\pi \in R(G)$ be of the form

$$\pi = n_1 v_1 + \ldots + n_k v_k$$

where $n_i \in \mathbb{Z}$, $i = \overline{1,k}$, are the multiplicities of the irreducible representations v_k.

We denote by χ_π the *character of* π which is defined by

$$\chi_\pi = n_1 \chi_{v_1} + \ldots + n_k \chi_{v_k}$$

(in view of A.7, Theorem A.9 and the definition of $R(G)$ χ_π is well defined).

A.22 For any fixed $g_0 \in G$ the number

$$\chi_{G\text{-ind}\,\tilde{b}}(g_0)$$

is called the *Lefschetz number* and is denoted by $L(g_0,\tilde{b})$. Thus

$$L(g,\tilde{b}) = \operatorname{Tr} \check{T}_g\Big|_{\operatorname{Ker}\tilde{b}} - \operatorname{Tr} \check{T}_g\Big|_{\operatorname{Coker}\tilde{b}} . \tag{A.28}$$

The Lefschetz numbers play an essential role in the index theory of the operators we are considering, and this is stated in the following.

Theorem A.23 *Let the operator $\tilde{b}$ considered in A.3 (or A.11) be Fredholm. Then*

(i) *the index of the operator b given by (A.7) (or (A.24)) is equal to the multiplicity of the trivial representation of the group G in the decomposition of G-ind $\tilde{b}$; and*

(ii) *it can be written via the formula*

$$\operatorname{ind} b = \frac{1}{|G|}\sum_g L(g,\tilde{b}) . \tag{A.29}$$

Proof The decomposition

$$\tilde{b} = \overset{p}{\underset{i=1}{\oplus}} \tilde{b}^i$$

(see Theorems A.17 and A.16) implies

$$\operatorname{Ker}\tilde{b} = \overset{p}{\underset{i=1}{\oplus}} \operatorname{Ker}\tilde{b}^i$$

and

$$\operatorname{Coker}\tilde{b} = \overset{p}{\underset{i=1}{\oplus}} \operatorname{Coker}\tilde{b}^i$$

and in view of the structure of $\tilde{b}^i$ both the subspaces Ker $\tilde{b}^i$ and Coker $\tilde{b}^i$ are invariant with respect to $\check{T}_g$. Thus (recalling (A.27))

$$G\text{-}\operatorname{ind}\tilde{b} = \sum_{i=1}^{p} G\text{-}\operatorname{ind}\tilde{b}^i .$$

Since in the subspace F_i (see A.15) the representation $\left\{\overset{\vee}{T}_g\right\}$ of the group G is a multiple of v_i it follows that

$$\left[\operatorname{Ker} \tilde{b}^i\right] = \frac{\dim\left(\operatorname{Ker} \tilde{b}^i\right)}{d_i}[v_i]$$

and

$$\left[\operatorname{Coker} \tilde{b}^i\right] = \frac{\dim\left(\operatorname{Coker} \tilde{b}^i\right)}{d_i}[v_i] .$$

So

$$G\text{-}\operatorname{ind} \tilde{b}^i = \frac{\operatorname{ind} \tilde{b}^i}{d_i}[v_i]$$

and

$$G\text{-}\operatorname{ind} \tilde{b} = \sum_{i=1}^{p} \frac{\operatorname{ind} \tilde{b}^i}{d_i}[v_i] . \tag{A.30}$$

Since by (A.26)

$$\operatorname{ind} b = \operatorname{ind} \tilde{b}^1 \tag{A.31}$$

and $d_1 = 1$ (A.30) proves statement (i) of the theorem.

To prove (ii) let us consider the character

$$\chi = \chi_{G\text{-ind}\, \tilde{b}}$$

(see A.21). Now (A.30) means that

$$\chi(g) = \sum_{i=1}^{p} \frac{\operatorname{ind} \tilde{b}^i}{d_i} \chi_i(g) \tag{A.32}$$

(where $\chi_i = \chi_{v_i}$).

Since the set $\{\chi_i\}_{i=\overline{1,p}}$ is an orthonormal system in $L^2(G)$ (see A.7) the coefficients of the expansion (A.32) can be found as the Fourier coefficients

$$\frac{\operatorname{ind} \tilde{b}^i}{d_i} = \int_G \chi \cdot \overline{\chi}_i \, d\mu = \frac{1}{|G|} \sum_g \chi(g) \overline{\chi_i}(g) . \tag{A.33}$$

Since

$$\chi_1(g) \equiv 1, \quad d_1 = 1$$

and by definition

$$\chi(g) = L(g, \tilde{b})$$

statement (ii) of the theorem follows from (A.31) and (A.33). □

Remarks A.24 (1) We emphasize once more that in general it may happen that b is Fredholm while $\tilde{b}$ is not. But if the ideal $\mathcal{J}$ (see A.1) consists of all the compact operators in H and the action $\left[\hat{T}_g\right]$, $g \in G$, on Prim$[A]$ is topologically free, then $\tilde{b}$ is Fredholm iff b is Fredholm (Corollary A.5).

(2) If b is Fredholm then (even if $\tilde{b}$ is not Fredholm) looking through the proof of the theorem we see that the index of b is equal to the multiplicity of the trivial representation of G in the decomposition of $[\text{Ker } \tilde{b}] - [\text{Coker } \tilde{b}]$.

Now we proceed from a finite group situation to the case $G = \mathbb{Z}$. Using the ideology of Section 20 along with that already developed in this section we shall obtain certain ($\mathbb{Z}$) analogues to the foregoing results.

A.25 Thus the situation we are considering hereafter is the following. Let $A \subset L(H)$ be a C^*-algebra, $\mathcal{J} \subset L(H)$ be a closed *-ideal and T be a unitary operator satisfying the condition

$$TAT^* = A, \quad T\mathcal{J}T^* = \mathcal{J}. \tag{A.34}$$

We denote by $\hat{T}$ the automorphism of A given by

$$\hat{T}(a) = TaT^*, \quad a \in A \tag{A.35}$$

and we denote by $[\hat{T}]$ the corresponding automorphism of

$$[A] = A/\mathcal{J}.$$

Recalling the remark in A.1 we can assume that

$$\mathcal{J} \subset A .$$

The main object we shall deal with is the operator

$$b = \sum_{k=0}^{m} a_k T^k, \ a_k \in A \tag{A.36}$$

where $m \geq 2$ is a certain fixed number.

A.26 (cf. A.3) Let us introduce the space

$$\tilde{H} = H^m \tag{A.37}$$

and the ideal $\tilde{\mathcal{J}} \subset L(\tilde{H})$ consisting of all the operator matrices $\{d_{j,k}\}_{\substack{j=\overline{1,m} \\ k=\overline{1,m}}}$ such that

$$d_{j,k} \in \mathcal{J} . \tag{A.38}$$

With the operator b given by (A.36) we shall associate the operator $\tilde{b} \in L(\tilde{H})$ having the form

$$\tilde{b} = c_0 + c_1 W \tag{A.39}$$

where

$$c_0 = \begin{bmatrix} a_0 & a_1 & \dots & a_{m-1} \\ 0 & \hat{T}(a_0) & \dots & \hat{T}(a_{m-2}) \\ \dots & \dots & \dots & \dots \\ 0 & 0 & \dots & \hat{T}^{m-1}(a_0) \end{bmatrix} \tag{A.40}$$

$$c_1 = \begin{bmatrix} a_m & 0 & \dots & 0 \\ \hat{T}(a_{m-1}) & \hat{T}(a_m) & \dots & 0 \\ \dots & \dots & \dots & \dots \\ \hat{T}^{m-1}(a_1) & \hat{T}^{m-1}(a_2) & \dots & \hat{T}^{m-1}(a_m) \end{bmatrix} \tag{A.41}$$

$$W = \begin{bmatrix} T^m & 0 & \dots & 0 \\ 0 & T^m & \dots & 0 \\ \dots & \dots & \dots & \dots \\ 0 & 0 & \dots & T^m \end{bmatrix}. \tag{A.42}$$

We have already observed in A.3 that in general the operator $\tilde{b}$ is not uniquely determined by b. But if the action $[\hat{T}]$ on Prim$[A]$ is topologically free then the coefficients a_k, $k=\overline{0,1}$, of b are uniquely determined (modulo $\mathfrak{J}$) and $\tilde{b}$ is uniquely determined (modulo $\tilde{\mathfrak{J}}$). Then the analogue of Lemma A.4 is

Lemma A.27 *Let $A, \mathfrak{J}, T$ be as mentioned in A.25 and let $[\hat{T}]$ act topologically freely on* Prim$[A]$. *Then the operator b (A.36) is $\mathfrak{J}$-Fredholm iff the operator $\tilde{b}$ (A.39) is $\tilde{\mathfrak{J}}$-Fredholm.*

Proof Apply Corollary 20.4 and the reasoning used in the proof of Lemma A.4. □

Remark Clearly it is enough to assume here not the unitary property of T but only the unitary property of $[T]$ that is

$$[T]^*[T] = [T][T]^* = [I] \tag{A.43}$$

and the property

$$\left[TAT^*\right] = [A] \tag{A.44}$$

instead of A.34.

Corollary A.28 *Let $\mathfrak{J}$ be the ideal of compact operators acting in H and let $[\hat{T}]$ act topologically freely on* Prim$[A]$. *Then the operator b (A.36) is Fredholm iff $\tilde{b}$ (A.39) is Fredholm.*

A.29 Let us now consider the situation when $C^*([A],[T])$ possesses the *property (*)*, that is when for any $p \geq 0$

$$\left\| \left[\sum_{k=0}^{p} a_k T^k \right] \right\| \geq \| [a_0] \| . \tag{A.45}$$

Recall that this is true when $[\hat{T}]$ acts topologically freely on Prim$[A]$ (in view of Theorem 12.14).

If b is the operator given by (A.36) then for any complex number λ we denote by $\varphi_\lambda(b)$ the operator

$$\varphi_\lambda(b) = \sum_{k=0}^{m} a_k (\lambda T)^k . \tag{A.46}$$

The next result shows the close interrelation between the Fredholm properties of b and $\varphi_\lambda(b)$.

Lemma A.30 *Let $C^*([A],[T])$ possess the property (*). Then for any λ, $|\lambda| = 1$:*

(i) *the operator b (A.36) is $\mathcal{J}$-Fredholm iff $\varphi_\lambda(b)$ is $\mathcal{J}$-Fredholm; and more over*

(ii) *if b is $\mathcal{J}$-Fredholm then the elements $[b]$ and $[\varphi_\lambda(b)]$ lie in one and the same linear connected component of the set of invertible elements of $L(H)/\mathcal{J}$; in particular if $\mathcal{J}$ is the ideal of compact operators and b $(\varphi_\lambda(b))$ is Fredholm then*

$$\text{ind}\, b = \text{ind}\, \varphi_\lambda(b) . \tag{A.47}$$

Proof In view of Proposition 13.5 the element $[b]$ is invertible iff $[\varphi_\lambda(b)]$ is invertible which means that (i) is true.

To obtain (ii) observe that for any $\lambda = \mathrm{e}^{\mathrm{i}\omega}$ the homotopy

$$b_t = \varphi_{\mathrm{e}^{\mathrm{i}\omega t}}(b),\ t \in [0,1]$$

connects $b = b_0$ and $\varphi_\lambda(b) = b_1$ and in view of (i) all the elements $[b_t]$ are invertible. □

A.31 In the space $\tilde{H}$ let us introduce the unitary operator

$$\check{T} = \begin{bmatrix} 0 & T^{-1} & 0 & \dots & 0 \\ 0 & 0 & T^{-1} & \dots & 0 \\ \dots & \dots & \dots & \dots & \dots \\ 0 & 0 & 0 & \dots & T^{-1} \\ T^{m-1} & 0 & 0 & \dots & 0 \end{bmatrix}. \tag{A.48}$$

Clearly $\check{T}^m = I$ which means that the operator $\check{T}$ defines a unitary representation of the group $\mathbb{Z}_m$.

Recall that $\mathbb{Z}_m$ has m inequivalent irreducible representations v_j, $j = \overline{0, m-1}$, which are one-dimensional and have the form

$$v_j(k) = \omega^{kj},\ k \in \overline{0, m-1},\ j = \overline{0, m-1} \tag{A.49}$$

where $\omega = \mathrm{e}^{\mathrm{i}\frac{2\pi}{m}}$.

For the representation

$$\pi : k \to \check{T}^k,\ k \in \mathbb{Z}_m$$

we denote by P_j the projection P_{v_j} given by (A.15). In view of (A.49) we have

$$P_j = \frac{1}{m} \sum_{k=0}^{m-1} \omega^{-kj} \check{T}^k. \tag{A.50}$$

If we set

$$F_j = \operatorname{Im} P_j$$

then the explicit form of P_j implies that F_j consists of vectors of the form

$$F_j = \left\{ \bar{u} \in \tilde{H} : \bar{u} = \left(u, \omega^j T u, \omega^{2j} T^2 u, \dots, \omega^{(m-1)j} T^{m-1} u \right), u \in H \right\} \tag{A.51}$$

and the mapping $S_j : F_j \to H$ given by

$$S_j \bar{u} = u \tag{A.52}$$

and having the inverse

$$S_j^{-1}u=\left(u,\omega^j Tu,\dots,\omega^{(m-1)j}T^{m-1}u\right) \tag{A.53}$$

establishes the isomorphism between F_j and H.

Lemma A.32 *Let $\tilde{b}$ be given by (A.39) and $\check{T}$ by (A.48). Then*

$$\tilde{b}\check{T}=\check{T}\tilde{b}\ .$$

Proof Analogous to that of Lemma A.13. □

Now the analogue of Theorem A.17 for this situation is

Theorem A.33 *Let b be the operator given by (A.36) and $\tilde{b}$ be the operator given by (A.39). Then $\tilde{b}$ is the direct sum of the operators $\tilde{b}^j$, $j=\overline{0,m-1}$, acting in the subspaces F_j described by (A.51).*

Moreover $\tilde{b}^j$ is conjugate to the operator

$$b_j=\sum_{k=0}^{m}a_k\left(\omega^j T\right)^k,\ \omega=\mathrm{e}^{\mathrm{i}\frac{2\pi}{m}} \tag{A.54}$$

and in particular $\tilde{b}^0$ is conjugate to the initial operator $b=b_0$. Thus if b (or $\tilde{b}^0$) is Fredholm then

$$\operatorname{ind} b=\operatorname{ind} b^0\ .$$

Proof Just follow the idea of the proof of Theorem A.16 and use the operators S_j (A.52) to establish the equality

$$b_jS_j\bar{u}=S_j\tilde{b}^j\bar{u}=S_j\tilde{b}\bar{u},\quad \bar{u}\in F_j\ .\ \square$$

A.34 (cf. A.20) Let $\tilde{b}$ (A.39) be Fredholm. Then the restrictions of $\check{T}$ (A.48) onto $\operatorname{Ker}\tilde{b}$ and $\operatorname{Coker}\tilde{b}$ generate the equivalence classes [$\operatorname{Ker}\tilde{b}$] and [$\operatorname{Coker}\tilde{b}$] of $R(\mathbb{Z}_m)$ and consequently

$$\mathbb{Z}_m\text{-ind}\,\tilde{b} = [\operatorname{Ker}\tilde{b}] - [\operatorname{Coker}\tilde{b}]\,. \tag{A.55}$$

Considering the corresponding Lefschetz numbers (see A.22) in this situation we have

$$L(k,\tilde{b}) = \operatorname{Tr}\check{T}^k\Big|_{\operatorname{Ker}\tilde{b}} - \operatorname{Tr}\check{T}^k\Big|_{\operatorname{Coker}\tilde{b}}\,, \tag{A.56}$$

$k = \overline{0, m-1}$.

The analogue of Theorem A.23 is

Theorem A.35 *Let the operator $\tilde{b}$ (A.39) be Fredholm. Then all the operators b_j (A.54) are Fredholm and*

(i) *the index of the operator b_j is equal to the multiplicity of the representation ν_j (A.49) of $\mathbb{Z}_m$ in the decomposition of $\mathbb{Z}_m$-ind $\tilde{b}$;*

(ii) ind b_j *can be written out via the formula*

$$\operatorname{ind} b_j = \frac{1}{m}\sum_{k=0}^{m-1} L(k,\tilde{b})\omega^{-jk}$$

and in particular

$$\operatorname{ind} b = \operatorname{ind} b_0 = \frac{1}{m}\sum_{k=0}^{m-1} L(k,\tilde{b})\,.$$

Proof Just follow the proof of Theorem A.23 bearing in mind that in this situation

$$d_j = 1, \quad j = \overline{0, m-1}$$

and by Theorem A.33

$$\operatorname{ind} b_j = \operatorname{ind} \tilde{b}^j$$

which means that here instead of (A.30) we have

$$\mathbb{Z}_m\text{-}\operatorname{ind}\tilde{b} = \sum_{j=0}^{m-1} \operatorname{ind} b_j [v_j] \tag{A.57}$$

thus proving (i).

Furthermore we have that (A.33) is transformed here into

$$\operatorname{ind} b_j = \frac{1}{m}\sum_{k=0}^{m-1} \chi(k)\overline{\chi_j(k)} = \frac{1}{m}\sum_{k=0}^{m-1} L(k,\tilde{b})\omega^{-kj}$$

thus proving (ii). □

In the case when $[\hat{T}]$ acts topologically freely on Prim[A] we can refine the result just obtained; namely, we have in mind the next

Theorem A.36 *Let $\mathcal{J}$ be the ideal of compact operators and A and T be those mentioned in A.25. If* $[\hat{T}]$ *acts topologically freely on* Prim[A] *then*:

(i) *b (A.36) is Fredholm iff $\tilde{b}$ (A.39) is Fredholm; and*

(ii) *if b is Fredholm then*

$$\operatorname{ind} b = \frac{1}{m}\operatorname{ind}\tilde{b}\ . \tag{A.58}$$

Proof (i) is the statement of Corollary A.28. To prove (ii) observe that if b (or $\tilde{b}$) is Fredholm then following the proof of the previous theorem we have

$$\operatorname{ind}\tilde{b} = \sum_{j=0}^{m-1} \operatorname{ind}\tilde{b}^j = \sum_{j=0}^{m-1} \operatorname{ind} b_j\ . \tag{A.59}$$

By Lemma A.30 (as $[\hat{T}]$ acts topologically freely on Prim[A] the algebra $C^*([A],[T])$ possesses the property (*) (Theorem 12.14)) we conclude that

$$\operatorname{ind} b = \operatorname{ind} b_0 = \operatorname{ind} b_j, \quad j = \overline{0, m-1}\ ,$$

which in view of (A.59) means that (A.58) is true. □

Remark A.37 We finish this section with the observation that an application of the ideas exploited in the proofs of Theorems A.16, A.17, A.35, A.36 makes it possible also to write out the corresponding index formulae for the operator b of the form

$$\sum_F a_g T_g, \quad |F| < \infty, \quad F \subset G$$

where

$$G = G_1 \oplus \mathbb{Z}$$

and G_1 is a certain finite group. We will not use such formulae in this book.

Appendix B
Quasicross-products and C^*-algebras associated with quasirepresentations of groups.

'Может быть, эту книгу вообще не следует читать.
Но иметь ее дома под рукой - необходимо.' И. Губерман 'Гарики на каждый день' (из надписи на обложке)*

In Section 16 we obtained a number of general results describing the structure of $C^*(A,T_g)$ in the case of an arbitrary group action. Of course in general one cannot suggest any concrete procedure of verification for say the equality (16.29) in advance (the level of complexity of such a verification corresponds to the level of complexity of the algebra $v_{g_0}(C^*(A,T_g))$ defined in 16.4). But in various natural situations (especially concerning the investigation of functional differential operators) we have the following property: if the set $\overset{\circ}{X}_g$ (see 16.1) is non-empty then the action of T_g on the corresponding ideal I_g coincides with the action of a certain element of A (or perhaps a certain multiplier of A). The development of the ideas arising from this observation is the essence of this section.

We shall manage (in particular) to derive a result which is an analogue to the isomorphism theorem (Corollary 12.17) for a rather wide and natural class of algebras $C^*(A,T_g)$ in the situation when the group action is very far from being topologically free. It is not surprising (in view of Corollary 12.16) that the C^*-algebraic structures arising in this way are not cross-products but have a somewhat more peculiar character. We call these algebras quasicross-products since they are associated with the so-called quasirepresentations of groups, that is group representations which are 'sploited' by means of cocycles generated by unitary elements of the algebra A.

---

* 'Maybe one should not read this book at all.
But having it at home at hand - is a necessity.' I. Guberman *Gariks for every day* (from the inscription on the cover)

In fact the structure of the quasicross-product is in many ways analogous to the structure of the cross-product but as we cannot refer to any source we present here a fairly thorough and self-contained exposition of this object (following the path of Sections 11, 12).

After this is done we proceed to develop (starting from B.22) the ideas of the proof of Theorem 16.21 for the situation considered here, and obtain results (Theorem B.28 and Corollary B.30) which, along with their own C^*-algebraic value, are convenient for applications. These results are very effective when constructing the symbolic calculus of functional differential operators (and are exploited mainly in Chapter 5) in the situation when the transformations of an argument are normally acting (see 40.11).

B.1 Let A be a C^*-algebra with identity and G be a discrete group. We fix a certain mapping (*cocycle*)

$$u: G\times G \to A$$

with the properties:

(1) $u(g,h)$ is unitary for any $(g,h)\in G\times G$;

(2) $u(g,e)=u(e,g)=u\left(g,g^{-1}\right)=I$.

We say that a family $\alpha=\{\alpha_g\}_{g\in G}$ of automorphisms forms a *quasirepresentation* of G (with cocycle u) if the following conditions hold:

(i) $\alpha_h\circ\alpha_g=\hat{u}(h,g)\circ\alpha_{h\cdot g}$, $(h,g)\in G\times G$ (recall that we denote by $\hat{u}(h,g)$ the automorphism of A given by $\hat{u}(h,g)a=u(h,g)au(h,g)^*$, $a\in A$);

(ii) $u\left(gh^{-1},h\right)=\alpha_{g\cdot h^{-1}}\left(u\left(h,g^{-1}\right)\right)$;

(iii) $\alpha_{g^{-1}}=\alpha_g^{-1}$;

(iv) $\alpha_e=I$.

Remark The naturalness of condition (ii) will be demonstrated in B.4.

B.2 (cf 11.1) For a given quasirepresentation described in B.1 we introduce the *quasicross-product* $A\underset{\alpha,u}{\times}G$ which is constructed as follows.

We define operations of involution and convolution on the set $K(G,A)$ of formal finite sums

$$b=\sum a_g \otimes g, \quad a_g \in A, \quad g \in G$$

by the equations

$$\left(a_g \otimes g\right)^* = \alpha_{g^{-1}}\left(a_g^*\right) \otimes g^{-1} \tag{B.1}$$

$$\left(a_g \otimes g\right) \cdot \left(a_h \otimes h\right) = a_g \cdot \alpha_g\left(a_h\right) \cdot u(g,h) \otimes g \cdot h \ . \tag{B.2}$$

If we take the operation of convolution in $K(G,A)$ as the product then these formulae (B.1) and (B.2) define the structure of a *-algebra in $K(G,A)$. With the norm

$$\|b\|_1 = \sum \left\|a_g\right\|$$

$K(G,A)$ becomes a normed algebra.

The quasicross-product $A \underset{\alpha,u}{\times} G$ is the enveloping algebra for $K(G,A)$.

B.3 Just as in the cross-product situation it is reasonable to use the regular representation of a quasicross-product. Let (π, H) be a representation of A in a Hilbert space H.

We define a quasicovariant representation $\tilde{\pi}$ of $A \underset{\alpha,u}{\times} G$ in the space $l^2(G,H)$ by the formulae (recall (15.3), (15.4))

$$\tilde{\pi}(a \otimes e)[g \otimes \xi] = g \otimes \alpha_g(a)\xi \ , \tag{B.3}$$

$$\tilde{\pi}(I \otimes h)[g \otimes \xi] = g \cdot h^{-1} \otimes \alpha_{g \cdot h^{-1}}\left(u\left(h, g^{-1}\right)\right)\xi \ , \tag{B.4}$$

$$h, g \in G, \ a \in A, \ \xi \in H \ .$$

Routine computations show that $\tilde{\pi}$ is indeed a representation of $K(G,A)$ and thus of $A \underset{\alpha,u}{\times} G$.

The representation $\tilde{\pi}$ is called the (*right*) *regular representation* of $A \underset{\alpha,u}{\times} G$ induced by the representation π of A.

Remark In an analogous manner one can construct the *left* regular representation of $A \underset{\alpha,u}{\times} G$ (recall 11.11).

B.4 Now let A be a C^*-subalgebra of the algebra $L(H)$ of linear continuous operators acting on a Hilbert space H and let $\left\{T_g\right\}_{g \in G}$ be a certain family of unitary operators.

We shall say that the family $\{T_g\}_{g\in G}$ forms a *quasirepresentation* of G (with cocycle u described in B.1) if for any $(h,g)\in G\times G$ we have

$$T_h\cdot T_g = u(h,g)T_{h\cdot g} \tag{B.5}$$

and

$$T_e = I \quad \text{and} \quad T_g^{-1} = T_{g^{-1}}\ . \tag{B.6}$$

This quasirepresentation of G by operators T_g is closely related to the quasirepresentation of G by automorphisms α_g described in B.1, since in the situation when

$$T_gAT_g^* = A,\ g\in G$$

(that is when the operators T_g generate the automorphisms $\hat{T}_g$ of A) the family $\hat{T}=\{\hat{T}_g\}_{g\in G}$ forms a quasirepresentation of G by automorphisms (that is, satisfies the conditions (i)-(iv) in B.1).

Note that condition (ii) for the family $\hat{T}$ is satisfied automatically. Indee

$$u(gh^{-1},h) = T_{gh^{-1}}\cdot T_h\cdot T^{-1}_{g\cdot h^{-1}\cdot h} = T_{gh^{-1}}\cdot T_h\cdot T_{g^{-1}}$$

$$= T_{gh^{-1}}\left(T_h\cdot T_{g^{-1}}\cdot T^{-1}_{hg^{-1}}\right)\cdot T^{-1}_{gh^{-1}} = \hat{T}_{gh^{-1}}\left(u(h,g^{-1})\right),$$

and the properties (iii), (iv) follow from (B.6).

B.5 If we consider the C^*-algebra $C^*(A,T_g)$ generated by the C^*-algebra A and the family $\{T_g\}_{g\in G}$ forming the quasirepresentation of G described in B.4 and satisfying the condition

$$T_gAT_g^* = A,\ g\in G\ ,$$

then (since $\hat{T}=\{\hat{T}_g\}_{g\in G}$ is a quasirepresentation of G by automorphisms of A) it follows that $C^*(A,T_g)$ forms a representation π of $A\underset{\hat{T},u}{\times} G$ generated by the mappings

$$\pi(a\otimes e) = a,\ a\in A$$

$$\pi(I\otimes g) = T_g,\ g\in G\ .$$

This representation will be called the *standard representation* of $A\underset{\hat{T},u}{\times} G$ (recall 12.6).

B.6 (cf. 12.2.) We shall say that the algebra $C^*(A,T_g)$ described in B.5 possesses the *property (*)* if for any finite sum $b=\sum a_g T_g$ we have the inequality

$$\|b\| = \left\|\sum a_g T_g\right\| \geq \|a_e\| .$$

This inequality naturally generates the mapping

$$N: C^*(A,T_g) \to A$$

with

$$N(b) = a_e .$$

B.7 The argument exploited in 12.3 shows that for the regular representation $\tilde{\pi}$ described in B.3 we have

$$\left\|\tilde{\pi}\left(\sum a_g \otimes g\right)\right\| \geq \|a_e\| .$$

This means that if we identify the elements $a \in A$ with the elements $a \otimes g \in A \underset{\alpha,u}{\times} G$ and define by T_g the element $I \otimes g$ then with this identification $A \underset{\alpha,u}{\times} G$ possesses the property (*).

B.8 In view of Theorem 11.14 we know that the structure of the positive functionals on the cross-product plays a significant role in the description of this algebra. For the same reason we shall now pass to consideration of the positive functionals on the quasicross-product $A \underset{\alpha,u}{\times} G$.

Let ϕ be an element of $\left[A \underset{\alpha,u}{\times} G\right]^*$. With the functional ϕ we shall associate the function

$$\Phi: G \to A^*$$

given by

$$\Phi(g)(a) = \phi(a \otimes g), \quad g \in G,\ a \in A . \tag{B.7}$$

We shall say that a *functional ϕ has finite support* if the corresponding function Φ has finite support.

We denote the set of functions Φ defined by (B.7) by $B\left(A \underset{\alpha,u}{\times} G\right)$. The subset of $B\left(A \underset{\alpha,u}{\times} G\right)$ of the functions corresponding to the positive functionals will be denoted by $B_+\left(A \underset{\alpha,u}{\times} G\right)$ and the elements of $B_+\left(A \underset{\alpha,u}{\times} G\right)$ will be called *positive definite* functions (with respect to α and u).

Let $\{\phi_i\}$ be a bounded net in $\left(A \underset{\alpha,u}{\times} G\right)^*$ and $\{\Phi_i\}$ be the corresponding net in $B\left(A \underset{\alpha,u}{\times} G\right)$. Since the set of elements of the form

$$\sum_{g\in F} a_g \otimes g, \quad |F| < \infty$$

is dense in $A \underset{\alpha,u}{\times} G$ it is clear that if the net $\{\Phi_i\}$ converges to $\Phi \in B\left(A \underset{\alpha,u}{\times} G\right)$ weak* uniformly on finite subsets of G then the net $\{\phi_i\}$ is weak* convergent to $\phi \in \left(A \underset{\alpha,u}{\times} G\right)^*$ (corresponding to Φ).

Remark In connection with the foregoing observation it is worth mentioning that if $\Phi \in B_+(A\times G)$ then

$$\|\phi\| = \|\Phi(e)\|.$$

The next result presents a description of the positive definite functions.

Lemma B.9 *The following conditions on a bounded function $\Phi: G \to A^*$ are equivalent:*

(i) *Φ is positive definite;*

(ii) $$\sum_{\tau,g\in F} \Phi(\tau^{-1}g)\left[\alpha_\tau^{-1}(a_\tau^* a_g)u(\tau^{-1},g)\right] \geq 0 \tag{B.8}$$

for all finite sets $F \subset G$ and $a_\tau \in A$;

(iii) *the matrix*

$$\left[\beta_{\tau,g}\right]_{\tau,g\in F}=\left[\Phi\left(\tau^{-1}g\right)\left(\alpha_{\tau}^{-1}\left(a_{\tau}^{*}a_{g}\right)u\left(\tau^{-1},g\right)\right)\right] \tag{B.9}$$

is positive (the matrix $\left[\gamma_{i,j}\right]_{i,j\in\overline{1,n}}$ *is called positive if it defines a positive form, that is for every* $\{t_i\}\in\mathbb{C}^n$ *we have*

$$\sum_{i,j}\gamma_{i,j}\bar{t}_i t_j\geq 0\ ). \tag{B.10}$$

Proof (i)⇒(ii)

Let Φ (associated with a functional ϕ) be positive definite. Then for any element

$$b=\sum_{F}a_g\otimes g,\quad |F|<\infty \tag{B.11}$$

we have

$$\phi\left(b^{*}b\right)\geq 0\ . \tag{B.12}$$

But (B.1) and (B.2) imply

$$\begin{aligned} b^{*}b &= \sum_{\tau,g\in F}\left(a_{\tau}\otimes\tau\right)^{*}\left(a_g\otimes g\right) \\ &= \sum_{\tau,g\in F}\alpha_{\tau}^{-1}\left(a_{\tau}^{*}a_g\right)u\left(\tau^{-1},g\right)\otimes\tau^{-1}g\ , \end{aligned}$$

which, along with (B.12), means that

$$\begin{aligned} \phi\left(b^{*}b\right) &= \phi\left(\sum_{\tau,g}\alpha_{\tau}^{-1}\left(a_{\tau}^{*}a_g\right)u\left(\tau^{-1},g\right)\otimes\tau^{-1}g\right) \\ &= \sum_{\tau,g}\Phi\left(\tau^{-1}g\right)\left(\alpha_{\tau}^{-1}\left(a_{\tau}^{*}a_g\right)u\left(\tau^{-1},g\right)\right)\geq 0 \end{aligned}$$

thus establishing the desired result.

(ii)⇔(iii)

Let us consider the matrix $\left[\beta_{\tau,g}\right]$ given by (B.9). If (B.8) holds for any $a_\tau \in A$ then taking instead of a_τ the elements $t_\tau a_\tau$ where t_τ are arbitrary complex numbers we obtain from (B.8) that

$$\sum_{\tau,g\in F} \Phi(\tau^{-1}g)\left[\alpha_\tau^{-1}\left(a_\tau^* a_g\right)u\left(\tau^{-1},g\right)\right]\bar{t}_\tau \cdot t_g \geq 0$$

which means that

$$\sum_{\tau,g} \beta_{\tau,g}\bar{t}_\tau t_g \geq 0$$

and consequently (in view of (B.10)) $\left[\beta_{\tau,g}\right]$ is positive.

On the other hand if $\left[\beta_{\tau,g}\right]$ is positive then taking $t_\tau = 1$, $\tau \in F$, we obtain (B.8).

(ii)⇒(i)

Let $b \in K(G,A)$ be the element given by (B.11). Then (B.8) means that

$$\sum_{h\in G} \Phi(h)\left[\left(b^* \cdot b\right)(h)\right] \geq 0$$

and consequently the functional ϕ defined on $K(G,A)$ via the formula

$$\phi(f) = \sum_{h\in G} \Phi(h)(f(h))$$

(where $f = \sum_{h\in S} a_h \otimes h$, $|S| < \infty$ and $f(h) = a_h$) is positive (on $K(G,A)$).

Since Φ is a bounded function this functional is bounded with respect to the norm

$$\|f\|_1 = \sum \|a_h\|$$

and it follows that it can be uniquely extended up to a (positive) functional on $A \underset{\alpha,u}{\times} G$.

The proof is complete. □

B.10 We shall say that a bounded function $\Psi: G \to \mathbb{C}$ is *positive definite* if for every finite set $F \subset G$ the matrix

$$\left[\gamma_{\tau,g}\right]_{\tau,g\in F} = \left[\Psi\left(\tau^{-1}g\right)\right]$$

is positive (that is satisfies (B.10)). The set of all such functions, Ψ will be denoted by $B_+(G)$.

Let us describe one rather useful interpretation of the set of positive definite functions. With every function $\Psi \in l^1(G)$ we shall associate the operator $\psi \in L\left(l^2(G)\right)$ defined by the formula

$$\psi = \sum_G \Psi(g) V_g \tag{B.13}$$

where V_g is the unitary shift operator in $l^2(G)$ given by

$$V_g(\tau \otimes t) = \tau g^{-1} \otimes t, \qquad \tau \in G,\ t \in \mathbb{C}\ .$$

Routine computation now shows that for any vector

$$\eta = \sum_{\tau\in G} \tau \otimes t_\tau \in l^2(G)$$

we have

$$\langle \psi\eta, \eta \rangle = \sum_g \sum_h \Psi\left(\tau^{-1}g\right) t_g \bar{t}_\tau\ , \tag{B.14}$$

which means in particular that a function $\Psi \in l^1(G)$ is positive definite iff the corresponding operator ψ (B.13) is positive.

Recalling now that in terms of the functions Ψ, multiplication of the operators of the form (B.13) is written out as a convolution, and the adjoint operator ψ^* corresponds to the function

$$\Psi^*(g) = \overline{\Psi}\left(g^{-1}\right), \tag{B.15}$$

we conclude that for any $\Psi \in l^1(G)$ the convolutions

$$\Psi^* \times \Psi, \text{ and } \Psi \times \Psi^*$$

are positive definite functions.

Our next step is the derivation of the following simple but rather important technical result.

Lemma B.11 *If* $\Phi \in B_+\left(A \underset{\alpha,u}{\times} G \right)$ *and* $\Psi \in B_+(G)$ *then*

$$\Psi \cdot \Phi \in B_+\left(G \underset{\alpha,u}{\times} A \right).$$

Proof As $(\Psi \cdot \Phi)(h) = \Psi(h) \cdot \Phi(h)$ the statement of the lemma follows from condition (iii) of Lemma B.9 and the observation that the pointwise product of positive matrices is again positive.

So let us check the latter assertion.

Recall that a matrix $\left[b_{ij} \right]$ is positive iff there exists a matrix $\left[c_{ij} \right]$ such that

$$b_{ij} = \sum_k \bar{c}_{ik} c_{jk} \; . \tag{B.16}$$

Now if $\left[a_{ij} \right]$ and $\left[b_{ij} \right]$ are two positive matrices then for any $\{t_i\} \in \mathbb{C}^n$ we have (applying (B.16) and (B.10)) that

$$\sum_{i,j} \left(a_{ij} b_{ij} \right) \bar{t}_i t_j = \sum_{i,j} a_{ij} \sum_k \bar{c}_{ik} c_{jk} \bar{t}_i t_j$$

$$= \sum_k \left[\sum_{i,j} a_{ij} \left(\overline{c_{ik} t_i} \right) \left(c_{jk} t_j \right) \right] \geq 0$$

which means that $\left[a_{ij} b_{ij} \right]$ is positive. □

The next result links positive definite functions on G and amenability.

Lemma B.12 *A group G is amenable iff there exists a net* $\{f_i\}$ *consisting of functions having finite supports and belonging to the unit sphere of* $l^2(G)$ *such that* $\left\{ f_i \times f_i^* \right\}$ *converges to* 1 *uniformly on finite sets of G (here* $f_i^*(g) = \bar{f}_i\left(g^{-1}\right)$ *(recall (B.15))).*

Proof By Theorem 11.5 (ii) for each finite set $K \subset G$ and $\varepsilon > 0$ there exists a positive function φ in the unit sphere of $l^1(G)$ such that

$$\left\| \varphi(g) - \varphi\left(s^{-1}g\right) \right\|_{l^1(G)} < \varepsilon \quad \text{for every } s \in K \ . \tag{B.17}$$

Clearly without loss of generality here we can consider only functions φ having finite support.

Now let φ be any such function. Set $f = \varphi^{\frac{1}{2}}$ (the pointwise square root). Then

$$1 = \|f\|^2_{l^2(G)} = \left(f \times f^*\right)(e) \ , \tag{B.18}$$

and, for each s in G we have

$$\begin{aligned} \left|1 - \left(f \times f^*\right)(s)\right|^2 &= \left|\left(f \times f^*\right)(e) - \left(f \times f^*\right)(s)\right|^2 \\ &= \left|\sum_G f(g)\left[f^*\left(g^{-1}\right) - f^*\left(g^{-1}s\right)\right]\right|^2 \\ &= \left|\sum_G f(g)\left(\overline{f(g)} - \overline{f\left(s^{-1}g\right)}\right)\right|^2 . \end{aligned} \tag{B.19}$$

Further (B.18) and (B.19) along with the Cauchy-Bunyakowskii-Schwartz inequality implies

$$\begin{aligned} \left|1 - \left(f \times f^*\right)(s)\right|^2 &\le \sum_G \left|f(g) - f\left(s^{-1}g\right)\right|^2 \\ &\le \sum_G \left|f(g) - f\left(s^{-1}g\right)\right| \left|f(g) + f\left(s^{-1}g\right)\right| \\ &= \sum_G \left|f^2(g) - f^2\left(s^{-1}g\right)\right| \\ &= \left\|\varphi(g) - \varphi\left(s^{-1}g\right)\right\|_{l^1(G)} . \end{aligned} \tag{B.20}$$

On the other hand, let f be any function lying in the unit sphere of $l^2(G)$ and having finite support. Set $\varphi = |f|^2$. Then φ has finite support and $\|\varphi\|_{l^1(G)} = 1$.

For each $s \in G$ we then have

$$\begin{aligned}
\left\|\varphi(g)-\varphi\left(s^{-1}g\right)\right\|_{l^1(G)} &= \sum_G \left| |f(g)|^2 - \left|f\left(s^{-1}g\right)\right|^2 \right| \\
&\le \sum_G \left| f(g)^2 - f\left(s^{-1}g\right)^2 \right| \\
&= \sum_G \left| f(g) - f\left(s^{-1}g\right)\right| \left| f(g) + f\left(s^{-1}g\right)\right| \\
&\le \left[\sum_G \left| f(g) - f\left(s^{-1}g\right)\right|^2 \cdot \sum_G \left| f(g) + f\left(s^{-1}g\right)\right|^2 \right]^{\frac{1}{2}} \\
&\le 2\left[\sum_G \left| f(g) - f\left(s^{-1}g\right)\right|^2 \right]^{\frac{1}{2}} \\
&= 2\left[\sum_G |f(g)|^2 - f(g)\overline{f\left(s^{-1}g\right)} - f\left(s^{-1}g\right)\overline{f(g)} + \left|f\left(s^{-1}g\right)\right|^2 \right]^{\frac{1}{2}} \\
&= 2\left[\left(1 - \sum_G f(g)\overline{f\left(s^{-1}g\right)}\right) + \left(1 - \sum_G \overline{f\left(s^{-1}g\right)}f(g)\right) \right]^{\frac{1}{2}} \\
&\le 2\left[2\left|1 - \sum_G f(g)\overline{f\left(s^{-1}g\right)}\right| \right]^{\frac{1}{2}} \\
&= 2\sqrt{2}\left|1 - \left(f \times f^*\right)(s)\right|^{\frac{1}{2}}.
\end{aligned} \tag{B.21}$$

Now (B.20) and (B.21) mean that

$$\left|1 - \left(f \times f^*\right)(s)\right|^2 \le \left\|\varphi(g)-\varphi\left(s^{-1}g\right)\right\|_{l^1(G)} \le 2\sqrt{2}\left|1 - \left(f \times f^*\right)(s)\right|^{\frac{1}{2}}$$

which, along with (B.17), proves the desired result. □

Corollary B.13 *If G is an amenable group then there exists a net* $\{\Psi_i\}$ *of positive definite functions with* $\Psi_i(e)=1$ *and having finite supports and such that* $\{\Psi_i\}$ *converges to* 1 *uniformly on finite subsets of G.*

Proof Just set

$$\Psi_i = f_i \times f_i^*$$

where f_i are those introduced in Lemma B.12 and recall B.10. □

Now the analogue of Theorem 11.14 is

Theorem B.14 *If G is amenable then the set of positive functionals with finite supports (see B.8) is weak* dense in the set of all positive functionals in* $\left(A \underset{\alpha,u}{\times} G\right)^*$.

Proof Let ϕ be any positive functional in $\left(A \underset{\alpha,u}{\times} G\right)^*$ and $\Phi \in B_+\left(A \underset{\alpha,u}{\times} G\right)$ be the corresponding function. Consider the net $\{\Psi_i\}$ mentioned in Corollary B.13. Bearing in mind Lemma B.11 we conclude that the net $\{\Phi_i\}=\{\Psi_i\Phi\}$ of positive definite functions having finite supports is weak* uniformly convergent to Φ on finite subsets. This means (recalling B.8) that the corresponding net $\{\phi_i\}$ of positive functionals with finite supports is weak* convergent to ϕ. □

The result just obtained gives us the possibility of proving analogues of the main results of Section 12 for the quasicross-product situation. Since the proofs in the quasicross-product situation are almost the same as in the cross-product case (one should simply follow the path of the proof of the corresponding assertion in Section 12 and use when necessary Theorem B.14 instead of Theorem 11.14) we shall simply sketch these results.

Theorem B.15 *Let* $\alpha=\{\alpha_g\}_{g\in G}$ *be a quasirepresentation of a discrete amenable group G by automorphisms of A (described in B.1). Let* π *be a faithful representation of A in a Hilbert space H and let a family* $\{T_g\}_{g\in G}$ *of unitary operators in H be a quasirepresentation of G (described in B.4 with cocycle* $\pi(u(g,h))$*) such that*

$$\pi(\alpha_g(a)) = \hat{T}_g(\pi(a)) \quad \textit{for any } a \in A .$$

Then the standard representation

$$A \underset{\alpha,u}{\times} G \xrightarrow{\ \pi\ } C^*(\pi(A), T_g)$$

*described in B.5 is a *-isomorphism between* $A \underset{\alpha,u}{\times} G$ and $C^*(\pi(A), T_g)$ *iff the algebra* $C^*(\pi(A), T_g)$ *possesses the property (*) (see B.6).*

Proof Analogous to that of Theorem 12.8. □

Corollary B.16 *Let* $C^*(A_1, T_g)$ *and* $C^*(A_2, V_g)$ *be two* C^**-algebras described in B.5 (with cocycles* u_1 *and* u_2*) and let* $\varphi: A_1 \to A_2$ *be an isomorphism such that*

$$\varphi(u_1(g,h)) = u_2(g,h), \qquad g,h \in G$$

and

$$\varphi(T_g a T_g^*) = V_g \varphi(a) V_g^*, \qquad g \in G .$$

If G *is amenable and both the algebras* $C^*(A_1, T_g)$ *and* $C^*(A_2, V_g)$ *possess the property (*) then* $C^*(A_1, T_g)$ *is *-isomorphic to* $C^*(A_2, V_g)$ *and the isomorphism is given by*

$$\sum a_g T_g \to \sum \varphi(a_g) V_g, \qquad a_g \in A_1 .$$

Remark B.17 Theorem B.15 along with B.7 means that in the case when G is amenable the regular representation of $A \underset{\alpha,u}{\times} G$ (described in B.3) is *-isomorphic to $A \underset{\alpha,u}{\times} G$.

B.18 We denote by

$$\tau_g : \hat{A} \to \hat{A}$$

and

$$t_g : \text{Prim } A \to \text{Prim } A$$

the mappings defined in 12.12.

Remark Since $\hat{u}(g,h)$ is an inner automorphism it follows that

$$t_g \cdot t_h = t_{gh}$$

and thus a quasirepresentation of G by automorphisms $\hat{T}_g$ induces a *representation*(!) of G by homeomorphisms t_g.

The *topologically free action* of the group G *for the quasirepresentation* $\{\hat{T}_g\}$ is defined in the same way as in 12.13 (or 12.13′).

Now following the reasoning of the proof of Theorem 12.14 we obtain

Theorem B.19 *If G acts topologically freely on A by a quasirepresentation $\{\hat{T}_g\}$ then the algebra $C^*(A,T_g)$ described in B.5 possesses the property (*).*

This result along with Theorem B.15 implies

Corollary B.20 *Let $C^*(A,T_g)$ be an algebra (described in B.5) such that G is amenable and acts topologically freely (by the quasirepresentation $\{\hat{T}_g\}$). Then $C^*(A,T_g)$ is *-isomorphic to $A \underset{\hat{T},u}{\times} G$.*

Corollary B.21 *(The isomorphism theorem for a quasirepresentation.) Let $C^*(A,T_g)$ and $C^*(A_1,V_g)$ be C^*-algebras (described in B.5 with cocycles u and u_1) and let $\varphi\colon A \to A_1$ be an isomorphism such that*

$$\varphi(u(g,h)) = u_1(g,h), \quad g,h \in G$$

and

$$\varphi\left(T_g a T_g^*\right) = V_g \varphi(a) V_g^*, \quad g \in G .$$

*If G is amenable and acts topoloically freely on A by the quasirepresentation $\{\hat{T}_g\}$ then $C^*(A,T_g)$ and $C^*(A_1,V_g)$ are *-isomorphic.*

Now we proceed to the derivation of a certain generalization of Theorem 16.21.

B.22 Let $\mathcal{J}$ be a non-degenerate C^*-subalgebra of $L(H)$. We say that an operator $x \in L(H)$ is a (two-sided) *multiplier* for $\mathcal{J}$ if

$$x\mathcal{J} \subset \mathcal{J}, \quad \mathcal{J}x \subset \mathcal{J} .$$

The set of all multipliers of $\mathcal{J}$ will be denoted by $M(\mathcal{J})$. Clearly $M(\mathcal{J})$ is a C^*-subalgebra of $L(H)$.

Since $x\mathcal{J} \subset \mathcal{J}$ implies $x\mathcal{J}'' \subset \mathcal{J}''$ and $1 \in \mathcal{J}''$ (in view of the non-degeneracy of $\mathcal{J}$) it follows that all the multipliers of $\mathcal{J}$ belong to $\mathcal{J}''$.

One could find more information about the multipliers, for example, in Pedersen [1], Section 3.12. For our further purposes we need only the following result which is an elementary corollary of Proposition 3.12.3 of Pedersen [1].

Lemma B.23 *Let $\mathcal{J}$ be a non-degenerate C^*-subalgebra of $L(H)$ and $\mathcal{J}_1$ be a non-degenerate subalgebra of $L(H_1)$.*

*If $\varphi: \mathcal{J} \to \mathcal{J}_1$ is a *-isomorphism then there exists a unique extension of this isomorphism up to an isomorphism*

$$\varphi: M(\mathcal{J}) \to M(\mathcal{J}_1)$$

which means in particular that for any operator $b \in M(\mathcal{J})$ there exists a unique operator $b_1 \in L(H_1)$ such that

$$\varphi(bj) = b_1\varphi(j) \quad \textit{for every}\; j \in \mathcal{J}$$

(and we set $\varphi(b) = b_1 \in M(\mathcal{J}_1)$).

B.24 Let α be an automorphism of a non-degenerate C^*-algebra $\mathcal{J} \subset L(H)$. The foregoing lemma means that α generates an automorphism of $M(\mathcal{J})$; namely, for any $b \in M(\mathcal{J})$ the element $\alpha(b)$ is defined via the equation

$$\alpha(bj) = \alpha(b)\alpha(j), \quad j \in \mathcal{J} .$$

B.25 Let $\{t_g\}_{g\in G}$ be a group of homeomorphisms of Prim A and G_x and G^x be the groups introduced in 16.16 and 16.19.

We say that *G acts normally* (on Prim A) if, for every $x \in \text{Prim}\, A$, G_x is normal and G^x acts topologically freely (recall Corollary 16.18 and 16.19).

B.26 Now let $C^*(A,T_g)$ be a C^*-algebra generated by a C^*-algebra A and a group $\{T_g\}_{g\in G}$ of unitary operators generating automorphisms $\{\hat{T}_g\}$ of A (and homeomorphisms $\{t_g\}$ (see 12.12) of Prim A).

We shall consider the situation when:

(i) G acts normally on Prim A, B.25;

and

(ii) for every I_g (defined in 16.1) there exists $u_g \in M(I_g)$ such that

$$T_g j = u_g j, \quad j \in I_g \,. \tag{B.22}$$

This situation is a natural generalization of that examined in Theorem 16.21. To investigate it we need to introduce the corresponding analogue of the representations described in 16.20.

For any $x \in \hat{A}$ consider the space $l^2(G^x, H_x)$ where $G^x = G^{\text{Ker}\, x}$ and H_x is the representation space for x. In this space we define the operators $\pi_x(a)$, $\pi_x(T_g)$ in the following way.

First in every equivalence class $[g] \in G^x$ we fix an element $\delta_{[g]} \in [g]$ in such a way that

$$\delta_{[g^{-1}]} = (\delta_{[g]})^{-1} \text{ and } \delta_{[e]} = e \,.$$

Then we set

$$\hat{T}_{[g]} = \hat{T}_{\delta_{[g]}} \,. \tag{B.23}$$

Now we define $\pi_x(a)$, $a \in A$, by the equation

$$\pi_x(a)([g] \otimes \xi) = [g] \otimes x(\hat{T}_{[g]}(a))\xi, \quad \xi \in H_x,\ [g] \in G^x \,. \tag{B.24}$$

Observe that, for every $g \in G_x$, $\pi_x(I_g)$ is a non-degenerate subalgebra in $l^2(G^x, H_x)$. Indeed, $x(I_g) \neq 0$ and Lemma 2.82 (i) along with the irreducibility of x implies that the essential subspace $K \subset H_x$ for $x(I_g)$ coincides with H_x which (in view of the structure of $\pi_x(A)$) proves the above assertion.

This observation together with Lemma B.23 and (B.22) means that for every $g \in G_x$ the representation $\pi_x(A)$ (B.24) determines uniquely the representation $\pi_x(T_g)$ (equal to $\pi_x(u_g)$).

Bearing this in mind we introduce the algebra

$$\overline{A}_x = C^*(A, T_g), \quad g \in G_x \tag{B.25}$$

and we shall denote by $\pi_x(\overline{A}_x)$ the representation, just described, of this algebra.

Remark Clearly this representation depends on the choice of the elements $\delta_{[g]} \in [g]$.

For any $[h],[g] \in G^x$ we set

$$\tau([h],[g]) = \delta_{[h]} \cdot \delta_{[g]} \cdot \left(\delta_{[h \cdot g]}\right)^{-1} . \tag{B.26}$$

Obviously

$$\tau([h],[g]) \in G_x . \tag{B.27}$$

Let

$$u([h],[g]) = T_{\tau([h],[g])} . \tag{B.28}$$

Now for every $[h] \in G^x$ we set

$$\pi_x\left(T_{\delta_{[h]}}\right) = \pi_x([h]) \tag{B.29}$$

and

$$\pi_x([h])([g] \otimes \xi) = \left[g \cdot h^{-1}\right] \otimes x\left(\hat{T}_{\left[g \cdot h^{-1}\right]}\left(u\left([h],[g]^{-1}\right)\right)\right)\xi . \tag{B.30}$$

Since by (B.28), (B.29) $u\left([h],[g]^{-1}\right) \in \overline{A}_x$ the formula (B.30) is well defined.

Observe that every element $g \in G$ could be written in the form $g = h \cdot \delta_{[g]}$ with some $h \in G_x$ and as $T_h \in \overline{A}_x$ the formula (B.30) along with the foregoing argument defines the representation $\pi_x(T_g)$ for any $g \in G$.

Remark B.27 Note that in fact the formulae (B.24), (B.26), (B.28), (B.30) describe the regular representation (see B.3) for the quasicross-product

$$\pi_x(\overline{A}_x) \underset{[\hat{T}_x],u_x}{\times} G^x$$

where $[\hat{T}_x]$ is the quasirepresentation of G^x by automorphisms $\{\widehat{\pi_x([h])}\}_{[h]\in G^x}$ induced by the quasirepresentation $\pi_x([h])$ (B.30) and u_x is the cocycle $\pi_x(u[h],[g])$.

Now we are ready to formulate the analogue of Theorem 16.21 for the situation considered here.

Theorem B.28 *Let A be a C^*-algebra such that the topology of* Prim A *has a countable base and let $C^*(A,T_g)$ be a C^*-algebra satisfying conditions (i) and (ii) in B.26. If, for every $y \in$ Prim A, G^y is an amenable countable group then*

$$C^*(A,T_g) \cong \underset{x\in\hat{A}}{\oplus} \pi_x(\overline{A}_x) \underset{[\hat{T}_x],u_x}{\times} G^x = \underset{x\in\hat{A}}{\oplus} \pi_x(C^*(A,T_g)) \tag{B.31}$$

where we use the notation introduced in B.26, B.27.

Finally, $b \in C^*(A,T_g)$ is invertible iff all $\pi_x(b)$, $x \in \hat{A}$, are invertible.

Proof The proof of this theorem is in fact the quasicross-product reinterpretation of the proof of Theorem 16.21. So we shall simply sketch it following the reasoning of the latter theorem and mentioning the differences.

It is enough to check that for any $x \in \hat{A}$ and any finite set $F \subset G$:

(i) $\left\|\sum_F a_g T_g\right\| \geq \left\|\pi_x\left(\sum_F a_g T_g\right)\right\|$;

(ii) for any irreducible representation ν of $C^*(A,T_g)$ there exists a point x such that ν is a representation for $\pi_x(C^*(A,T_g))$; and

(iii) $\pi_x(\overline{A}_x) \underset{[\hat{T}_x],u_x}{\times} G^x \cong \pi_x(C^*(A,T_g))$.

Note that (iii) is true in view of Remarks B.27 and B.17.

Now let us prove (ii).

Note that for any point $y \in \operatorname{Prim} A$ we have

$$\overline{\{t_g(y)\}}_{g\in G} = \overline{\{t_g(y)\}}_{g\in G^y} . \tag{B.32}$$

Bearing this in mind and using the countable base of the topology on Prim A we (by an argument analogous to that used in the proof of Theorem 16.21) obtain that for any irreducible representation $\nu \in \widehat{C^*(A,T_g)}$ there exists a y_0 such that

$$\operatorname{supp}\operatorname{Ker}[\nu(A)] = \overline{\{t_g(y_0)\}}_{g\in G^{y_0}} .$$

Now we have to consider the following cases.

(a) $y_0 \in X_\infty$, that is $G_{y_0} = \{e\}$, $G^{y_0} = G$.

By the same argument as in the proof of Theorem 16.21 for any $x_0 \in \hat{A}$ with $\operatorname{Ker} x_0 = y_0$ we have here

$$\pi_{x_0}\left(C^*\left(A,T_g\right)\right) \cong \nu(A) \underset{\hat{T}}{\times} G$$

thus proving (ii) (for this case).

(b) $y_0 \,\overline{\in}\, X_\infty$ that is $G_{y_0} \neq \{e\}$.

As in the previous situation here

$$\pi_{x_0}(A) \cong \nu(A) . \tag{B.33}$$

For any finite set $N \subset G_{y_0}$ let us consider the ideal

$$\mathcal{J} = \bigcap_N I_g .$$

The argument used in B.26 (when checking the non-degeneracy of $\pi_x\left(I_g\right)$) shows that $\pi_{x_0}(\mathcal{J})$ is non-degenerate while the argument used in the proof of Theorem 16.21 (when checking the non-degeneracy of $\nu\left(I_h\right)$) shows that $\nu(\mathcal{J})$ is non-degenerate.

Now this observation, along with Lemma B.23 (and bearing in mind that $\overline{A}_{x_0}$ is the closure of *finite* sums $\sum a_g T_g$, $g \in G_{x_0}$), means that

$$\pi_{x_0}\left(T_g\right) \cong \nu\left(T_g\right), \quad g \in G_{x_0}$$

and (B.34)

$$\pi_{x_0}\left(\overline{A}_{x_0}\right) \cong \nu\left(\overline{A}_{x_0}\right),$$

which in turn implies

$$\pi_{x_0}\left(C^*\left(A, T_g\right)\right) \cong \nu\left(\overline{A}_{x_0}\right) \underset{\left[\hat{T}_{x_0}\right], u_{x_0}}{\times} G^{x_0}$$

thus proving (ii).

To finish the proof now we have to verify (i).

Again the only non-trivial situation is $y = \mathrm{Ker}\ x \,\overline{\in}\, X_\infty$.

Recalling Theorem B.14 and bearing in mind (iii) we conclude that to prove (i) it is enough to verify that for any state $f \in \left(\pi_x\left(\overline{A}_x\right) \underset{\left[\hat{T}_x\right], u_x}{\times} G^x\right)^*$ with a finite support and any finite sum $b = \sum a_g T_g$ there exists a state $\varphi \in \left[C^*\left(A, T_g\right)\right]^*$ such that

$$\varphi(b^* b) \geq f\left(\pi_x(b^* b)\right) . \tag{B.35}$$

But the state φ could be constructed from the state f by means of reasoning similar to that exploited in the proof of Lemma 12.7, provided $C^*\left(A, T_g\right)$ possesses the property

$$\left\| \sum_F a_g T_g \right\| \geq \left\| \sum_{[g]=[e]} \pi_x\left(a_g T_g\right) \right\| \tag{B.36}$$

for any finite set $F \subset G$ (here $[g] = [e]$ means $g \in G_y$, $y = \mathrm{Ker}\ x$ (recall (16.43))). Therefore we proceed to the verification of the latter inequality.

Set

$$\{G(y)\} = \{t_g(y)\}_{g\in G}$$

$$M = \{g \in F : [g] = [e]\} ,$$

$$\mathcal{J} = \bigcap_{h\in M} I_h , \qquad \text{(B.37)}$$

$$W = \bigcap_{h\in M} \overset{\circ}{X}_h$$

$$\beta = \left\| \sum_{g\in M} \pi_x(a_g T_g) \right\| = \sup_{\substack{\eta\in\hat{A} \\ \text{Ker } \eta\in G(y)}} \left\| \eta\left(\sum_{g\in M} a_g T_g \right) \right\| .$$

The equality in the latter expression follows from the construction of π_x and the fact that

$$\eta(a_g T_g) \in \eta(A)'', \quad g \in M, \ \text{Ker } \eta \in G(y)$$

(recall once more Lemma B.23).

The last equality in (B.37) implies in particular that for any $\varepsilon > 0$ there exists a point $z_0 \in \{G(y)\}$ and a representation η_0 with Ker $\eta_0 = z_0$ such that

$$\left\| \eta_0\left(\sum_{g\in M} a_g T_g \right) \right\| > \beta - \varepsilon . \qquad \text{(B.38)}$$

The only difference in the verification of (B.36) in comparison with (16.43) is that $\pi_x(a_g T_g)$, $g \in M$, does not belong to $\pi_x(A)$ but to $\pi_x(A)''$. But in fact this 'obstacle' can be easily overcome.

For the ideal $\mathcal{J}$ described by (B.37) we set

$$H_{\mathcal{J}} = \bigoplus_{\substack{x\in\hat{A} \\ \text{Ker } x\in W}} H_x$$

and consider the corresponding faithful representation of $\mathcal{J}$ in $L(H_{\mathcal{J}})$:

$$\varphi = \bigoplus_{\substack{x\in\hat{A} \\ \operatorname{Ker} x \in W}} x .$$

By construction $\varphi(\mathcal{J})$ is non-degenerate and (recalling B.22) we have

$$\varphi\left(\sum_{g\in M} a_g T_g\right) \in M(\varphi(\mathcal{J})) \subset \varphi(\mathcal{J})'' . \tag{B.39}$$

In view of this and Kaplansky's density theorem (Theorem 2.97) the inequality (B.38) implies the existence of an element $j \in \mathcal{J}$ with $\|j\| \le 1$ and such that

$$\left\|\eta_0\left(j\sum_{g\in M} a_g T_g\right)\right\| = \left(\widecheck{\sum_{g\in M} j a_g T_g}\right)(z_0) > \beta - \varepsilon . \tag{B.40}$$

On the other hand, for the same j we have

$$\left\|j\sum_F a_g T_g\right\| \le \left\|\sum_F a_g T_g\right\| . \tag{B.41}$$

Now (B.40) and (B.41) mean that the inequality (B.36) will be proved if we show that

$$\left\|\sum_F j a_g T_g\right\| \ge \left(\widecheck{\sum_{g\in M} j a_g T_g}\right)(z_0) . \tag{B.42}$$

But the verification of (B.42) can be carried out by an argument absolutely analogous to that used in the proof of Theorem 16.21 when checking (16.43), only instead of the inequality (16.49) here we have the inequality

$$\left\|\pi\left(\sum_{g\in M} j a_g T_g\right)\xi\right\| > \beta - \varepsilon \tag{B.43}$$

and instead of the deduction (16.52) we have respectively

$$\left\|\nu\left(\sum_F ja_gT_g\right)\right\|^2 \geq \left\|\nu(j_\lambda)\nu\left(\sum ja_gT_g\right)\xi\right\|^2$$

$$=\left\|\sum_F \nu\left(j_\lambda ja_gT_g\right)\xi\right\|^2$$

$$=\left\|\sum_M \nu\left(j_\lambda ja_gT_g\right)\xi\right\|^2+\left\|\sum_{F\setminus M} \nu\left(j_\lambda ja_gT_g\right)\xi\right\|^2$$

$$\geq\left\|\sum_M \nu\left(j_\lambda ja_gT_g\right)\xi\right\|^2$$

$$=\left\|\nu(j_\lambda)\sum_M \nu\left(ja_gT_g\right)\xi\right\|^2$$

$$=\left\|\nu(j_\lambda)\sum_M \pi\left(ja_gT_g\right)\xi\right\|^2$$

which, along with (B.43) and the arbitrariness of ε, proves (B.42), thus completing the proof of (B.31).

Finally the equivalence of the invertibility of $b \in C^*\left(A, T_g\right)$ and the invertibility of all $\pi_x(b)$, $x \in \hat{A}$, can be proved in the manner presented in the proof of Theorem 21.2. □

B.29 Since $C^*\left(A, T_g\right)$ is the uniform closure of the finite sums

$$\sum_F a_gT_g, \quad |F| < \infty$$

it is reasonable to consider for every finite set $F \subset G$ the group G_F generated by F. Clearly G_F is countable.

A useful observation is to note that

if G is normally acting then, for every finite F, G_F is normally acting as well.

Indeed, for any $x \in \text{Prim}\, A$ we have

$$(G_F)_x = G_x \cap G_F \tag{B.44}$$

which means in particular that $(G_F)_x$ is a normal subgroup of G_F.

On the other hand, (B.44) implies that for any $g \in G_F$ if its equivalence class $[g] \in (G_F)^x = G_F / (G_F)_x$ is not equal to $[e]$ (in $(G_F)^x$) then its equivalence class $[g] \in G^x$ is also not equal to $[e]$ (in G^x), and consequently if G^x acts topologically freely at x then $(G_F)^x$ acts topologically freely as well.

Bearing all this in mind we obtain an immediate corollary of Theorem B.28.

Corollary B.30 *Let A be a C^*-algebra with such that the topology of* Prim A *has a countable base and $C^*(A, T_g)$ be a C^*-algebra satisfying conditions (i) and (ii) in B.26 and such that, for every $y \in$* Prim A *and any finite $F \subset G$, $(G_F)^y$ is amenable.*

*Let $C^*(A_1, V_g)$ be another C^*-algebra such that there exists a *-isomorphism $\varphi: A \to A_1$ having the properties*

$$\varphi(T_g a T_g^*) = V_g \varphi(a) V_g^*, \quad g \in G$$

and

$$\varphi(T_g j) = V_g \varphi(j), \quad j \in I_g .$$

*Then $C^*(A, T_g)$ is *-isomorphic to $C^*(A_1, V_g)$ and the isomorphism is given by*

$$\sum a_g T_g \to \sum \varphi(a_g) V_g, \quad a_g \in A .$$

B.31 In fact we can relax the condition on the topology of Prim A in the latter result; namely, looking through the proof of Theorem B.28 one sees that we used the countable base of the topology of Prim A only to establish, for every $\nu \in \widehat{C^*(A, T_g)}$, the existence of a point $y \in \text{Prim}\, A$ such that $\overline{\{t_g(y)\}_{g \in G}} = \text{supp Ker}[\nu(A)]$. Note that the set $\Omega = \text{supp}[\text{Ker}\, \nu(A)]$ possesses the following property: for any open set $u \subset \Omega$ (in the topology induced from Prim A) we have

$$\overline{\left[\bigcup_{g\in G} t_g(u)\right]} = \Omega\ . \tag{B.45}$$

We shall call any closed set Ω possessing the property (B.45) *irreducible* (with respect to a group of homeomorphisms $\{t_g\}_{g\in G}$).

We shall say that a *pair* (A,G) (A is a C^*-algebra, G is a group of homeomorphisms of Prim A) is *separably compatible* if for any finite $F\subset G$ every irreducible set Ω, with respect to G_F, has a countable base of topology.

Looking through the proof of Theorem 16.21 we see that if (A,G) is separably compatible then for any finite F and any irreducible representation $\nu\in \widehat{C^*\left(A,T_g\right)_{g\in G_F}}$ there exists a point $y_0\in \text{Prim}\, A$ such that

$$\overline{\{t_g(y_0)\}_{g\in G_F}} = \text{supp Ker}[\nu(A)]\ ,$$

which means that the statement of Corollary B.30 will be true if instead of the condition on the topology of Prim A that it has a countable base we assume only that (A,G) is separably compatible.

Remark The idea of the foregoing argument shows that there could be some other reasonable ways of relaxing the condition on the topology of Prim A but we will not treat this theme further.

Our final observation in this section is a generalization of Theorem 16.23.

Theorem B.32 *Let $C^*\left(A,T_g\right)$ be a C^*-algebra satisfying conditions (i) and (ii) in B.26 and such that (A,G) is separably compatible and for every $y\in \text{Prim}\, A$ and any finite $F\subset G$ $(G_F)^y$ is amenable. If $\mathcal{J}\neq\{0\}$ is a closed *-ideal in $C^*\left(A,T_g\right)$ then $\mathcal{J}\cap A\neq\{0\}$.*

Proof The argument here is analogous to that used in the proof of Theorem 16.23; namely, let $A\cap\mathcal{J}=\{0\}$. Consider the canonical mapping

$$\pi: C^*\left(A,T_g\right)\to C^*\left(A,T_g\right)/\mathcal{J}\ .$$

We have

$$\pi(A) = A/\mathcal{J} \cong A$$

$$\pi\left(T_g a T_g^*\right) = \pi\left(T_g\right)\pi(a)\pi\left(T_g\right)^*, \quad a \in A,\ g \in G$$

and for every $[j] \in \pi\left(I_g\right)$

$$\pi\left(T_g\right)[j] = \pi\left(T_g\right)\pi(j) = \pi\left(T_g j\right) = \pi\left(u_g j\right) = \left[u_g j\right]$$

which means (by Corollary B.30) that

$$C^*\left(A, T_g\right)/\mathcal{J} = \pi\left(C^*\left(A, T_g\right)\right) = C^*\left(\pi(A), \pi\left(T_g\right)\right) \cong C^*\left(A, T_g\right)$$

and consequently $\mathcal{J} = \{0\}$. □

Notes and Remarks

A. The results of A.1-A.24 present the general C^*-algebraic essence of the results obtained by Antonevich [2,5]. In connection with these results we should also mention Gordeev [1].

The results of A.25-A.35 are borrowed from Antonevich, Lebedev [5].

Theorem A.36 is a natural generalization of Theorem 20.4 from Antonevich [18].

B. As we have already mentioned in B.1-B.21 we follow the path laid in Sections 11 and 12. The description of certain properties of the quasicross-product arising from this approach has much in common with that presented for the cross-product situation in, say, Pedersen [1], Chapter 7 (one can compare for example Lemma B.9 with Proposition 7.6.8 of the latter source, Lemma B.11 with Corollary 7.6.9 and note that Lemma B.12 is for our further purposes, simply a more convenient *discrete* analogue to Proposition 7.3.8).

In B.1-B.21 we follow Lebedev [19].

It is worth mentioning that in Antonevich [18], §7, the isomorphism theorem for admissible groups was proved by means of subexponential groups and C^*-algebras associated with quasirepresentations of groups.

In B.22-B.32 we follow Lebedev [20].

Bibliography

M.S. Agranovich

[1] Elliptic singular integro-differential operators. *Uspekhi mat. nauk*, 1965, v.20, N5, p.3-120 (Russian).

M.S. Agranovich, A.S. Dynin

[1] General boundary value problems for elliptic systems in a multidimensional domain. *DAN SSSR*, 1962, v.146, N3, p.511-514 (Russian).

A.B. Antonevich

[1] On the normal solvability of a general boundary value problem with a shift for elliptic type equations. *Trudy 2 resp. kouf. matematikov BSSR*, Minsk, 1969, p.253-255 (Russian).

[2] On the index of a pseudodifferential operator with a finite group of shifts. *DAN SSSR*, 1970, v.190, N4, p.751-752 (Russian).

[3] On the index and normal solvability of a general elliptic boundary value problem with a finite group of shifts on the boundary. *Diff. uravnenija*, 1972, v.8, N2, p.309-317 (Russian).

[4] The Gelfand representation of a certain ring of operators. *Vestn. Belorus, univ. Ser. I*, 1973, N2, p.3-7 (Russian).

[5] Elliptic pseudodifferential operators with a finite group of shifts. *Izv. AN SSSR, Ser. Mat.*, 1973, v.37, N3, p.663-675 (Russian).

[6] On a class of pseudodifferential operators with deviating argument on the torus. *Diff. Uravnenija*, 1975, v.11, N9, p.1550-1557 (Russian).

[7] On variations of spectrum under small perturbations of an operator. *Vestn. Belorus univ. Ser. I*, 1976, N3, p.60-62 (Russian).

[8] On convolution type operators with oscillating coefficients. *Vestsi Akad. Navuk BSSR, Ser. Fiz.-Mat. Navuk*, 1976, N2, p.42-46 (Russian).

[9] On operators generated by linear extensions of diffeomorphisms. *DAN SSSR*, 1978, v.243, N4, p.825-828 (Russian).

[10] On a certain factorization of a difference operator. *Vestsi Akad Navuk BSSR, Ser. Fiz.-Mat. navuk*, 1978, N3; p.10-15 (Russian).

[11] Operators with a shift generated by the action of a compact Lie group. *Sibirsk. Mat. Zh.*, 1979, v.20, N3, p.467-478 (Russian).

[12] The conditions of invertibility of operators with a convex rationally independent system of shifts. *DAN SSSR*, 1981, v.256, N1, p.11-14 (Russian).

[13] Algebras generated by weighted shift operators and pseudodifferential operators with a shift. *DAN SSSR*, 1981, v.256, N6, p.1293-1296 (Russian).

[14] On the spectrum of a weighted shift operator in the space $W_p^1(X)$. *DAN SSSR,* 1982, v.264, N5, p.1033-1035 (Russian).

[15] On two methods of studying the invertibility of operators in C^*-algebras induced by dynamical systems. *Mat. Sborn.,* 1984, v.124 (166), N1, p.3-23 (Russian).

[16] Operators with a convex rationally independent system of shifts in vector-function spaces. *IX Shkola poeorii operat. v funkts. prostr., Ternopol'*, 1984, p.7 (Russian).

[17] Two-point problem for the string oscillations equation and related to it functional equations. *Diff. uravnenija,* 1985, v.21, N3, p.426-436 (Russian).

[18] *Linear Functional Equations. Operator approach.* Minsk, Universitetskoe, 1988 (Russian).

[19] The symbol of a functional pseudodifferential operator on a compact manifold. *Funkts. anal. i egho prilozhen.*, 1988, v.22, N4, p70-71 (Russian).

[20] Boundary value problems with strong nonlocalness for elliptic equations. *Izv. AN SSSR*, 1989, v.53, N1, p.3-24 (Russian).

[21] The index of a system of singular integral functional equations with a shift having an arbitrary set of periodic points. *International conference 'Boundary value problems, special functions and fractional calculus'*, Minsk, 1996, p.12 (Russian).

A.B. Antonevich, M.V. Belousov

[1] On the index formula of a pseudodifferential operator with a finite group of shifts. *Voprosy funkts. analiza (sb. statej), Petrozavodsk*, 1980, p.3-7 (Russian).

A.B. Antonevich, V.V. Brenner

[1] On algebras of operators with a finite group of shifts. *Mat. issled., Karaghanda*, 1977, v.4, p.3-10 (Russian).

[2] On symbol of pseudodifferential operator with locally independent shifts. *Dokl. AN BSSR*, 1980, v.24, N10, p.884-887 (Russian).

A.B. Antonevich, A. Buraczewski

[1] Dynamics of a linear mapping and invariant measures on sphere. *Demonstratio Math.*, 1996, v.29, N4, p.817-824.

A.B. Antonevich, Yu. A. Bykadorov

[1] On operators $(Sy)(t) = a(t)y(\alpha(t))$ related to functional-differential equations. *Dep. VINITI,* 1985, N2043-85, 53p. (Russian).

A.B. Antonevich, V.A. Dobrushkin

[1] A general boundary value problem for elliptic systems with a deviating argument. *Dokl. AN BSSR*, 1973, v.17, N6, p.491-493 (Russian).

A.B. Antonevich, M.B. Gankin

[1] On the homotopic classification of singular integral functional operators with a shift. *Izv. Vuzov. Matem.*, 1987, N2, p.9-14 (Russian).

A.B. Antonevich, J. Hodzaj

[1] On a certain class of integral difference equations with oscillating coefficients. *Diff. uravnenija*, 1989, v.26, N11, p.1997-1999 (Russian).

A.B. Antonevich, A.V. Lebedev

[1] Extension of operator algebras by unitary operators generating automorphisms. *Dokl. AN BSSR*, 1980, v.24, N5, p.404-407 (Russian).

[2] On Noetherity of functional-differential operator with partial derivatives containing a linear transformation of an argument. *Diff. uravnenija*, 1982, v.18, N6, p.987-996 (Russian).

[3] On spectral properties of operators with a shift. *Izv. AN SSSR. Ser. Mat.*, 1983, v.47, N5, p.915-941 (Russian).

[4] The symbols of functional pseudodifferential operators in the scales of Sobolev spaces. (To appear, *Dokl. AN Belarusi*)

[5] On the index of operators in algebras associated with automorphisms. (To appear, *Dokl. AN Belarusi*)

A.B. Antonevich, V.B. Ryvkin

[1] On the normal solvability of a problem on periodic solutions of a linear differential equation with a deviating argument. *Diff. uravnenija*, 1974, v.10, N3, p.1347-1353 (Russian).

[2] On conditions for a certain singular integral operator with a shift to be Fredholm. *Dokl. AN BSSR*, 1975, v.19, N1, p.5-7 (Russian).

[3] Operators generated by homeomorphisms of the circle conjugate to a rotation. *Mat. zametki*, 1982, v.31, N5, p.773-783 (Russian).

A.B. Antonevich, Serigne Aliou Lo

[1] Sur les mesures, qui apparaissent lors de l'etude d'une classe d'equations differentielles fonctionnelles. (To appear.)

A.B. Antonevich, Yu. V. Vol'pert

[1] On the spectrum of a weighted diagonal shift operator on the torus. *Vestsi Akad. Navuk BSSR Ser. Fiz.-Mat. navuk*, 1976, N3, p.48-53 (Russian).

W.B. Arveson

[1] A theorem on the action of abelian unitary groups. *Pacif. J. Math.*, 1966, v.16, N2, p.205-212.

[2] Operator algebras and measure preserving automorphisms. *Acta. Math.*, 1967, v.118, p.95-109.

[3] *An Invitation to C^*-algebras*. Springer-Verlag, 1976.

W.B. Arveson, K.B. Josephson

[1] Operator algebras and measure preserving automorphisms. II. *J. Funct. Anal.*, 1969, v.4, N1, p.100-134.

M.F. Atiyah
[1] *K-theory*. Harvard Univ., Cambridge, Mass., 1965.

M.F. Atiyah, G.B. Segal
[1] The index of elliptic operators. II. *Ann. of Math.*, 1968, v.87, p.531-545.

M.F. Atiyah, I.M. Singer
[1] The index of elliptic operators. I, III. *Ann. of Math.*, 1968, v.87, p.484-530, 546-604.

N.V. Azbelev, L.F. Rahmatullina
[1] Functional-differential equations. *Diff. uravnenija*, 1978, v.14, N5, p.771-797 (Russian).

Ch. Babbage
[1] An essay towards the calculus of functions. *Philos. Trans. London*, 1816, v.106, N2, p.179-256.

S. Banach
[1] *Functional Analysis Course*. Kiiv, Radjan'ska shk., 1948 (Ukrainian).

S. Banach, A. Tarski
[1] Sur le decomposition des ensembles de points en partes respectivement congruentes. *Fund. Math.*, 1924, N6, p.244-277.

A.B. Batkhin
[1] On Noetherity of two dimensional singular integral operators with discrete groups of shifts in $L_2(\mathbb{R}^2)$. *Linejnye operatory v funktsional'nykh prostranstvah: Tez. dokl. Sev.-kavkaz. reg. konf. Groznyi*, 1989, p.17 (Russian).
[2] An isomorphism theorem for C^*-algebras of multidimensional convolution type operators with discrete groups of shifts. *Dokl. rasch. zased. sem. Inst. prikl. mat. im. I.N. Vekua, Tbilisi*, 1990, v.5, N1, p.26-29 (Russian).
[3] Multidimensional convolution type operators with piece-wise continuous symbol and shifts, generating discrete group. The Noetherity in algebras of operators. *Dep. v Ukr. Zh. INTEI*, 1992, N654-UK92 (Russian).
[4] The Noether theory of multidimensional operators of convolution type with discrete groups of shifts in the space $L_2(\mathbb{R}^n)$. Ph.D. diss., Odessa, 1992 (Russian).

R. Beals
[1] Non-local boundary value problem for elliptic operators. *Amer. J. Math.*, 1965, v.87, N2, p.315-352.

E.F. Beckenbach, R. Bellman
[1] *Inequalities*. Springer-Verlag, 1961.

R. Bellman, K. Cooke

[1] *Differential Difference Equations*. Academic Press, 1963.

M.V. Belousov

[1] The algebra of the Boutet de Monvel operators with a finite group of shifts on the boundary. *IV Shkola po teorii operat. v funkts. prostr., Tez. dokl Minsk.*, 1978, p.19 (Russian).

[2] On a certain boundary value problem for an elliptic operator. *5 resp. konf. matematikov Belorussii, Toz. dokl. chast' 2, Grodno*, 1980, p.103 (Russian).

[3] BMOs with a finite group of shifts on the boundary. *Shkola po teorii operat. v funkts. prostr., Tez. dokl. Minsk.*, 1982, p.215-216 (Russian).

[4] The generalized Green operators with a finite group of shifts on the boundary. *Diff. uravnenija*, 1982, v.18, N8, p.1369-1377 (Russian).

[5] On the theory of the generalized Green operators. *Diff. uravnenija*, 1983, v.19, N4, p.693-694 (Russian).

[6] The index of general boundary value problems with a finite group of shifts. *11 Vsesoyuznaya shkola po teorii operat. v funkts. prostr., Tez. dokl. chast' 3., Chelyabinsk*, 1986, p.17 (Russian).

[7] On the Fredholmity of boundary value problems containing the shift operators on the boundary. *Mekhanika i prikladnaya matematika. Materialy Vsesoyuzn. konf. "Sovremennye problemy informatiki, vychislitelnoi tekhniki i avtomatizatsii", Tula*, 1988, p.123-125 (Russian).

[8] Certain elliptic nonlocal boundary value problems. Ph.D. diss., Minsk, 1989 (Russian).

M.V. Belousov, A.V. Lebedev

[1] The symbolic calculus of nonlocal boundary value problems with the shifts on the boundary. (To appear, *Dokl. AN Belarusi*)

L.M. Berezanskij

[1] On the question of conditions of reducibility of functional differential equations to the canonical form. *Diff. uravnenija,* 1977, v.13, N10, p.1749-1756 (Russian).

[2] Linear functional equations and functional differential equations with a deviating argument. Ph.D., diss., Alma-Ata, 1979 (Russian).

A. Beurling

[1] Sur les integrales de Fourier absolument convergentes et leur application a une transformation functionelle. *IX Congres des Math. Scand., Helsingfors,* 1938, p.345-366.

R.A. Biktashev

[1] Spectra of pseudodifferential operator over C^*-algebras. *Vestnik Moskov Univ. Ser. I. Mat. Mekh.,* 1982, N4, p.36-38 (Russian).

R.A. Biktashev, A.S. Mischenko
[1] Spectra of elliptic unbounded pseudodifferential operators over C^*-algebras. *Vestnik Moskov Univ. Ser. I., Mat. Mekh.*, 1980, N3, p.56-58 (Russian).

A.V. Bitsadze
[1] *Certain Classes of Partial Differential Equations.* M. Nauka, 1981 (Russian).

A.V. Bitsadze, A.A. Samarskij
[1] On certain simplest generalizations of the linear elliptic boundary value problems. *DAN SSSR,* 1969, v.185, N4, p.739-740 (Russian).

S. Bochner, R.S. Phillips
[1] Absolutely convergent Fourier expansions for non-commutative rings. *Ann. Math.*, 1942, v.43, N3, 409-418.

Ju. G. Borisovic, B.D. Gel'man, A.D. Myškis, V.V. Obuhovskii
[1] Topological methods in the theory of fixed points of multivalued mapping. *Uspekhi Mat. Nauk,* 1980, v.35, N1, p.59-126 (Russian).

V.M. Borok
[1] On a boundary value problem on a stripe for certain differential functional equations. *Izv. AN SSSR. Ser. Mat.*, 1971, v.35, N1, p.13-19 (Russian).

V.M. Borok, Ya. I. Zhitomirskij
[1] On the Cauchy problem for partial differential equations with a deviating argument. *Differential equations with a deviating argument. Kiev, Nauk. Dumka*, 1977, p.32-62 (Russian).

A. Böttcher, Yu. I. Karlovich, B. Silbermann
[1] Singular integral equations with PQC coefficients and freely transformed argument. *Math. Nachr.*, 1994, 166, p.113-133.

A. Böttcher, B. Silbermann
[1] *Analysis of Toeplitz Operators.* Berlin, Akademie, 1989.

N. Bourbaki
[1] *Integration (Eléments de mathématique,* LIVRE VI). Hermann, 1952.
[2] *Theoreies Spectrales* (Eléments de mathématique, Fascicule XXXII). Hermann, 1967.
[3] *Varietes Differentielles et Analytiques* (Eléments de mathématique, Fasc. XXXIII, XXXVI). Hermann, 1967, 1971.

L. Boutet de Monvel
[1] Boundary problems for pseudo-differential operators. *Acta. Math.*, 1971, v.126, N1-2, p.11-51.

M. De Brabanter
[1] The classification of rational rotation C^*-algebras. *Arch. Math.*, 1984, v.43, N1, p.79-83.
[2] A classification theorem for C^*-algebras generated by weighted shifts. *Simon Stevin*, 1984, v.58, N3, p.267-270.

M. De Brabanter, H. Zettl
[1] C^*-algebras associated with rotation groups and characters. *Manuscr. math.*, 1984, v.47, N1-3, p.153-174.

B. Brebken
[1] Representations and automorphisms of the irrational rotation algebra. *Pacif. J. Math.*, 1984, v.111, N2, p.257-282.

G.E. Bredon
[1] *Introduction to Compact Transformation Groups*. Academic Press, 1972.

V.V. Brenner
[1] On the spectral radius of operators with locally independent shifts. *Izv. AN BSSR*, 1981, N3, p.48-55 (Russian).
[2] Operators with locally independent shifts. Ph.D. diss., Minsk, 1981 (Russian).
[3] On spectral properties of operators with a shift. *Uspekhi Mat. Nauk.*, 1982, v.37, N1, p.139-140 (Russian).
[4] On symbolic sequences I, II. *Dep. VINITI*, 1983, N4399-83; *Dep. VINITI*, 1983, N4400-83 (Russian).

I.U. Bronstein
[1] *Nonautonomous Dynamical Systems*. Kishinev. Shtiintsa, 1984 (Russian).

F.F. Browder
[1] Non-local elliptic boundary value problems. *Amer. J. Math.*, 1964, v.86, N4, p.735-758.

M. Burnat
[1] Die Spectraldarstellung einiger differential Operatoren mit periodischen Koeffizienten in Raum der fastperiodischen Functionen. *Studia Math.*, 1964, v.25, p.33-64.

B.F. Bylov, R.E. Vinoghrad, D.M. Grobman, V.V. Nemytskij
[1] *The Lyapunov Exponents Theory and its Applications to Problems of Stability*. M. Nauka, 1966 (Russian).

S.L. Campbele
[1] Linear operators for which T^*T and TT^* commute. I. *Proc. Amer. Math. Soc.*, 1972, 34, N1, p.177-180; II. *Pacif. J. Math.*, 1974, 53, N2, p.355-361.

T. Carleman

[1] Sur la theorie des equations integrales et ses applications. *Verhandlungen des Internationalen Matematiker. Kongresses, Zürich*, 1932, Bd. 1, p.138-151.

C. Chicone, R. Swanson

[1] The spectrum of the adjoint representation and the hyperbolicity of dynamical systems. *J. Diff. Eqns.*, 1980, v36, p.28-39.

J.M. Cohen

[1] Cogrowth and amenability of discrete groups. *J. Funct. Anal.*, 1982, v.48, p.301-309.

P.E. Conner, E.E. Floyd

[1] *Differentiable Periodic Maps*. Springer-Verlag, 1964.

H.O. Cordes

[1] *Elliptic Pseudo-differential Operators - an Abstract Theory*. Lect. Notes Math. 756, Springer, 1979.

[2] On some C^*-algebras and Fréchet-*-algebras of pseudodifferential operators. *Proc. Symp. Pure Math., AMS*, 1985, v.43, p.79-104.

[3] *Spectral Theory of Linear Differential Operators and Comparison Algebras*. Lond. Math. Soc. Lect. Notes Ser. 76, 1987.

C.C. Cowen

[1] Composition operators on Hilbert spaces of analytic functions: A status report. *Oper. Th.: Oper. Algebras and Appl.: Proc. Summer Res. Inst., Durham, July 3-23, 1988. Providence (R.I), 1990, p.131-145.*

Yu. L. Daletskij, M.G. Krein

[1] *Stability of Solutions of Differential Equations in Banach Space*. M. Nauka, 1970 (Russian).

Dangh Suan Tkhan'

[1] C^*-algebras generated by local type operators and their automorphisms. Ph.D. diss., Minsk, 1992 (Russian).

J. Dauns, K.H. Hofmann

[1] Representations of rings by sections. *Mem. Amer. Math. Soc.*, 1968, N83, 180p.

A.M. Davie

[1] Invariant subspaces for Bishop's operator. *Bull. London Math. Soc.*, 1974, N6, p.343-348.

A.A. Dezin

[1] The simplest solvable extensions for the ultrahyperbolic and pseudoparabolic operators. *DAN SSSR*, 1963, v.148, N5, p.1013-1016 (Russian).

[2] The operators with the first 'time' derivative and non-local boundary conditions. *Izv. AN SSSR. Ser. Mat.*, 1967, v.31, N1, p.61-86 (Russian).

[3] *The General Problems of Boundary Value Problems Theory.* M. Nauka., 1980 (Russian).

J. Dixmier

[1] *Les C^*-algébras et Leurs Representations,* 2nd edition. Paris, Gauthier-Villars, 1969.

[2] *Les Algebres d'operateurs l'espace Hilbertien,* 2nd edition. Paris, Gauthier-Villars, 1969.

D.P. O'Donovan

[1] Weighted shifts and covariance algebras. *Trans. Amer. Math. Soc.,* 1975, v.208, p.1-25.

R.G. Douglas

[1] *Banach Algebras Technique in Operator Theory.* Academic Press, 1972.

[2] Banach algebra techniques in the theory of Toeplitz operators. *Lectures given at a GBMS Regional Conference, Series in Mathematics*, 1973.

B.A. Dubrovin, S.P. Novikov, A.T. Fomenko

[1] *Modern Geometry: Methods and Applications.* M. Nauka, 1979 (Russian).

R.V. Duduchava

[1] Integral convolution equations with discontinuous presymbols, singular integral equations with fixed singularities and their applications to mechanics problems. *Tr. Tbilissk. mat. in-ta,* 1979, v.60, p.5-135 (Russian).

N. Dunford, Ja. T. Schwartz

[1] *Linear Operators: General Theory.* Interscience, New York, v.I, 1958.

L.E. El'sgholts, S.B. Norkin

[1] *Introduction to the Theory and Applications of Differential Equations with a Deviating Argument* (2nd edition). M. Nauka, 1971 (Russian).

G.I. Éskin

[1] *Boundary Value Problems for Elliptic Pseudodifferential Equations.* M. Nauka, 1972 (Russian).

G.M. Fihtengol'ts

[1] *Differential and Integral Calculus Course. v.II*, M. Nauka, 1970 (Russian).

Fr. Forelli

[1] The isometries of H^p. *Canad. J. Math.*, 1964, 16, p.721-728.

F.I. Frankl'
[1] *Selected Memoirs on Gas Dynamics*. M. Nauka, 1973 (Russian).

B. Fuglede, R.V. Kadison
[1] On determinants and a property of the trace in finite factors. *Proc. Nat. Acad. Sc. USA*, 1951, 37, p.425-431.
[2] Determinant theory in finite factors. *Ann. Math.*, 1952, 55, p.520-530.

F.D. Gahov
[1] *Boundary Value Problems* (3rd edition). M. Nauka, 1977 (Russian).

F.D. Gahov, Yu. I. Cherskij
[1] *Convolution Type Equations*. M. Nauka, 1978 (Russian).

Th. W. Gamelin
[1] *Uniform Algebras*. Prentice-Hall, Englewood Cliffs, N.J., 1969.

F.R. Gantmakher
[1] *The Theory of Matrices*. M. Nauka, 1988 (Russian).

L.M. Glazman
[1] *The Direct Methods of the Qualitative Spectral Analysis of Singular Differential Operators*. M., Physmatghiz., 1963 (Russian).

A.V. Glebov, V.M. Deundyak
[1] The homotopic classification of the Fredholm systems of singular and bisingular integral operators with discontinuous coefficients. *Funkts. anal. i egho prilozhen.*, 1978, v.12, N3, p.76-77 (Russian).
[2] The homotopic classification of the Fredholm systems of singular and bisingular integral operators with discontinuous coefficients. *Matem. anal. i egho prilozhen. Rostov-na-Dony, Izd. Rostov. univ.* 1981, p.33-41 (Russian).

J. Glimm
[1] A Stone-Weierstrass theorem for C^*-algebras. *Ann. Math.*, 1960, v.72, p.216-244.

B. Godunov, P.P. Zabreiko
[1] The successful approximations convergence character for equations with smooth operators. *Dokl AN BSSR*, 1989, v.33, N7, p.583-586 (Russian).

I. Ts. Gohberg
[1] On an application of the theory of normed rings to singular integral equations. *Uspekhi Mat. Nauk.*, 1952, v.7, N2, p.149-156 (Russian).

I. Ts. Gohberg, I.A. Feldman
[1] *Convolution Equations and Projection Methods of their Solution*. M. Nauka, 1971 (Russian).

I. Ts. Gohberg, N. Ja. Krupnik

[1] On the algebra generated by one-dimensional singular integral operators with piecewise continuous coefficients. *Funkts. anal i egho prilozhen*, 1970, v.4, N3, p.27-38 (Russian).

[2] Singular integral operators with piecewise continuous coefficients and their symbols. *Izv AN SSSR*, Ser. Mat., 1971, v.35, N4, p.940-964 (Russian).

[3] *An Introduction to the Theory of One-dimensional Singular Integral Operators.* Shtiintsa, Kishinjev, 1973 (Russian).

[4] *One-dimensional Linear Singular Integral Equations. I. Introduction.* Birkhäuser Verlag, 1992.

V. Ja. Golodets

[1] Cross products of Neumann algebras. *Uspekhi Mat. Nauk*, 1971, v.26, N5, p.3-50 (Russian).

I. Ja. Gol'shejd, G.A. Margulis

[1] Lyapunov exponents of the product of random matrixes. *Uspehi Mat. Nauk,* 1989, v.44, N5, p.13-60 (Russian).

S.R. Gordeev

[1] The homotopic classification of families of Fredholm operators with a shift. *Dep. VINITI*, 1985, N8846-13, 51p (Russian).

B. Gramsh

[1] Relative Inversion on der Störungstheorie von Operatoren und Ψ-Algebren. *Mat. Ann.*, 1984, v.269, p.27-71.

B. Gramsh, W. Kaballo

[1] Decompositions of meromorphic Fredholm resolvents and Ψ^*-algebras. *Integral Eq. and Operat. Th.*, 1989, v.12, p.23-41.

B. Gramsh, K.G. Kalb

[1] *Pseudo-locality and Hypoellipticity of Operator Algebras.* Semesterbericht Funktionalanalysis 51-61, Tübingen, Sommersemester, 1985.

Fr. P. Greenleaf

[1] *Invariant Means on Topological Groups and Their Applications.* Van Nostrand-Reinhold Comp., 1969.

R.I. Grigorchuk

[1] Symmetrical wanderings on discrete groups. *Mnoghokompponent. slutchain. sistemy., M. Nauka*, 1978, p.132-152 (Russian).

[2] The growth degrees of finitely generated groups and the invariant means theory. *Izv. AN SSSR, Ser. Mat.*, 1984, v.48, N5, p.939-985 (Russian).

A.L. Gromov

[1] On invariant subspaces of weighted composition operators. *Funkts. anal. i egho prilozhen.*, 1988, v.22, N2, p.75-76 (Russian).

[2] On spectral properties of unitary composition operators with a weight. *Funkts. anal. i egho prilozhen.,* 1990, v.24, N3, p.74-75 (Russian).

[3] Spectral classification of certain types of unitary composition operators with a weight. *Alg. i anal.*, 1991, v.3, N5, p.62-87 (Russian).

[4] Spectral analysis of operators of an irrational rotation with a weight in $L^2(T)$. Ph.D. diss., Sankt-Peterbourgh, 1991 (Russian).

G. Grubb

[1] A characterisation of the non-local boundary value problems associated with elliptic operators. *Ann. Scuola norm. super. Pisa Sci. fis e math.* 1968, v.22, N3, p.433-442.

[2] *Functional Calculus of Pseudo-differential Boundary Problems.* Birkhäuser, 1986.

D.K. Gupta, B.S. Komal

[1] Fredholm composition operators on weighted sequence spaces. *Indian J. pure and appl. Math.*, 1983, v.14, N3, 293-296.

J. Hale

[1] *Theory of Functional Differential Equations.* Springer-Verlag, 1977.

P.R. Halmos

[1] *Lectures in Ergodic Theory.* Math. Soc., Japan, Tokyo, 1956.

[2] *A Hilbert Space Problem Book.* D. Van Nostrand Comp. Inc, Princeton, N.J., 1967.

P. de la Harpe, G. Scandalis

[1] Déterminant associé á une trace sur unt algébre de Banach. *Ann. Inst. Fourier,* 1984, 34, N1, p.241-26.

F. Hausdorff

[1] *Grundznge der Mengenlehre.* Leipzig, 1914.

H. Helson

[1] Cocycles on the circle. *J. Oper. Theory*, 1986, v.16, p.183-199.

[2] The spectral theorem. *Lect. Notes Math.,* Springer-Verlag, 1986, v.1227, p.87-91.

A.B. Hevelev

[1] On invertibility of equations with two shifts on the circle. *Dep. VINITI,* N1305-78, 1978, 7p (Russian).

[2] On invertibility and Noetherity of certain classes of operators with non Carleman shift. Ph.D. diss., University Rostov-na-Donu, 1979 (Russian).

E. Hewitt, K.A. Ross

[1] *Abstract Harmonic Analysis* v.1. Springer-Verlag, 1963.

[2] *Abstract Harmonic Analysis* v.2. Springer-Verlag, 1970.

M.W. Hirsch

[1] *Differential Topology*. Springer-Verlag, 1976.

K. Hoffman

[1] *Banach Spaces of Analytic Functions*. Prentice-Hall, Englewood Hills, N.J., 1962.

K.H. Hofmann

[1] Representations of algebras by continuous sections. *Bul. Amer. Math. Soc.*, 1972, N78, p.291-373.

W. Höppner

[1] Einige Abschätzungen für die Symbolrechung von Pseudodifferentialoperatoren im Halbraum. I, II. *Math. Nachr.* 1979, v.90, p.267-278; 1980, v.93, p.33-52.

L. Hörmander

[1] *The Analysis of Linear Partial Differential Operators* v.I-IV. Springer-Verlag, 1983, 1983, 1985, 1985.

S.-T. Hu

[1] *Homotopy Theory*. Academic Press, N.Y., 1959.

W. Hurewicz, H. Wallman

[1] *Dimension Theory*. Princeton University Press. Princeton, 1941.

D. Husemoller

[1] *Fibre Bundles*. McGraw-Hill Book Company, 1966.

N.I. Ionkin

[1] The solution of a certain heat conductivity boundary value problem with non-classical boundary conditions. *Diff. uravnenija*, 1977, v.13, N2, p.294-304 (Russian).

R.A. Johnson, K. Palmer, G.R. Sell

[1] Ergodic properties for linear differential systems. *SIAM J. Math. Anal.*, 1987, v.18, N1, p.1-33.

R.V. Kadison

[1] Strong continuity of operator functions. *Pacif. J. Math.*, 1968, v.26, p.121-129.

R.V. Kadison, J.R. Ringrose

[1] *Fundamentals of the Theory of Operator Algebras* v.I, II. Academic Press, 1986.

G.A. Kamenskij, A.L. Skubachevskij

[1] *Linear Boundary Value Problems for Differential Difference Equations.* Izdatel'stvo MAI., 1992 (Russian).

H. Kamowitz

[1] On compact multipliers of Banach algebras. *Proceeding of AMS,* 1981, v.81, N1, p.79-80.

N.K. Karapetyants

[1] On a class of discrete convolution operators with oscillating coefficients. *DAN SSSR*, 1974, v.216, N1, p.28-31 (Russian).

[2] On a certain class of shift operators. *Izv. Sev.-Kavkaz. nautch. tzentr. vysch. shk.*, 1976, N3, p.11-12 (Russian).

N.K. Karapetyants, S.G. Samko

[1] On functional equation $\psi(x+\alpha)-b(x)\psi(x)=g(x)$. *Izv. AN Arm. SSR.*, 1971, v.5, N5, p.441-445 (Russian).

[2] On discrete Wiener-Hopf operators with oscillating coefficients. *DAN SSSR*, 1971, v.200, N1, p.17-20 (Russian).

[3] *Equations with Involutive Operators and their Applications.* Izdatel'stvo Rostovskogho univ., 1988 (Russian).

Yu. I. Karlovich

[1] Singular integral operators with a shift of the curve of integration into a domain and their applications. Ph.D. diss., Odessa, 1975 (Russian).

[2] The Fredholm theory of a certain class of convolution type operators with a shift. *DAN SSSR*, 1987, v.295, N1, p.24-29 (Russian).

[3] C^*-algebras of convolution type operators with discrete groups of shifts and oscillating coefficients. *DAN SSSR*, 1988, v.302, N3, p.535-540 (Russian).

[4] Locally-trajectorial method of investigation of invertibility in C^*-algebras of operators with discrete groups of shifts. *DAN SSSR*, 1988, v.299, N3, p.546-550 (Russian).

[5] On algebras of singular integral operators with discrete groups of shifts in L_p spaces. *DAN SSSR*, 1989, v.304, p.274-280 (Russian).

[6] Algebras of convolution type operators with discrete groups of shifts and oscillating coefficients. Doctoral diss., Odessa, 1990 (Russian).

Yu. I. Karlovich, V.G. Kravchenko

[1] The Noether theory of a singular integral operator with a shift having periodic points. *DAN SSSR*, 1976, v.231, p.277-280 (Russian).

[2] On systems of functional and integro-functional equations with a non-Carleman shift. *DAN SSSR*, 1977, v.236, N5, p.1064-1067 (Russian).

[3] On a certain algebra of singular integral operators with a non-Carleman shift. *DAN SSSR*, 1978, v.239, N1, p.38-41 (Russian).

[4] Systems of singular integral equations with a shift. *Mat. Sborn.*, 1981, N1, p.87-110 (Russian).

[5] An algebra of singular integral operators with piecewise continuous coefficients and piecewise smooth shift on a composite curve. *Izv. AN SSSR, Ser. Mat.*, 1983, v.47, N5, p.1030-1077 (Russian).

Yu. I. Karlovich, V.G. Kravchenko, G.S. Litvinchuk
[1] The Noether theory of singul integral operators with a shift. *Izv. Vuzov. Matem.*, 1983, N4, p.3-27 (Russian).

Yu. I. Karlovich, G.S. Litvinchuk
[1] On certain classes of semi-Noetherian operators. *Izv. Vuzov. Matem.*, 1990, N2, p.3-16 (Russian).

Sh. Kawamura, J. Tomiyama
[1] Properties of topological dynamical systems and corresponding C^*-algebras. *Tokyo J. Math.*, 1990, v.13, N2, p.251-257.

A.K. Kitover
[1] On the spectrum of an automorphism with a weight, and Kamowitz-Sheinberg theorem. *Funktsion. Anal. i. Prilozhen*, 1979, v.13, N1, p.70-71 (Russian).

Sh. Kobayashi, K. Nomizu
[1] *Foundations of Differential Geometry.* v.I. Interscience Publ., 1963.

V.B. Komanovskij, V.R. Nosov
[1] *Stability and Periodic Regimen of Controlled Systems with an Aftereffect.* M. Nauka, 1981 (Russian).
[2] Systems with an aftereffect of neutral type. *Avtomatica i telemehanika*, 1984, N1, p.5-35 (Russian).

R.A. Kordzadze
[1] A general boundary value problem with a shift for a second order elliptic equation. *DAN SSSR*, 1964, v.155, N4, p.739-742 (Russian).

I.P. Kornfel'd, Ja G. Sinaj
[1] The general ergodic theory of transformation groups with an invariant measure. *Dinamicheskie sistemy. Sovr. probl. matem. Fundam. inapravl.*, 1985, v.2, p.5-106 (Russian).

I.P. Kornfel'd, Ja G. Sinaj, S.V. Fomin
[1] *Ergodic Theory.* M. Nauka, 1980 (Russian).

Yu. V. Kostarchuk
[1] The isomorphism theorems for elliptic boundary value problems with discontinuous coefficients and boundary transmission conditions which are not normal and their applications. *Ukrainsk. Mat. Zh.*, 1976, v.28, N4, p.532-535 (Russian).

A.V. Kozak
[1] Local principle in the theory of projection methods. *DAN SSSR*, 1973, v.212, N6, p.1287-1289 (Russian).

S.M. Kozlov, M.A. Shubin
[1] On the coincidence of spectra of random elliptic operators. *Mat. sb.*, 1984, v.123, N4, p.460-476 (Russian).

V.G. Kravchenko
[1] On a singular integral operator with a shift. *DAN SSSR*, 1974, v.215, N6, p.1301-1304 (Russian).

V.G. Kravchenko, G.S. Litvinchuk
[1] *Introduction to the Theory of Singular Integral Operators with Shift.* Kluwer Academic Publishers, 1994.

V. Krengel
[1] Ergodic theorems. 1985.

N. Ja. Krupnik, I.A. Feldman
[1] On invertibility of certain Fredholm operators. *Izv. AN MSSR. Ser. fiz.-tehn. i mat. nauk.*, 1982, N2, p.8-14 (Russian).

A. Kumar
[1] Fredholm composition operators. *Proc. AMS,* 1980, v.79, N2, p.233-236.

K. Kuratowski
[1] *Topology* v.I. Academic Press, 1966.

V.G. Kurbatov
[1] On estimate of the spectral radii of delaying operators in the space of continuous and bounded on the axis functions. *Funkts. analis i egho prilozhen.*, 1975, v.9, N3, p.56-60 (Russian).
[2] Linear functional-differential equations of neutral type and the delaying spectrum. *Sibirsk. Mat. Zh.,* 1975, v.16, N3, p.536-548 (Russian).
[3] On invertibility of differential-difference operators in various topologies. *Dep. VINITI*, 1984, N764 (Russian).
[4] *Linear Differential-Difference Equations.* Voronegsh, Izdat. Voronezhskogo Univer., 1990 (Russian).

D.A. Kveselava
[1] Some boundary value problems of the theory of functions. *Trudy matem. inst. AN Gruz. SSR*, 1948, v.16, p.39-80 (Russian).

M.B. Landstad

[1] Duality theory for covariant systems. *Trans. Amer. Math. Soc.*, 1979, v.248, p.223-267.

B.V. Langhe, V.S. Rabinovich

[1] Algebra of generalized discrete convolution operators with oscillating coefficients. *Dep. VINITI*, 1982, N3661-82 (Russian).

Yu. D. Latushkin, A.M. Stepin

[1] Weighted shift operators, spectral theory of linear extensions and multiplicative ergodic theorem. *Mat. Sborn*, 1990, v.181, N6, p.723-792 (Russian).

[2] Weighted shift operators and linear extensions of dynamical systems. *Uspekhi Mat. Nauk.*, 1991, v.46, N2, p.85-143 (Russian).

A.V. Lebedev

[1] On the spectrum of weighted shift operators. *IV Shkola po teorii operat. v. funkts. prostr., Tez. dokl., Minsk.*, 1978, p.78 (Russian).

[2] On the invertibility of elements in C^*-algebras generated by dynamical systems. *Uspekhi Mat. Nauk.,* 1979, v.34, N4, p.199-200 (Russian).

[3] On the spectrum of weighted shift operators generated by a Morse-Smale diffeomorphism. *Vestsi Akad Navuk BSSR, Ser. Fiz-Mat. Navuk*, 1979, N3, p.24-28 (Russian).

[4] On certain operator algebras. *Grodnensk. konfer. molod. utchen., Tez. dokl part III Grodno*, 1979, p.25-27 (Russian).

[5] On the spectrum of operators which can be decomposed into a direct integral. *Vestsi Akad. Navuk BSSR, Ser. Fiz-Mat. Navuk,* 1979, N5, p.57-62 (Russian).

[6] On the decomposition of weighted shift operators into integrable families of operators. *Vestski Akad. Navuk BSSR, Ser. Fiz.-Mat. Navuk*, 1980, N3, p.25-32 (Russian).

[7] Weighted shift type operators. Ph.D. diss., Minsk, 1980 (Russian).

[8] On the spectrum of intransitive weighted shift operator on the circle. *Voprosy funkts. analiza (sb. statej), Petrozavodsk*, 1980, p.8-18 (Russian).

[9] On the Fredholmity of the Toeplitz operators with a shift. *Dep. VINITI*, 1980, N1461-80 (Russian).

[10] Invertibility of elements in algebras of operators with a shift. I. *Vestsi Akad Navuk BSSR Ser. Fiz.-Mat. Navuk*, 1982, N6, p.36-43 (Russian).

[11] On invertibility of operators with a shift. *Dokl. AN BSSR*, 1983, v.27, N9, p.773-775 (Russian).

[12] On Wiener type theorems in cross-products and algebras associated with automorphisms. *Dokl. AN BSSR*, 1986, v.30, N6, p.493-495 (Russian).

[13] On invertibility and compactness of elements in algebras associated with automorphisms. *Dokl. AN BSSR*, 1986, v.30, N7, p.589-592 (Russian).

[14] On certain C^*-methods used while investigating algebras associated with automorphisms and endomorphisms. *Dep. VINITI*, 1987, N5351-B87 (Russian).

[15] On Szego-Kolmogorov-Krein theorem and Fuglede-Kadison determinant. *Dep. VINITI*, 1990, N2188-B90 (Russian).

[16] On certain algebras related to C^*-dynamical systems. *Dep. VINITI*, 1991, N1281-B91 (Russian).

[17] On certain ideas arising when searching for the conditions of invertibility of elements in algebras associated with automorphisms. *Dep. VINITI*, 1991, N3290-B91 (Russian).

[18] On hyperbolicity of operators in algebras associated with automorphisms. *Dokl. AN Belarusi*, 1994, v.38, N6, p.21-24 (Russian).

[19] The quasicross-products and the isomorphism theorem for C^*-algebras associated with the quasirepresentations of discrete groups. *Dokl. AN Belarusi*, 1996, v.40, N5, p.40-43 (Russian).

[20] C^*-algebras associated with the normally acting groups of automorphisms. *Dokl. AN Belarusi*, 1996, v.40, N6, p.29-33 (Russian).

[21] Non-local Belousov boundary value problem operators. C^*-scheme. *VII Belorusskaja matem. kouf., Tez. dokl., Part II., Minsk*, 1996, p.131-132 (Russian).

[22] The Kveselava-Vekua type theorems and the index calculation for G-singular integral operators with piecewise continuous coefficients. *International conference 'Boundary value problems, special functions and fractional calculus', Minsk*, 1996, p.61 (Russian)

[23] C^*-algebras associated with automorphisms acting on a singled out ideal. *Dokl. AN Belarusi*, 1997, v.41, N1, p.40-44 (Russian).

A.V. Lebedev, Dangh Suan Tkhan

[1] Local method - C^*-scheme. *Dep. VINITI*, 1991, N1796-B1, 25p. (Russian).

[2] C^*-algebras generated by local type operators and their automorphisms. *Dep. VINITI*, 1991, N2449-B91, 29p. (Russian).

[3] C^*-algebra generated by the Toeplitz operators and a group of shifts. *Dokl. AN BSSR*, 1991, v.35, N11, p.983-985 (Russian).

G.S. Litvinchuk

[1] The Noether theory for a system of singular integral equations with the Carleman shift and complex conjugate unknowns. *Izv. AN SSR. Ser. mat.,* 1967, v.31, N3, p.563-586 (Russian).

[2] *Boundary Value Problems and Singular Integral Equations with a Shift.* M. Nauka, 1977 (Russian).

A.A. Lyapunov, G.P. Bagrinovskaya, Yu. M. Sviretev and others

[1] *Mathematical Simulation in Biology.* M. Nauka, 1975 (Russian).

J.L. Massera, J.J. Sheffer

[1] *Linear Differential Equations and Function Spaces.* Academic Press, 1966.

J. Mather

[1] Characterization of Anosov diffeomorphisms. *Indag. Math.*, 1968, v.30, p.479-483.

S.G. Michlin, S. Pröbdorf

[1] *Singulare Integral Operatoren.* Akademic-Verlag, 1980.

A.S. Mischenko

[1] *Vector Bundles and their Applications.* M. Nauka, 1984 (Russian).

A.S. Mischenko, F. Sharipov

[1] Independence of spectrum of an elliptic operator with random coefficients. *Vestn. Mosk. Univ., Ser. I, Matem. Mekh.,* 1983, N6, p.51-56 (Russian).

Yu. A. Mitropol'skij, D.I. Martynjuk

[1] *Periodic and Quasiperiodic Oscillations of Systems with Delay.* Kiev, Vischa. shk., 1979 (Russian).

Ae. Mukhamadiev, B.N. Sadovskij

[1] On a spectral radius estimate for a certain operator related to neutral-type equations. *Mat. zametki*, 1973, v.13, N1, p.61-78 (Russian).

P.S. Muhly

[1] Imprimitive operators. Preprint, 1972.

N.I. Muskhelishvili

[1] *Singular Integral Equations.* M. Nauka, 1968 (Russian).

F.J. Murray, J. von Neumann

[1] On rings of operators. I. *Ann. Math.*, 1936, v.37, p.116-229; II. *Trans. Amer. Math. Soc.*, 1937, v.41, p.208-248.

A.G. Myasnikov, L.I. Sazonov

[1] On singular integral operators with a non-Carleman shift. *DAN SSSR*, 1977, v.237, N6, p.1289-1292 (Russian).

[2] On singular integral operators with non-Carleman shifts and piecewise continuous coefficients. *DAN SSSR*, 1979, v.245, N6, 1304-1307 (Russian).

[3] On singular operators with a non-Carleman shift and their symbols. *DAN SSSR*, 1980, v.254, N5, p.1076-1080 (Russian).

[4] On singular integral operators with a non-Carleman shift on a non-closed curve. *Funkts. analis i egho prilozhen.*, 1980, v.14, N1, p.71-72 (Russian).

A.D. Myshkis

[1] *Linear Differential Equations with a Delay in an Argument.* M. Nauka, 1972 (Russian).

[2] On certain problems of the theory of equations with a deviating argument. *Uspekhi Mat. Nauk.,* 1977, v.32, N2, p.173-202 (Russian).

P.A. Myshkis

[1] On an algebra generated by conjugate pseudodifferential operators on a manifold. *Uspekhi Mat. Nauk.*, 1976, v.31, N4, p.269-270 (Russian).

M.A. Naimark

[1] *Normed Rings*. M. Nauka, 1968 (Russian).

A.M. Nakhushev

[1] On certain approximate method of the solution of boundary value problems for differential equations and its applications to the dynamics of the soil moisture and subsoil waters. *Diff. uravnenija*, 1982, v.18, N1, p.72-81 (Russian).

J. Von Neumann

[1] On rings of operators. III. *Ann. Math.*, 1949, v.41, p.94-161.

N.K. Nikolskij

[1] *Lectures on the Shift Operator*. M. Nauka, 1980 (Russian).

Z. Nitecki

[1] *Differentiable Dynamics*. M.I.T. Press, 1971.

E.A. Nordgren

[1] Composition operators. *Canad. J. Math.*, 1968, v.20, p.442-449.

[2] Composition operators in Hilbert spaces. *Lect. Notes Math.*, Springer-Verlag, 1978, v.693, p.37-64.

E.A. Nordgren, P. Rosental, F.S. Wintrobe

[1] Composition operators and the invariant space problem. *Math. Repts. Acad. Sci. Can.*, 1984, v.6, N5, p.279-283.

O.A. Oleinik

[1] The solution of the basic boundary value problems for the second order equations with discontinuous coefficients. *DAN SSSR*, 1959, v.124, N6, p.1219-1222 (Russian).

[2] Boundary value problems for linear equations of elliptic and parabolic type with discontinuous coefficients. *Izv. AN SSSR, Ser. Mat.*, 1961, v.25, N1, p.3-20 (Russian).

D. Olesen, G.K. Pedersen

[1] Applications of the Connes spectrum to C^*-dynamical systems I, II, III. *J. Funct. Anal.*, 1978, v.30, p.179-197; 1980, v.36, p.18-32; 1982, v.45, p.357-390.

V.I. Oseledets

[1] Multiplicative ergodic theorem. Lyapunov characteristic exponents. *Trudy Mosk. mat. obschest.*, 1968, v.19, p.179-210 (Russian).

R.S. Palais

[1] *Seminar on the Atiyah-Singer Index Theorem.* Princeton University Press, 1965.

S. Parrot

[1] Weighted translation operators. Ph.D. diss., Univ. of Michigan, Ann Arbor, Mich., 1965.

L.T.A. Paterson

[1] Amenability. *Math. Surv. and Monogr. N29, Amer. Math. Soc.*, 1988.

G.K. Pedersen

[1] *C^*-algebras and their Automorphisms Groups.* Academic Press, 1979.

Ja. B. Pesin

[1] General theory of smooth hyperbolic dynamical systems. *Itogi nauki i tehniki Sovr. problemi matematiki. Fund. naprovl.*, 1985, v.2, p.123-171 (Russian).

K. Petersen

[1] The spectrum and commutant of a certain weighted translation operator. *Math. Scand.*, 1975, v.37, N2, p.295-306.

R.R. Phelps

[1] *Lectures on Choquet's Theorem.* Van Nostrand, Princeton, N.J., 1966.

M. Pimsner, D. Voiculescu

[1] Imbedding the irrational rotation C^*-algebras into AF-algebras. *J. Oper. Theory*, 1980, N4, p.93-118.

B.A. Plamenevskii

[1] *Algebras of Pseudodifferential Operators.* M. Nauka, 1986 (Russian).

L.S. Pontryagin

[1] *Continuous Groups* (3rd edition). M. Nauka, 1973 (Russian).

D. Przeworska-Rolewicz

[1] *Equations with Transformed Argument: an Algebraic Approach.* Warsawa, PWN, 1973.

I. Putnam, K. Schmidt, Ch. Skau

[1] C^*-algebras associated with Denjoy homeomorphisms of the circle. *J. Oper. Theory.*, 1986, 16, N1, p.99-126.

V.S. Rabinovich

[1] On the solvability of differential difference equations on $\mathbb{R}^n$ and the half-space. *DAN SSSR*, 1978, v.243, N5, p.1134-1137 (Russian).

[2] On the algebra generated by the pseudodifferential operators on $\mathbb{R}^n$, the operators of multiplication by almost-periodic functions and the shift operators. *DAN SSSR*, 1982, v.263, N5, p.1066-1069 (Russian).
[3] The limit operators method in the problems of solvability of pseudodifferential equations and convolution type equations. Doctoral diss., Har'kov, 1993 (Russian).

N.I. Radbel'
[1] Decomposition of spectrum and normal points of linear bundle of operators in *B*-spaces. *Vestu. Khar'kov. univ. Ser. Mekh. Mat.*, 1974, N39, p.17-20 (Russian).

S. Rempel
[1] The index theorem for elliptic transmission problems. *Sem. Analisis, Inst. Math., Berlin*, 1984, p.141-162.

S. Rempel, B.-W. Schulze
[1] *Index Theory of Elliptic Boundary Problems*. Akademie-Verlag, Berlin, 1982.
[2] Parametrices and boundary symbolic calculus for elliptic boundary problems without transmission property. *Math. Nachr.*, 1982, N105, p.45-149.
[3] A theory of pseudo-differential boundary value problems with discontinuous conditions. I, II. *Ann. Global Anal. and Geometry*, 1984, v.2, N2, p.163-251; 1984, v.2, N3, p.289-384.

C.E. Rickart
[1] *General Theory of Banach Algebras*. Van Nostrand Company Inc., 1960.

W.C. Ridge
[1] Composition operators. Thesis, Indiana University, 1969.
[2] Spectrum of a composition operator. *Proceedings AMS*, 1973, v.37, N1, p.121-127.

N. Riedel
[1] Classification of C^*-algebras associated with minimal rotations. *Pacif. J. Math.*, 1982, v.101, N1, p.153-161.
[2] Almost Mathieu operators and rotation C^*-algebras. *Proc. London Math. Soc.*, 1988, v.56, N8, p.284-302.

M. Rieffel
[1] C^*-algebras associated with irrational rotations. *Pacif. J. Math.*, 1981, v.93, N2, p.415-429.

F. Riesz, B. Sz.-Nagy
[1] *Lecons d'analyse Fonctionnele* (2nd ed.). Budapest, Akad. Kiado, 1972.

Ya. A. Roitberg, Z.G. Sheftel'

[1] General boundary value problems for elliptic equations with discontinuous coefficients. *DAN SSSR*, 1963, v.148, N5, p.1034-1037 (Russian).

E. Yu. Romanenko, A.N. Sharkovskij

[1] An asymptotics of solutions of linear differential-functional equations. *Asimpt. poved, reshen. differents.-funktsion, uravn., Kiev Inst. Mat. AN USSR*, 1978, p.3-39 (Russian).

W. Rudin

[1] *Functional Analysis*. McGraw-Hill Book Company, 1973.

D. Ruelle

[1] Characteristics exponents and invariant manifolds in Hilbert space. *Ann. Math.*, 1982, v.115, p.243-290.

R.J. Sacker, G.R. Sell

[1] Existence of dichotomies and invariant splitting for linear differential systems. I. *J. Diff. Eqns.*, 1974, v.15, p.429-458.

[2] A spectral theory for linear differential systems. *J. Diff. Eqns.*, 1978, v.27, p.320-358.

S. Sakai

[1] *C^*-algebras and W^*-algebras*. Springer-Verlag, 1971.

A.A. Samarskij

[1] On certain problems of the theory of differential equations. *Diff. uravnenija*, 1980, v.17, N11, p.1925-1938 (Russian).

A.M. Samoilenko

[1] *Elements of Mathematical Theory of Multifrequency Oscillations. Invariant Tori*. M. Nauka, 1987 (Russian).

M. Schehter

[1] Non-local elliptic boundary value problems. *Anal. Scuola norm. super. Pisa Sci. fis. e math.*, 1966, v.20, N2, p.421-441.

E. Schrohe

[1] The symbols of an algebra of pseudodifferential operators. *Pacif. J. Math.*, 1986, v.125, N1, p.211-224.

[2] A ψ^*-algebra of pseudodifferential operators on noncompact manifolds. *Arch. Math.*, 1988, v.51, p.81-86.

B.-W. Schulze

[1] Pseudodifferential operators and asymptotics on manifolds with corners I-IX. *Rept. Akad. Wiss DDK. Karl-Weierstrass-Inst. Math.*, 1989, N7, I-IV, p.1-78; 1990, N2, I-IV, p.1-60; 1990, N4, I-V, p.1-70; 1990, N6, I-IV, p.1-61; 1990, N7, I-VI, p.1-70; 1990, N8, I-V, p.1-71; 1991, N1, I-VI, p.1-123; 1991, N2, I-VII, p.1-91; 1991, N3, p.1-121.

R.T. Seeley

[1] Integro-differential operators on vector bundles. *Trans. AMS*, 1965, v.117, N5, p.167-204.

G.R. Sell

[1] Linearization and global dynamics. *Proc. Intern. Congress of Math.*, Aug. 16-24, 1983, Warsawa, v.2, p.1283-1296.

V.N. Semenjuta

[1] On singular operator equations with a shift on the circle. *DAN SSSR*, 1977, v.237, N6, p.1301-1302 (Russian).

[2] Local approach to problems with a shift. *Izv. Sev.-Kavkaz. nautch. tzentra vysch. schk. Estestv. n.*, 1985, N4, p.28-31 (Russian).

B.V. Shabat

[1] *An Introduction to Complex Analysis, Part I, Functions of One Variable*. M. Nauka, 1976 (Russian).

F. Sharipov

[1] Independence of spectrum of an elliptic operator over C^*-algebra. *Vestn. Mosk. Univ., Ser. I, Matem. Mekh.*, 1985, N1, p.87-88 (Russian).

Z.G. Sheftel'

[1] The general theory of boundary value problems for elliptic systems with discontinuous coefficients. *Ukrainsk. Mat. Zh.*, 1966, v.18, N3, p.132-136 (Russian).

A.L. Shields

[1] Weighted shift operators and analytic function theory. *Top. oper. theory, Mathem. Surv.*, 1974, v.13, p.49-128.

M. Kh. Shkhanukov

[1] On certain problems for a third order equation arising when simulating the filtration of liquid in porous media. *Diff. uravnenija*, 1982, v.18, N4, p.689-699 (Russian).

M.A. Shubin

[1] *Pseudodifferential Operators and Spectral Theory*. M. Nauka, 1978 (Russian).

[2] Pseudodifference operators and their inversion. *DAN SSSR*, 1984, v.276, N3, p.567-570 (Russian).

[3] Pseudodifference operators and their Green function. *Izv. AN SSSR. Ser. mat.*, 1985, v.49, N3, p.652-671 (Russian).

I.B. Simonenko

[1] A new general method of investigation of linear operator integral equations. I. II. *Izv. AN SSSR, Ser. mat.*, 1965, v.29, N3, p.567-586; N4, p.775-782 (Russian).

[2] On multidimensional discrete convolutions. *Mat. issled. Kishinev*, 1968, v.3, N1, p.108-122 (Russian).

I.B. Simonenko, Chin'Ngok Min'

[1] *Local Method in the Theory of One-dimensional Singular Integral Equations with Piecewise Continuous Coefficients. Noetherity.* Izdatel'stvo Rostovskogho univ., 1986 (Russian).

R.K. Singh

[1] Compact and quasinormal composition operators. *Proceedings AMS,* 1974, v.45, N1, p.80-82.

R.K. Singh, N.S. Dharmdhikari

[1] Compact and Fredholm composite multiplication operators. *Acta. sci. math.*, 1988, 52, N3-4, p.437-441.

R.K. Singh, A. Kumar

[1] Compact composition operators. *J. Austral. Math. Soc.*, 1979, v.28, N3, p.309-314.

R.K. Singh, R. Kumar

[1] Compact weighted composition operators on L_2. *Acta Sci. Math.*, 1985, v.49, p.339-344.

[2] Weighted composition operators on functional Hilbert spaces. *Bull. Austral. Math. Soc.*, 1985, v.31, p.117-126.

R.K. Singh, R. Kumar, David Chandra

[1] Compact weighted composition operators. *Acta sci. math.*, 1985, v.49, N1-4, p.339-344.

R.K. Singh, S.D. Sharma

[1] Non-compact composition operators. *Bull. Austral. Math. Soc.*, 1980, v.21, p.125-130.

R.K. Singh, T. Veluchamy

[1] Spectra of normal composition operators on L^2 spaces. *Indian J. pure and appl. Math.*, 1985, v.16, N10, 1123-1131.

[2] Non-atomic measure spaces and Fredholm-composition operators. *Acta. sci math.*, 1987, v.51, N3-4, p.461-465.

A.L. Skubachevskij
[1] On the spectrum of some non-local elliptic boundary value problems. *Mat. Sborn.*, 1982, v.117, N1, p.548-558 (Russian).
[2] Elliptic problems with non-local conditions near the boundary. *Mat. Sborn.*, 1986, v.129, N2, p.279-302 (Russian).

S.L. Sobolev
[1] An example of a correct boundary value problem for the string oscillation equation with the conditions set on the whole of the boundary. *DAN SSSR*, 1956, v.109, N4, p.707-709 (Russian).

A.P. Soldatov
[1] Algebra B of singular operators. *DAN SSSR*, 1985, v.282, N6, p.1312-1316 (Russian).
[2] *One Dimensional Singular Operators and Boundary Value Problems of the Theory of Functions.* M. Vysch. schk., 1991 (Russian).

V.A. Steklov
[1] *The Basic Problems of Mathematical Physics.* M. Nauka, 1983 (Russian).

M.E. Taylor
[1] *Pseudodifferential Operators.* Princeton Univ. Press, Princeton, N.J., 1981.

P. Thieullen
[1] Fibres dynamiques asymptotiquement compact, exposants de Lyapunov. Entropie. Dimension. *Ann. Inst. H. Poincare*, 1987, v.4, N1, p.49-97.

F. Treves
[1] *Introduction to Pseudodifferential and Fourier Integral Operators.* v.1-2. Plenum Press, New York and London, 1982.

M. Troisi
[1] Sulla regularizzazioni delle soluzioni di taluni problemi di transmissione. *Ricerche di Math.*, 1964, v.13, p.281-315.

K. Vala
[1] On compact sets of compact operators. *Ann. Acad. Sci. Fenn. A.I.,* 1964, N351, p.2-9.
[2] Sur les éléments compact d'une algébre normée. *Ann. Acad. Sci. Fenn. A.I.*, 1967, N407, p.2-8.

J. Varela
[1] Duality of C^*-algebras. *Mem. Amer. Math. Soc.*, 1974, N148, p.97-108.

N.L. Vasilevskii

[1] Local principles in operator theory. *Lineinyie operatory v funktzionalnyikh prostranstvah. Tez. dokl. Severo-Kavkaz. reg. konf., Groznii*, 1989, p.32-33 (Russian).

N.P. Vekua

[1] *Systems of Singular Integral Equations.* M. Nauka, 1970 (Russian).

A.M. Vershik

[1] Trajectorial theory ('Dynamical systems' 2, ch. 5). *Sovrem. problemy matematiki. Fundament. napravl,* v.2. *VINITI*, 1985 (Russian).

A.M. Vershik, A.L. Fedorov

[1] Trajectorial theory. *Sovrem. problemy matematiki. Noveisch. dostizh.,* v.26. *VINITI*, 1985, p.171-212 (Russian).

A.M. Vershik, S.V. Kerov

[1] Locally semisimple algebras. Combinatorial theory and K_0-functor. *Sovrem. problemy matematiki Noveisch. dostizh,* v.26. *VINITI*, 1985, p.3-56 (Russian).

A.M. Vershik, I.P. Kornfel'd

[1] Periodic approximations and their applications. Ergodic theorems, spectral and entropy theory for general group actions ('Dynamical systems' - 2, ch.4). *Sovrem. problemy matematiki. Fundament. napravl,* v.2. *VINITI*, 1985 (Russian).

M.I. Vishik, G.I. Éskin

[1] Boundary value problems for general singular equations in a bounded domain. *DAN SSSR*, 1964, V.155, N1, p.24-27 (Russian).

[2] Convolution equations in a bounded domain. *Uspekhi Mat. Nauk.*, 1965, V.20, N3, p.89-152 (Russian).

[3] Elliptic convolution equations in a bounded domain and their applications. *Uspekhi Mat. Nauk*, 1967, V.22, N1, p.15-76 (Russian).

[4] Convolution equations of varying order. *Trudy Mosk. mat. obsch.*, 1967, V.16, p.25-50 (Russian).

[5] Normally solvable problems for elliptic systems of convolution equations. *Mat Sborn.,* 1967, V.74, N3, p.323-356 (Russian).

P. Walters

[1] *An Introduction to Ergodic Theory*. Graduate texts in Math., Springer 1982, v.79.

H. Weyl

[1] Uber die Gleich-verteilung Von Zahlen mod. Eins. *Math. Ann.*, 1916, Bd. 77S, p.313-352.

N. Wiener

[1] *The Fourier Integral and Certain of its Applications.* Dover publ., N.Y., 1956.

K. Ylinen

[1] Compact and finite-dimensional elements of normed algebras. *Ann. Acad. Sci. Fenn. A.I.*, 1968, N428, p.6-37.

N.V. Zhitrashu

[1] A note on general boundary value problems for elliptic and parabolic systems of equations with discontinuous coefficients. *Mat. issled., Kishinev*, 1978, v.46, p.53-60 (Russian).

N.V. Zhitrashu, S.D. Èidel'man

[1] On non-local boundary value problems for elliptic equations. *Mat. issled., Kishnev*, 1971, v.6, N2(20), p.63-73 (Russian).

Index

O

P

List of Notation

Certain standard algebras, ideals and sets of operators

Standard operators and automorphisms

The objects related to algebras and automorphisms

The objects related to the elements of algebras

Special algebras and sets of operators

Symbols of various operators

Algebras of symbols

Special operators

Spaces

Special mappings

Special representations

Some special objects and notations

Contents of Volume I

Contents of Volume II (Part 2)

Contents of Volume III